Principles of Hydrology
Third Edition

Principles of
Hydrology

Third Edition

R. C. Ward
Professor of Geography
University of Hull

M. Robinson
Senior Scientific Officer
Institute of Hydrology
Wallingford

McGRAW-HILL BOOK COMPANY

London · New York · St Louis · San Francisco · Auckland
Bogotá · Guatemala · Hamburg · Lisbon · Madrid · Mexico
Montreal · New Delhi · Panama · Paris · San Juan · São Paulo
Singapore · Sydney · Tokyo · Toronto

Published by
McGraw-Hill Book Company (UK) Limited
MAIDENHEAD · BERKSHIRE · ENGLAND

British Library Cataloguing in Publication Data
Ward, R. C. (Roy Charles), 1937–
Principles of hydrology.—3rd ed
1. Hydrology
I. Title II.   Robinson, M. (Mark)
551.48
ISBN 0-07-707204-9

Library of Congress Cataloging-in-Publication Data
Ward, R. C.
    Principles of hydrology / R. C. Ward, M. Robinson.—3rd ed.
        p.   cm.
    Bibliography: p.
    Includes index.
    ISBN 0-07-707204-9
    1. Hydrology.   I. Robinson, M. (Mark).   II. Title.
GB661.2.W35   1989
551.4—dc 19   88-32409

1234 WL9109

Typeset by Eta Services (Typesetters) Ltd, Beccles, Suffolk
Printed and bound in Great Britain by
Whitstable Litho Ltd. Whitstable, Kent

To Kay and Mary

Contents

CHAPTER 3 INTERCEPTION

CHAPTER 4 EVAPORATION

CHAPTER 5 SOIL WATER

CHAPTER 6 GROUNDWATER

CHAPTER 7 RUNOFF

CHAPTER 8 WATER QUALITY

Preface

Perhaps the most dramatic of the early photographs taken in space showed the earth rising over the moon's horizon—a white-swirled disk of blue and grey contrasting vividly with the barren aridity of the moon's surface and the blackness of space. Significantly the colour of the 'blue planet' reflects the fact that it is the only body in the solar system on which water exists freely. Indeed, almost three-quarters of the earth's surface is covered by water. And yet, although ours is a water world, we are still unable to determine precisely the amounts and distributions of water, to define its movements and fluxes, and to understand fully the hydrological processes at work on and near the earth's surface.

Even so, great progress has been made since the appearance of the First Edition of *Principles of Hydrology* in 1967. By that date, shortly after the beginning of the International Hydrological Decade (IHD) in 1965, attempts were already in train to consolidate a large number of international programmes of data collection and water resource assessment and some of this was reflected in the 'Water for Peace' Conference in that year.

Further considerable progress had been made, especially in quantifying the global water balance, by the time the Second Edition appeared in 1975. The IHD had been superseded by the International Hydrological Programme (IHP), and a number of large-scale programmes of data collection, such as FREND (Flow Regimes from Experimental and Network Data), involving extensive international cooperation, had been initiated.

The appearance of this Third Edition coincides with a better understanding of physical hydrology, coupled with a much greater emphasis on applying the principles of physical hydrology to the solution of human problems. In both respects there is still some way to go but at no time in the history of mankind have the potential rewards for success been greater. For example, as we approach the end of the International Drinking Water Supply and Sanitation Decade (IDWSSD), 45 per cent of the world's population do not have access to adequate and safe water for their daily needs, 30 per cent are at frequent risk from floods and it is estimated that 80 per cent of all disease is directly related to inadequate water supplies.

Long-term, reliable solutions will not be found without a proper understanding of the processes responsible for the occurrence, distribution, movement and quality of water on and below the earth's surface. It is these processes, and particularly our greatly improved understanding of their operation, which form the subject matter of this discussion of the principles of hydrology. Although the structure of the book remains essentially unchanged, with chapters devoted to the major components of the hydrological cycle, the subject matter has been fully updated and, indeed, apart from the retention of some of the diagrams, the text has been completely rewritten. In addition, a new chapter on water quality reflects an increasing awareness of the importance of considering both quality and quantity in the approach to most major hydrological problems. Also, in order to ensure an appropriately fresh approach, and in recognition of the growth in scope and complexity of hydrology over the past 21 years, this task has been accomplished with the help of a co-author.

As in previous editions we have tried to give a wide range of modern and up-to-date references for readers who wish to follow up specific topics in more detail than our allocation of space allows us here. Where appropriate, however, we have also given original references, some of them quite ancient, in order to make clear to the reader how the understanding of a particular aspect of hydrology has developed over time.

Both of us would wish to thank the many friends and colleagues who have given willingly of their time to read and comment on earlier drafts of chapters. Particular thanks are due to Ann Calver, David Cooper, John Gash, Colin Neal, Mike Price, John Rodda and Brian Smith, and we are also sincerely grateful to Richard Harding, Terry Marsh, Ian Nicholson, Duncan Reed and Gareth Roberts. Their valuable advice has done much to improve this text; for the failings that remain, however, we take full responsibility. We also thank our wives, Kay and Mary, for their forbearance, encouragement and help in the preparation of this book.

Roy Ward and Mark Robinson

Preface to First Edition

This book presents a non-mathematical treatment of 'pure' as opposed to 'applied' hydrology. It is intended to fill a long-felt need at university level, among geographers in particular and earth scientists in general, for a straightforward, systematic analysis of the distribution and movement of water in the physical environment. It is, therefore, the first British textbook covering the general field of hydrology that is not aimed exclusively at the engineer. Nevertheless, it will probably be found useful as an elementary text for students in engineering faculties in which hydrology forms a compulsory part of the course, and for technical staffs in water boards and river authorities. In addition, the book is written in a way that should make it quite acceptable to the intelligent layman seeking background information on the pressing problems of water supply, and on the factors underlying the initiation in 1965 of the International Hydrological Decade.

Inevitably, its general scope and framework are similar to those of existing engineering texts. The essential differences lie in the selection, emphasis, and treatment of material and, particularly, in the inclusion of numerous examples from the British Isles. A conscious effort has been made to reduce engineering and other 'applied' aspects to a minimum, although in many cases natural events have been so modified by man's activities that references to the latter are unavoidable. On the other hand, since many of the problems of hydrology are geographical problems of spatial distribution and of climatic and regional differences, the application of geographical methods and techniques should contribute positively to their solution.

The arrangement of the text follows the conventional, although logical, systematic approach, whereby the concept of the continuous natural movement of water in the hydrological cycle is very briefly introduced in Chapter 1, and, in the remaining seven chapters, the main phases of the cycle, i.e., precipitation, interception, evaporation, evapotranspiration, soil moisture, groundwater and runoff, are examined in detail. There was a temptation to write a ninth, concluding chapter but it was felt that the discussion of runoff in Chapter 8 provides a natural conclusion, emphasizing, as it does, the effects and interactions of many of the components of the hydrological cycle that have been previously discussed. In view of this arrangement, I have necessarily drawn quite heavily in places on the existing standard hydrological texts, and I hope that this debt is adequately acknowledged in the bibliographies. These have been placed at the end of each chapter, despite the fact that this has resulted in some obvious repetitions, in order to avoid a large and cumbersome bibliography at the end of the book. It will be noted that reference has been made to a large number of articles and other publications throughout this book in the hope that the interested reader will be encouraged to extend his study of particular topics. For this reason, the articles referred to are not necessarily always the normally accepted 'classics'. They have been selected largely because they represent a modern statement of ideas and facts; or they themselves contain an extensive bibliography that will lead the reader back through the development of thought on a given topic; or the journals in which they appear are comparatively easily accessible in this country.

I am sincerely grateful to all who have helped in the preparation of this book. Particular thanks are due to the numerous individuals and publishers for permission to copy or adapt a large number of the many diagrams; individual acknowledgements are made in each case. I would also like to thank Professor H. R. Wilkinson, of the Department of Geography at Hull University, and other colleagues for reading parts of the manuscript, although they are in no way responsible for the deficiencies that still remain; Mr R. R. Dean and the staff of the Geography Department drawing office for the preparation of all the diagrams; Miss J. M. Bailey and Miss P. A. Ashcroft, who bore the brunt of the typing; finally, my wife, Kay Ward, for her commendable assistance during many hours of proofreading and checking.

R. C. WARD
Hull, 1967

1. Introduction

1.1 Water—facts and figures

Hydrology is the science of water. This book focuses on the *principles* of hydrology which are concerned with the basic physical processes governing both the occurrence, distribution and movement of water over, on and under the surface of the earth and also its variation of quality from time to time and from one place to another. In this context of *environmental hydrology*, water, no less than soil, vegetation, climate or rock, is an element of the landscape demanding rigorous, scientific quantification and analysis. Water, the subject matter of hydrology, is both commonplace and unique. It is found everywhere in the earth's ecosystem and taken for granted in much of the developed world. It is, however, the only naturally occurring inorganic liquid and is the only chemical compound that occurs in normal conditions as a solid, a liquid and a gas. Its distribution over the globe is amazingly uneven.

About 97 per cent (depending on the method of calculation) occurs as saline water in the seas and oceans. Of the remaining, fresh, water considerably more than one-half is locked up in the ice sheets and glaciers and another substantial volume occurs as deep groundwater. The really mobile fresh water which contributes frequently and actively to rainfall, evaporation and streamflow thus represents only about 0.3 per cent (again depending on the method of calculation) of the global total. Estimated values of global water storage and flux, compiled by Speidel and Agnew (1988), are shown in Table 1.1 together with a range of values found in the recent literature.

Not surprisingly, it is on this relatively small amount of fresh water occurring as lakes, rivers, soil moisture and shallow groundwater, and in the vegetation cover and in the atmosphere, that hydrologists have concentrated their attention. It is important, however, not to underestimate the dominant role played by the oceans in the global water and energy budgets. It is equally important to recognize that the small volume of mobile fresh water is itself distributed unevenly in both space and time. Wetland and prairie, forest and scrub, snowfield and desert, each exhibits different regimes of precipitation, evaporation and streamflow, each offers different challenges of understanding for the hydrologist and of water management for the planner and engineer, and each poses different benefits and threats to human life and livelihood as between the developed and the developing world.

1.2 The changing nature of hydrology

Although hydrology is concerned with the study of water, especially atmospheric and terrestrial fresh water, its emphases have changed from time to time and from one practitioner to another. Some have discerned in such changes identifiable historical patterns or 'eras' of hydrology (cf. Kundzewicz *et al.*, 1987).

The origins of hydrology have been variously attributed to Greece, Egypt or South America. Jiaqi (1987) traced its history from the ancients of China and postulated three

1

Table 1.1 Global water reservoirs and fluxes. (Based on a table in D. H. Speidel and A. F. Agnew, Water, in *The natural geochemistry of our environment*, Boulder, Colo. © 1982 by Westview Press, Inc.)

	Values ($km^3 \times 10^3$)	Percentage of total	Range of values in recent literature ($km^3 \times 10^3$)
Reservoir			
Ocean	1 350 000.0	97.403	$1.32–1.37 \times 10^6$
Atmosphere	13.0	0.000 94	10.5–14.0
Land	35 977.8	2.596	
Rivers	1.7	0.000 12	1.0–2.1
Freshwater lakes	100.0	0.007 2	30.0–150.0
Inland seas, saline	105.0	0.007 6	85.4–125.0
Soil water	70.0	0.005 1	16.5–150.0
Groundwater	8 200.0	0.592	$7.0–330.0 \times 10^3$
Ice caps/glaciers	27 500.0	1.984	$16.5–29.2 \times 10^3$
Biota	1.1	0.000 08	1.0–50.0
Annual flux			
Evaporation	496.0		446.0–577.0
Ocean	425.0		383.0–505.0
Land	71.0		63.0–73.0
Precipitation	496.0		446.0–577.0
Ocean	385.0		320.0–458.0
Land	111.0		99.0–119.0
Runoff to oceans	41.5		33.5–47.0
Rivers	27.0		27.0–45.0
Groundwater	12.0		0.0–12.0
Glacial meltwater	2.5		1.7–4.5

stages of development. The stage of *geographical* hydrology saw the establishment of the hydrological cycle and the concepts of the water balance. The stage of *engineering* hydrology was brought about because the design of control structures demanded quantitative analysis of hydrological phenomena. Finally, social development leading to increased water demand and larger-scale use of water resources, coupled with the development of new techniques, has, argued Jiaqi, brought about a new stage of *water resources* hydrology involving resource analysis and management.

Far more convincing, however, is the evidence that understanding, engineering and large-scale resource development have progressed simultaneously and interdependently and from very early times, especially in areas where water was a 'problem' either because of its shortage or its abundance. In Egypt, for example, the Nile flood formed the basis of successful, large-scale, agricultural irrigation for more than 5000 years and the remains of what is possibly the world's oldest dam, built between 2950 and 2750 BC, may be found near Cairo (Biswas, 1970). Again, field systems, dating from before the discovery of the New World, have been found on Columbian floodplains, where construction must have required large-scale operational effort and a high degree of social organization to take advantage of the seasonal floods (Parsons and Bowen, 1966). Indeed, the fact that water is essential to life and that its distribution and availability are intimately associated with the development of human society means that it was almost inevitable that some development of water resources *preceded* a real understanding of their origin and formation.

Archaeological discoveries and later documentary evidence emphasize the significant part played by the location and magnitude of water supplies in the lives of, for example, the Old Testament peoples, the ancient Egyptians and later the Greeks and Romans. During these periods, throughout the Middle Ages and indeed until comparatively recent times, the search continued for an explanation of springs, streamflow and the occurrence and movement of groundwater. However, the hypotheses put forward were either based on guesswork or mythology or else were biased by religious convictions; few, if any, were based on the scientific measurement of the relevant hydrological factors. And yet some of the ideas developed by the ancient writers were remarkably close to the truth as we now know it. Aristotle (384–322 BC) explained the mechanics of precipitation, Vitruvius, three centuries later, believed in the pluvial origin of springs, da Vinci (1452–1519) had somewhat confused ideas about the hydrological cycle, but a much better understanding of the principles of flow in open channels than either his predecessors or contemporaries, and Palissy (1510–1590) stated categorically that rainfall was the only source of springs and rivers (Biswas, 1970).

It was not until near the end of the seventeenth century, however, that plausible theories about the hydrological cycle, based on experimental evidence, were put forward. The greatest advances came largely through the work of three men: Pierre Perrault and Edme Mariotte, whose work on the Seine drainage basin in northern France demonstrated that, contrary to earlier assumption, rainfall was more than adequate to account for river flow; and the English astronomer, Edmund Halley, who showed that the total flow of springs and rivers could be more than accounted for by evaporation from the oceans. Because Perrault, Mariotte and Halley undertook hydrological research of the modern scientific type they may well be regarded as the founders of hydrology (UNESCO/WMO/IAHS, 1974).

In this context hydrology is still a young science in which progress has been largely dependent on technological advances in *other* sciences (McCulloch, 1988) and in which the development of theory to underpin the broad field of endeavour has been hampered by the diversity and complexity of the terrestrial portion of the hydrological system (Collins, 1987).

1.3 The hydrological cycle and system

The interdependence and continuous movement of all forms of water provide the basis for the concept of the *hydrological cycle* (Fig. 1.1), which has long been a central concept in hydrology. Water vapour in the atmosphere condenses and may give rise to precipitation. In the terrestrial portion of the cycle not all of this precipitation will reach the ground surface because some will be intercepted by the vegetation cover or by the surfaces of buildings and other structures, and will from there be evaporated back into the atmosphere.

The precipitation that reaches the ground surface may then follow one of three courses. First, it may remain as surface storage in the form of pools, puddles and surface moisture which are eventually evaporated back into the atmosphere. Second, it may flow over the surface into streams and lakes, from where it will move either by evaporation back into

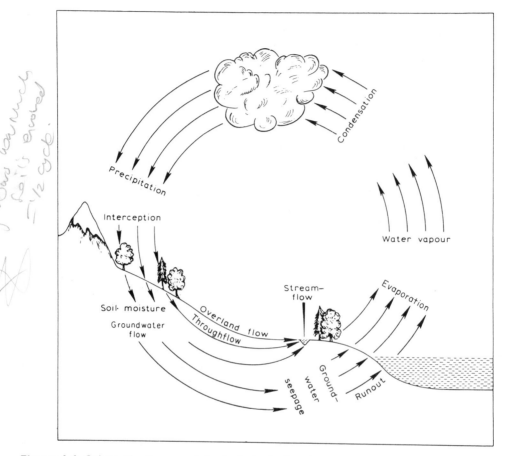

Figure 1.1 *Schematic diagram of the hydrological cycle.*

the atmosphere, or by seepage towards the groundwater, or by further surface flow into the oceans. Following the third course, precipitation may infiltrate through the ground surface to join existing soil moisture. This may be removed either by evaporation from the soil and vegetation cover, or by throughflow towards stream channels, or by downward percolation to the underlying groundwater, where it may be held for weeks or months or even longer. The groundwater component will eventually be removed either by upward capillary movement to the soil surface or to the root zone of the vegetation cover, whence it will be returned by evaporation to the atmosphere or by groundwater seepage and flow into surface streams and into the oceans.

Inevitably, the simplifications and generalizations involved in the broad concept of the hydrological cycle may be misleading unless treated with caution. Thus, the implication of a smooth, uninterrupted, sequential movement of water is belied by the complexity of natural events. The cycle is short-circuited when, for example, water precipitated from the atmosphere and falling to the ground surface is immediately returned to the atmosphere by evaporation without becoming involved in streamflow, soil moisture or groundwater movement, or the oceans. Similarly, precipitation may fall upon a lake and be evaporated from there without touching the land surface at all, or it may fall upon the land and percolate down to the main groundwater body within which it moves slowly towards a discharge point such as a spring, which it may not reach for a thousand years or more. Also, increasingly, the cycle is interrupted and modified by human activities.

The irregularity of water movement within the cycle is illustrated further by conditions in hot deserts and subpolar regions. In the former, rainfall is spasmodic, occurring perhaps once in a period of years. Other phases of the cycle, such as evaporation and streamflow, which can take place only for a short period during and after rainfall, are therefore equally spasmodic, so that a short burst of hydrological activity for a week or so may be followed by a long period of virtual inactivity, apart from a slow redistribution of groundwater at some depth below the surface. In cold climates, where most of the precipitation is in the form of snow, there may be an interval of several months between precipitation and the active involvement of the precipitated moisture, after melting, in the subsequent phases of the hydrological cycle.

The hydrological cycle provides a useful introductory concept and permits the relationship between precipitation and streamflow to be expressed in a very general way. It is, however, of little practical value to the hydrologist concerned with understanding and quantifying the occurrence, distribution and movement of water in a *specific* area, whether a small experimental plot of a few square metres or a large continent but which, for most hydrological purposes, will comprise a river basin or group of basins. In this context it is helpful to recognize that in natural conditions most rivers and streams receive water only from their own topographic drainage basin or catchment area. Each drainage basin can therefore be regarded as an individual system (Fig. 1.2) receiving quantifiable inputs of precipitation and transforming these, via various flows and storages, into quantifiable outputs of evaporation and streamflow. In some cases leakage from deeper subsurface water may represent either an additional input or (as shown in Fig. 1.2) an additional output from the drainage basin system. Clearly each of the five storages shown also has the qualities of a system, and may therefore be regarded as subsystems of the drainage basin hydrological system.

Drainage basin hydrological processes rarely operate completely uninfluenced by human activity (Fig. 1.3). It is, therefore, important to recognize that human modifications may be made to virtually every component of the system. At the present time the most important of these relate to:

(a) large-scale modifications of channel flow and storage, e.g. by means of surface changes such as afforestation, deforestation, urbanization, etc., which affect surface runoff and the incidence or magnitude of flooding,
(b) the widespread development of irrigation and land drainage, and
(c) the large-scale abstraction of groundwater and surface water for domestic and industrial uses.

Other important modifications include artificial recharge of groundwater and interbasin transfers of surface and groundwater. In the future other modifications, such as the artificial stimulation of precipitation or the use of transpiration suppressants, may become more generally important.

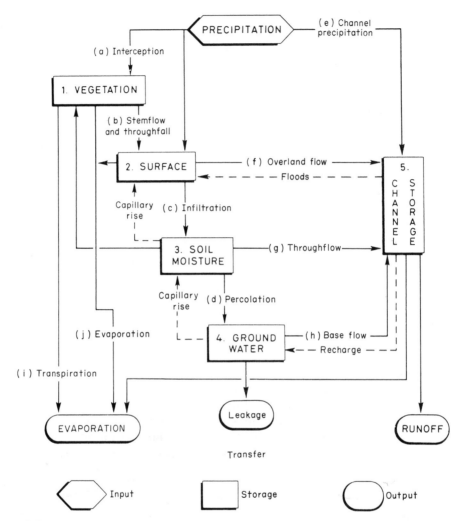

Figure 1.2 *The drainage basin hydrological system (from an original diagram by Professor J. Lewin).*

1.4 The nature of hydrological processes

Although frequent reference has been made in the preceding section to the drainage basin system, hydrological processes can in fact be investigated over a very wide range of spatial and temporal *scales*. At one extreme *microscale* investigations are exemplified by studies of the movement of soil solution through the interstices of the soil matrix or of the evaporation characteristics of individual plants growing in controlled-environment chambers. At the other extreme some of the emerging problems of environmental change associated with large-scale forest clearance or desertification, for example, will be resolved only by a better understanding of the hydrological system at the *macroscale*, i.e. global or regional, rather than at the *mesoscale* level of the catchment.

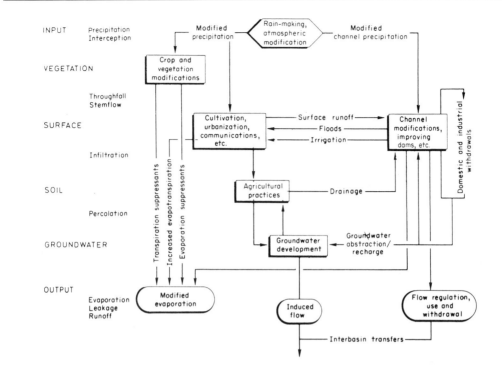

INPUT   Precipitation
                Interception

VEGETATION

               Throughfall
               Stemflow

SURFACE

               Infiltration

SOIL

               Percolation

GROUNDWATER

OUTPUT
               Evaporation
               Leakage
               Runoff

Figure 1.3 *Principal areas of human intervention in the hydrological system (from an original diagram by Professor J. Lewin).*

Such large-scale studies demand international as well as interdisciplinary collaboration and depend increasingly on large data networks and in recent years on the growing use of data derived from remote sensing techniques. For example, in the mid-1980s a series of large-scale investigations of the interrelations of climate and vegetation formed part of the World Climate Research Programme (WCRP). Similarly, the International Satellite Land Surface Climatology Project (ISLSCP) aimed to improve our ability to relate satellite data to surface hydrological and climatological conditions. As a final example, the study by eight national scientific groups of the dynamics of desertification in the Sahel region of West Africa using remote sensing techniques was supported by EC funding.

Implicit in these larger-scale studies is a recognition that hydrological processes form part of an *interactive* earth surface–atmosphere system. Hydrologists must deal with an enormous *spatial diversity* both at the earth surface and in the overlying atmosphere, and this diversity exists across the entire range from micro- to macroscale occurrences. Not only do soils, geology and land use vary spatially but so also do the components of the water balance. Especially crucial is the relation between precipitation and potential evaporation, which varies enormously from one area to another *and* over time, both seasonally and in the long term (cf. the Sahel). Accordingly, hydrologists are increasingly taking a long-term, as well as a large-scale, view in order to meet the considerable challenges posed by, for example, the impact of climatic change on water resources.

Within this framework of spatial diversity it is virtually impossible to argue from the particular to the general using a simple deterministic notion of causality. Quoting Max Born, 'nature is ruled by laws of cause and laws of chance in a certain mixture', Matalas (1982) remarked:

> Our experience suggests that hydrological events are of a recurrent kind and are the realizations of hydrologic processes. And our intuition tells us that the events are chance outcomes of experiments, the collection of processes themselves, and that chance has something to do with the experimental arrangements, particularly with their initial conditions. However, knowing the initial conditions is not tantamount to knowing what the outcome of the experiment is to be. Thus, intuitively, probability is an intrinsic empirical property of the system.

In other words, hydrological processes may be *deterministic* (i.e. chance independent) or *stochastic* (i.e. chance dependent). All hydrological phenomena are subject to laws that govern their evolution and behaviour so that at the microscale all hydrological processes may be deterministic. However, the macroscale processes which comprise the combined effect of the individual microscopic events may be stochastic in nature. Thus the evolution of rainfall phenomena in time is stochastic in the sense that, although each microevent is deterministic, at the macroscale present conditions do not uniquely determine future conditions; only *probabilities* of future conditions may be predicted from the present situation (Todorovic and Yevyevich, 1969).

Beven (1987) suggested that our perception of the hydrological system, which is necessarily personal and will have been conditioned by teacher, textbooks and experience in hydrological modelling and fieldwork, is rarely set down in print, perhaps because it so often conflicts with current hydrological theory. He then continued:

> My own perceptual model is one of complex spatial and temporal variability of input rates, flow paths and nonlinear dynamic responses resulting from the effects of spatial variability in rainfalls, vegetation canopy, soil structure, and topography. My model embodies the idea that preferential pathways are important in flow processes at all scales, from the micro-scale in soil physics, through the hillslope scale in overland and subsurface flows, to the catchment scale in expanding networks of small channels.
>   Inherent in this perceptual model is an element of unknowability of the system resulting from heterogeneity. . . . Unknowability implies uncertainty and yet hydrologists persist in using deterministic models for predicting catchment responses.

Beven argued the need for two developments in hydrology. First, hydrological prediction must be associated with a realistic estimate of uncertainty. Levels of uncertainty could then be used directly in the decision-making process, thereby permitting the consideration of hydrological processes that are not properly understood as well as those that are. Second, an inherently stochastic, macroscale theory will be needed which can accommodate the spatial integration of heterogeneous non-linear interacting processes, including preferential pathways, in order to provide a rigorous basis for both 'lumped' and 'physically based' predictions.

The discussion of hydrological systems and processes in the ensuing chapters of this book will show that, although there is some way to go in both respects, hydrology has, nevertheless, made enormous advances since the first edition of this book was published in 1967. That it is likely to continue to do so was suggested by the stimulating report of an IAHS Working Group of young hydrologists on prospects for hydrology to the end of this century under the title *Hydrology 2000* (Kundzewicz *et al.*, 1987). This made it all the more surprising that Dooge (1988) should have come to the rather negative conclusion that:

> If we take a broad enough viewpoint, hydrology can indeed establish itself as a respectable science and as a useful art before the year 2000 and therefore provide a sound foundation for continuing development into the next century.

It is hoped that the concentration, in this book, on basic physical processes will demonstrate that those factors identified by Dooge as having hampered the development of hydrological theory, e.g. fragmentation of approach to the subject, failure of communication between those using different techniques and lack of proper structures and procedures at national and international level, are rapidly receding in importance. Thus, although stochastic and deterministic approaches have been considered as rivals in the past, they are now seen to be complementary; it *is* recognized that if hydrology is to be applied effectively and economically to human needs then 'hydrologists must learn to listen to people from other disciplines and hopefully to understand them' (Dooge, 1988). Finally, the degree of international cooperation now in evidence, as a result of the Hydrological Decade and the Hydrological Programme, and of the efforts of bodies such as the International Council of Scientific Unions (ICSU), the Council on Water Research (COWAR), together with the long-established contribution of the International Associa-tion of Hydrological Sciences (IAHS), is most encouraging.

Hydrology is already both 'respectable' and 'useful'. After all, Chow's massive *Handbook of Applied Hydrology* was published a quarter of a century ago (Chow, 1964). Since then great strides forward have been taken in our ability to collect hydrological data, including remote sensing, weather radar and microprocessor-based data loggers, and recent instrument developments, cf. the Institute of Hydrology's 'Hydra', now permit direct measurements of actual evaporation. Equally impressive improvements have been made in our ability to process and analyse these data because of information technology advances in computers and geographical information systems. Hydrologists now have the means, and the confidence, to work with 'real' rather than 'ideal' problem situations, and have recognized that water is important for its quality as well as for its quantity.

This is a scientific and not an engineering textbook, but we should be ever mindful that hydrology is more than just an intellectual exercise. The practical and theoretical advances in hydrology, which are reflected in this book, will provide an essential basis for meeting the growing challenges posed, for example, by increasing environmental pressures resulting from population growth and climatic changes, such as the 'greenhouse effect', which have the potential to greatly affect water resources. Events in the Sahel have made us uncomfortably aware that recent hydrological records may not be a good guide to the future unless we fully understand the hydrological processes at work. Water is our most precious resource—the original elixir of life—and for the majority of the global population a better understanding of the principles of hydrology may, literally, be a matter of life and death.

References

Beven, K. (1987) Towards a new paradigm in hydrology, in *Water for the future: hydrology in perspective*, IAHS Publ. 164: 393–403.

Biswas, A. K. (1970) *History of hydrology*, North-Holland, Amsterdam.

Chow, V. T. (ed) (1964) *Handbook of applied hydrology*, McGraw-Hill, New York, 1467 pp.

Collins, D. N. (1987) Hydrology and hydrologists, in *Hydrology 2000*, Z. W. Kundzewicz, L. Gottschalk and B. Webb (eds), IAHS Publ. 171: 91–100.

Dooge, J. C. I. (1988) Hydrology in perspective, *Hydrology Science Journal*, **33**: 61–85.

Jiaqi, C. (1987) The new stage of development of hydrology—water resources hydrology, in *Water for the future: hydrology in perspective*, IAHS Publ. 164: 17–25.

Kundzewicz, Z. W., L. Gottschalk and B. Webb (eds) (1987) *Hydrology 2000*, IAHS Publ. 171.

McCulloch, J. S. G. (1988) Hydrology—science or just technology?, *Research Report 1984–87*, Institute of Hydrology, Wallingford.

Matalas, N. C. (1982) *Reflections on hydrology*, 1st Chester C. Kisiel Memorial Lecture, University of Arizona, Department of Hydrology and Water Resources, 16 pp.

Parsons, J. J. and W. A. Bowen (1966) Ancient ridged fields of the San Jorge River floodplain, Colombia, *Geographical Review*, **61**: 317–43.

Speidel, D. H. and A. F. Agnew (1988) The world water budget, in *Perspectives on water*, D. H. Speidel, L. C. Ruedisili and A. F. Agnew (eds), Oxford University Press, New York, 27–36.

Todorovic, P. and V. Yevyevich (1969) Stochastic process of precipitation, *Colorado State University, Hydrology Papers*, Fort Collins, Colo., Paper 35, 61 pp.

UNESCO/WMO/IAHS (1974) *Three centuries of scientific hydrology*, 110 pp.

2. Precipitation

2.1 Introduction and definitions

Precipitation is a major factor controlling the hydrology of a region—it is the main input of water to the earth's surface and a knowledge of rainfall patterns in space and time is essential to an understanding of soil moisture, groundwater recharge and river flows. Data are more readily available, for more sites and for longer periods, than for other components of the hydrological cycle. In some countries precipitation data may constitute the only available hydrological record (Wiesner, 1970). The study of precipitation is thus of fundamental importance to the hydrologist, but detailed investigation of the mechanisms of its formation is the domain of the meteorologist and climatologist. This chapter will concentrate on those aspects that are of direct relevance to the hydrologist, and it will be assumed that the reader will refer to standard meteorological and climatological texts for a more systematic treatment of the subject.

The total amount of water vapour in the atmosphere represents only a minute proportion of the world's water budget. It has been estimated that atmospheric water accounts for less than 0.001 per cent of the world's total supply of land, oceanic and atmospheric water, and yet this small amount serves as a continuing source of supply in the form of precipitation. The atmospheric vapour amounts to about 25 mm of liquid water which, given the average annual precipitation over the whole globe of about 1000 mm, represents about 10 days' average supply. While the *mean* residence time is 10 days, this figure hides a great variation since some water may be carried up into the stratosphere where it could remain for up to 10 years, but at the other extreme, some water that is evaporated into the lower layers of a thunderstorm cloud may be precipitated out within an hour (Lamb, 1972).

In large storms the amount of precipitation may be several times greater than the average water content of a column of atmosphere (although in fact it could never all be precipitated out, even in the greatest storms), indicating that large-scale lateral movements of moist air must play a key role in the distribution of precipitation. The meteorologist is concerned to analyse and explain the mechanisms responsible for this distribution, an interest ceasing when the precipitation reaches the ground. The hydrologist is interested in the distribution itself, in how much precipitation occurs and in when and where it falls. Thus the hydrological aspects of precipitation studies are concerned with the form in which precipitation occurs, its variations in both space and time, and the problems of the correct use and interpretation of the measured data.

Precipitation occurs in a number of forms, and a simple but fundamental distinction can be made between liquid and solid forms. Liquid precipitation principally comprises *rainfall* and *drizzle*, the latter having smaller drop sizes and lighter intensity. In contrast to these forms, which may play an immediate part in the movement of water in the hydrological cycle, solid precipitation, comprising mainly *snow*, may remain upon the ground surface for some considerable time until the temperature rises sufficiently for it to melt. For this reason solid precipitation, particularly snow, is discussed separately in Section 2.7. Snow may be the predominant form of precipitation in colder latitudes and at higher altitudes. Hail is a rather special case since, although falling to the ground as a solid, it normally does so in temperature conditions that favour rapid melting, and so it tends to act hydrologically like a heavy shower of rain.

Other types of precipitation may be important locally. For example, in semi-arid areas the main source of moisture may be dew, formed by cooling of the air and condensation of water vapour by cold ground surfaces at night. In practice, although it is not strictly correct, the terms precipitation and rainfall are often applied indiscriminately and interchangeably to any or all of these forms.

2.2 Precipitation mechanisms

Precipitation takes place when a body of moist air is cooled sufficiently for it to become saturated and, if condensation nuclei are present, for water droplets or ice crystals to form. These processes are discussed in detail elsewhere (e.g. Mason, 1971), and the following is a brief summary. Air may be cooled in a number of ways, for example by the meeting of air masses of different temperatures or by coming into contact with a cold object such as the ground. The most important cooling mechanism, however, is due to the uplift of air. As air is forced to rise its pressure decreases and it expands and cools. As it cools its ability to hold water is reduced until at a certain point (the dew point temperature) the air becomes saturated and condensation occurs. The formation of *clouds* does not in itself result in precipitation as there must be a mechanism to provide a source of inflow of moisture. The fact that some types of clouds are associated with, for example, dry weather or only light rain while other cloud types are indications of heavy intense rainfall has long been used for weather forecasting (Flohn, 1969; Mason, 1975).

Only when water droplets or ice crystals grow to a certain size are they able to fall through the rising air currents as precipitation, and, depending upon the temperature, they may reach the ground as rain, hail or snow. Since uplift is the major cause of cooling and precipitation, the following three-fold division may be used in a very general way to distinguish precipitation according to the meteorological conditions causing the vertical air motion.

2.2.1 *Cyclonic precipitation*

This comprises both frontal and non-frontal types. The non-frontal type results from the convergence and uplift of air within a low-pressure area. In extratropical areas this usually produces moderate rainfall of fairly long duration while in tropical areas, due to the greater heating, the resulting precipitation may be much more intense, and of short duration (Eagleson, 1970). In the case of frontal precipitation, warm moist air is forced to rise up and over a wedge of denser cold air. This may be either at a warm or cold front and, in broad terms, the two may be distinguished in terms of the resulting precipitation. Cold fronts normally have steep frontal surface slopes that give rise to rapid lifting and heavy rain of short duration. In contrast, warm frontal surfaces are usually much less steep, giving more gradual lifting and cooling, leading to less intense rainfall of longer duration.

2.2.2 *Convectional precipitation*

This results when heating of the ground surface causes convectional upcurrents of thermally unstable air. Thus it is dependent on heating and moistening of the air from below and is most common in tropical regions, although it occurs widely in other areas, too, especially in the summer. The precipitation is characteristically intense, but of limited duration and areal extent. Spatially, the occurrence of these convection *cells* may be organized or disorganized. The former occurs, for example, where a cold, moist airstream moves over a warmer surface and the cells tend to travel with the wind, producing a streaky distribution of precipitation parallel to the wind direction (Bergeron, 1960). In tropical cyclones, cumulonimbus cloud cells may form spiralling bands about the vortex, giving rise to heavy, prolonged rain affecting large areas (Barry and Chorley, 1987). Disorganized convection cells are typified by summer heating of the ground surface, resulting in scattered heavy showers of short duration but high intensity.

2.2.3 *Orographic precipitation*

This results from the mechanical lifting of moist air over barriers such as mountain ranges or islands in oceans. It may not be as efficient in producing precipitation as convective or cyclonic storms, but the lifting may induce convectional instability which may be more important than the orographic uplift itself (Miller, 1977). Frequently more rain falls on the windward than on the leeward slopes. This so-called 'rainshadow' effect can be seen, for example, in the Western Cordillera of North America, with high rainfall amounts along the western coast, and in the northern and western highland areas of the British Isles. The intensity of orographic precipitation tends to increase with the depth of the uplifted layer of moist air. The vertical enhancement of precipitation has been studied using weather radar, and maps have been produced for Britain showing the amounts of enhancement of precipitation for given wind speeds and directions (Browning and Hill, 1981).

It will be clear from this discussion that the three-fold division of precipitation into cyclonic, convectional and orographic types is rather simplistic as they do not necessarily occur independently of each other. It is therefore very difficult to relate rainfall characteristics to the causative mechanism except in the generalized manner presented here.

2.3 General spatial patterns of precipitation

The large variations in the amount of precipitation, both in time and in space, are of considerable interest to the hydrologist. There is, for example, a great contrast between some of the driest deserts of the world which receive rainfall perhaps only once in 20 years and places such as Bahia Felix in Chile which on average has rain on 325 days per year (Todd, 1970). The average annual precipitation over the land areas of the globe has been estimated to be about 720 mm, and may be contrasted with places such as Mount Waialeale in the Hawaiian Islands which receives about 12 000 mm annually and Cherrapunji in Assam, India, where over 26 400 mm were recorded in one year and more than 3800 mm fell in one five-day period. The magnitude of these falls emphasizes the crucial role of large-scale horizontal and vertical movements within the atmosphere in transferring large masses of moist air from areas of high evaporation to areas of high precipitation.

While the behaviour and pattern of individual storms may be complex and variable, broad areal patterns of precipitation exist when averaged over long periods. This is the essential difference between 'weather' as the day-to-day state of the atmosphere and 'climate' which is the normal or average course of the weather.

2.3.1 *Global pattern of precipitation*

The average water vapour content of the atmosphere, expressed as a precipitation equivalent, is about 25 mm. Values decline systematically from the equator to the polar regions, and also vary seasonally, increasing in summer with greater heating and evaporation. The overall distribution of atmospheric moisture over the globe is well related to the areal pattern of evaporation and transport by winds (Lamb, 1972). On the other hand, the pattern of world precipitation is not well related, being instead closely dependent on the processes causing precipitation, generally a vertical motion in the atmosphere which produces condensation. A review of the historical development of world precipitation maps and global water balance estimates was given by Jaeger (1983).

In broad terms, the areas of maximum rainfall are in equatorial areas associated with converging trade wind systems and monsoon rainfall regimes. The lowest rainfalls are in (a) high latitude polar areas, due to descending air masses and the low water content of the extremely cold air, and (b) subtropical areas, which include many of the world's largest deserts, where high-pressure cells give rise to descending, drying air.

This simple general pattern is then modified by a number of other factors in addition to, as yet, unexplained or random variations in the global atmospheric circulation. A detailed account of the water fluxes in the atmosphere was given by Peixoto and Oort (1983) who noted that most of the water vapour for precipitation over land is supplied by evaporation from the oceans (especially subtropical oceans) and that evaporation from continents provides only a small proportion of precipitation over land (see also Table 1.1). In coastal areas, precipitation is generally greater over land than over the nearby sea due to the greater mechanical and thermal overturning of the air. As evaporation from the oceans is the main source of atmospheric moisture, precipitation will tend to decrease with distance from the sea, resulting in areas of extremely low rainfall near the centres of most of the major landmasses. Mountain ranges tend to accentuate precipitation amounts, particularly in areas where the prevailing air movement is onshore.

2.3.2 *Regional precipitation*

When smaller areas such as the United States or the British Isles are considered in detail, the orographic influence is far more apparent, dominating the annual and, to a lesser extent, the seasonal distributions. In the British Isles (Fig. 2.1) the predominant direction of rain-bearing winds is from the west. Where mountains force them to rise over parts of Wales, the Lake District and Scotland, rain may exceed 2000 mm and may fall on 200–250 days per year. In North America (Fig. 2.2), the influence of the Western Cordillera and of the Appalachian Mountains on the annual distribution of precipitation can be distinguished, but because of the greater size of the country other major influences such as latitude are also apparent. The Rocky Mountains are such a barrier to moisture from the Pacific

Figure 2.1 *Average annual precipitation (mm) over the British Isles, 1931–60 (from an original diagram by B. W. Atkinson and P. A. Smithson in Chandler and Gregory, 1976, based on data from the UK Meteorological Office and the Eire Meteorological Service.*

Ocean that the main source of precipitation over much of North America is supplied by evaporation from the warm waters of the Gulf of Mexico.

2.4 Precipitation measurement

At this point it is appropriate to review the different means of measuring and recording precipitation, and to discuss briefly some of their problems and limitations before dealing

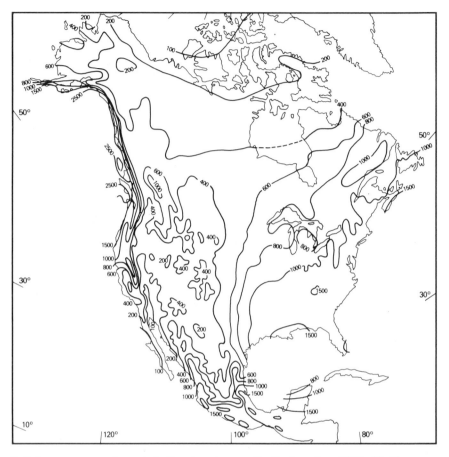

Figure 2.2 *Average annual precipitation (mm) over North America, 1931–60 (from an original diagram by R. G. Barry and R. J. Chorley, 1987).*

in detail with studies of its variation in time and space and methods of analysing aspects of its magnitude and frequency.

Of the different forms of precipitation (rain, hail, snow, etc.) only rainfall is extensively measured with any degree of certainty (Rodda, 1969). For this reason the following sections deal primarily with rainfall. Methods specifically for measuring snow are dealt with in a separate section. The measurement of rainfall comprises two aspects: first, the point measurement of rainfall at a gauge and, second, the use of the catches at a number of gauges to estimate areal rainfall.

2.4.1 *Point measurement*

A raingauge is basically a can to catch vertically falling raindrops or snowflakes (Miller, 1977). The aim of a raingauge is to collect rain over a known area bounded by the raingauge rim and to measure the amount of rain so collected. This could be by manually emptying and recording the amount of water accumulated in a storage gauge, usually at daily or longer intervals, or else by using a recording raingauge which automatically registers the *intensity*, or rate of accumulation of rainfall (Meteorological Office, 1982). The main types of recording raingauges are either 'continuous' recorders which register changing water levels on a chart (e.g. the Dines siphoning raingauge) or tipping bucket raingauges which record increments of rainfall, the increment depending on the size of the bucket, but typically holding from 0.1 to 0.5 mm of rain. Short-period rainfall data are necessary to understand rainfall interception losses, limits to soil infiltration rates and catchment runoff hydrographs. For runoff studies from urban catchments rainfall depths over only a few minutes' duration are necessary (Folland and Colgate, 1978). Some of the problems and procedures in processing and checking recording raingauge data were discussed by Kelway (1975).

The long history of rainfall measurement goes back over 2000 years in India, and the first raingauges were reported in Europe from about the seventeenth century (Biswas, 1969, 1970). In the 1670s Pierre Perrault used a raingauge to prove for the first time that the magnitude of rainfall was adequate to account for streamflow. Nevertheless, many problems remain in the collection and accuracy of rainfall data. In either type of raingauge (storage or recording) the major problem of accuracy is due to wind turbulence around the gauge,which usually results in underestimates (Sevruk, 1982). It has been known for some time (Kurtyka, 1953; Weiss and Wilson, 1958; Rodda, 1973) that the higher the rim of the gauge is above the ground, then the greater is this underestimate. This effect is much more pronounced in the case of snowfall. Controlled experiments in wind tunnels show how the raingauge acts as an obstacle to the wind flow, leading to turbulence and an increase in wind speed above the gauge orifice. The result is that precipitation particles that would have entered the gauge tend to be deflected and carried further downwind (Robinson and Rodda, 1969; Green and Helliwell, 1972). Errors due to turbulence increase with wind speed and with reducing drop size. Thus it may be expected that errors would be greater in Britain than in some tropical areas due to smaller drop sizes and higher wind speeds. Comparisons between countries are complicated by the fact that raingauge designs vary considerably. Thus, for example, while the standard gauge in Britain has a 127 mm diameter orifice at 305 mm (1 ft) above the ground, in the United States gauge diameters of 203 and 305 mm are normally used. In some European countries prone to heavy snowfall, rim heights of 1500 or even 2000 mm are adopted. Clearly, then, what is recorded as, say, 20 mm of precipitation on one side of a national frontier *could* be registered as something different on the other side, so that the existing global and continental precipitation maps are not as meaningful as they would be with uniform measurement techniques (Rodda, 1969). In order to try to provide a basis for comparison the WMO interim reference precipitation gauge was introduced, but it, too, suffers errors from the effect of wind. Furthermore, many countries have been reluctant to change to this type of gauge because of the danger of introducing inhomogeneity into rainfall records, leading to problems in the use of such records for studies of climatic variations and in evaluating long-term average values.

Before the effect of wind on catches was recognized many early raingauges were installed on house roofs (Craddock, 1976). This led to more serious undercatches in exposed upland areas than in adjacent valleys and resulted in the impression that rainfall decreased with altitude. For example, in 1838 this belief led to the abandonment of plans

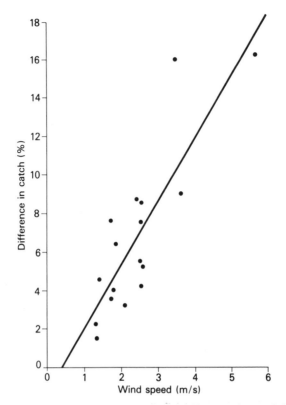

Figure 2.3 *Relation between average wind speed and difference in catch between standard and ground-level raingauges at 17 long-term sites in Britain. (Reprinted with permission from* Atmospheric Environment, *vol. 20, J. C. Rodda and S. W. Smith, The significance of the systematic error in rainfall measurement for assessing atmospheric deposition,* © *1986 Pergamon Journals Ltd.)*

to build an upland reservoir to supply water to the town of Oldham in northern England (Binnie, 1981). One early attempt to solve the problem of exposure at windy sites was to build a circular turf wall around the raingauge with its crest at the same height as the rim of the gauge (Hudleston, 1934; Meteorological Office, 1982). Attempts to use wind shields around the raingauge to reduce turbulence have only been partially successful (Weiss and Wilson, 1958). The most satisfactory solution is to install the raingauge with its rim at ground level surrounded by a grid to prevent splash into the gauge (Koschmieder, 1934; Rodda, 1968). Rodda and Smith (1986) presented the results of a long-term comparison of a network of standard (305 mm height) and ground-level gauges in the United Kingdom. Using average rainfall and wind speed values for sites with at least five years of data, they demonstrated a clear average relation between catch deficiency and wind (Fig. 2.3). While annual total catches differ by about 3–10 per cent, this may increase to 30–40 per cent or more in individual storms. The loss of catch also varies between seasons, being largest in winter when drop sizes are smaller and wind velocities are higher (Rodda, 1968; Green and Helliwell, 1972). Consideration of such wind effects must be made when, for example, interpreting historical rainfall records. For example, the first continuous record of daily rainfall in Britain was made by Richard Townley in the seventeenth century, using a gauge on the roof of his house (Biswas, 1970; Craddock, 1976). When measuring the rainfall onto forest areas it has been found that catches very similar to ground-level gauge values can be achieved using special raingauges with their rims set at the level of the top of the tree canopy (Newson and Clarke, 1976).

Measuring extreme rainfall events
Of special importance to the hydrologist is the measurement of extreme rainfall events which may be responsible for rare floods of great magnitude. Under such severe conditions, the performance of the instruments, rather than wind effects, may be the major problem (Sevruk and Geiger, 1981), e.g. exceeding the capacity of storage gauges, causing recording gauge mechanisms to jam, or simply loss of accuracy due to the finite time taken for the tipping buckets to tip or the float gauges to siphon empty.

In addition, since extreme storm events may be quite localized and with steep rainfall gradients, they may not be recorded adequately by a raingauge network and it may be necessary to make use of so-called 'bucket catches'. This involves measuring the depth of water caught in chance-placed open containers, e.g. buckets, cans, water tanks and barrels, if their contents before the storm are known or can be assumed zero. For example, following severe flooding of Louth in Lincolnshire by the River Ludd on 29 May 1920, the daily rainfall from the nearest gauge was 116.6 mm, but measurements from rainfall accumulated in milk cans indicated that up to 153.4 mm (nearly a third higher) had fallen (Crosthwaite, quoted in Newson, 1975a). Extreme care must be taken in interpreting such point rainfall measurements since their accuracy is likely to be considerably less than that of conventional raingauges. Not only will the shape of these receptacles influence the catch, but their chance siting in relation to larger-scale obstructions, including local topography and buildings, may be such that there are considerably greater errors due to wind effects than for a standard raingauge.

Nevertheless, information on such catches may be of great use when trying to reconstruct an extreme rainfall event for which there was insufficient information from raingauge stations due to the limited coverage of raingauge networks. In some cases bucket catches may provide the *only* source of information on very localized bursts of rainfall not recorded by official networks.

2.4.2 *Areal rainfall*

Hydrologists often need to know the volume of rainfall over a catchment area and require an adequate number of measurements in order to assess the areal variation and estimate the total rainfall. This may be achieved with a network of raingauges alone, or by using additional information from remote sensing by radar or satellites.

Design of raingauge networks
A network of raingauges represents a finite number of point samples of the two-dimensional pattern of rainfall depths. In estimating the areal pattern of rainfall from a given gauge network, errors will occur due to the random nature of storms and their passage between gauges (Wiesner, 1970). The accuracy will depend on the spatial

variability of precipitation; thus more gauges would be required in steeply sloping terrain (Hutchinson, 1970) and in areas prone to localized thunderstorms rather than frontal rainfall (Browning *et al.*, 1977). The density of gauges required also depends upon the time scale of interest; shorter period rainfall intensities (e.g. hourly) are much more spatially variable than daily or annual totals.

In general, estimates of areal precipitation will increase in accuracy as the density of the gauging network increases, but a dense network is difficult and expensive to maintain. A number of general guidelines for the density required have been produced. The US Weather Bureau (1947) carried out a statistical analysis of raingauge data for an area of relatively flat terrain and found that for a given level of accuracy, a less dense network was needed as larger areas are considered. Of course, other factors than the size of the area of interest are also likely to be important, including type of topography and climate characteristics. The World Meteorological Organization (WMO, 1981) gave the following broad guidelines for the minimum gauge density of precipitation networks in various geographical regions: 25 km² per gauge for small mountainous islands with irregular precipitation; 100–250 km² per gauge for mountainous areas (and 600–900 km² per gauge in flat areas) in temperate, Mediterranean and tropical climates and 1500–10 000 km² per gauge for arid and polar climates. In Great Britain there are about 4500 gauges (May, 1986), giving an average density of about one gauge per 60 km², but for the United States this figure rises to about 600 km² per gauge (Linsley *et al.*, 1975). In other parts of the world, areas of over 2500 km² per gauge are quite common (Todd, 1970).

The accuracy of the network depends not only on its overall density but also on the distribution of the gauges. An improvement in the accuracy of areal rainfall estimation may be achieved by making allowance for the spatial pattern of rainfall over the area. This can be done in a number of ways. Raingauges may be sited *a priori* within a classification of 'domains' representing classes with different ranges of topographic characteristics such as altitude, slope and aspect (Clarke *et al.*, 1973; Catterall, 1972). If the main purpose of precipitation measurement is for runoff studies, then one approach to network design would be to locate gauges in those areas that contribute most to runoff (McKay, 1965; Moore, 1987).

If an area to be instrumented has no existing gauges to indicate the spatial distribution of precipitation, then it would be necessary to transpose information on rainfall variations in time and space from a similar area in order to derive a preliminary network. This is often the case for developing countries. In developed countries, in contrast, networks usually exist already but have probably evolved in an arbitrary manner. The available data could be used to quantify the statistical structure of rainfall patterns and enable a more effective network design to be identified (Huff, 1970). A large number of studies have used data from existing networks to investigate the effect of different network densities and configurations. Stol (1972), for example, used correlation analysis to evaluate the relative efficiency of different gauge networks in the Netherlands. Caffey (1965) and Hendrick and Comer (1970) showed that storm size and direction were important in determining spatial variations in precipitation, and concluded that the most effective gauging network would have a strong directional component reflecting the predominant direction of moisture flux across the area.

A comprehensive account of the steps involved in assessing and redesigning an existing network was presented for an area of nearly 10 000 km² in southern England by O'Connell

Table 2.1 Survey of the rainfall data required for various purposes by water authorities and government agencies in the UK. (Based on Nicholass *et al.*, 1981, by permission of the Controller of Her Majesty's Stationery Office.)

Purpose	Daily	Recording	Telemetry
Design of flood alleviation scheme	×	×	
Reservoir yield studies	×		
River regulation	×		×
	(design)		(operation)
Aquifer recharge calculation	×		
Urban drainage design	×	×	
Flood forecasting			×
Soil moisture deficit maps	×		×
Climatic change	×		
Weather radar research	×	×	×
Irrigation requirements	×		
Waste tip leaching	×	×	
Field drainage design	×		
General research purposes	×	×	×

et al. (1978) and Nicholass *et al.* (1981). As a first step they produced a useful survey of *users* of precipitation data (Table 2.1). Users were asked to specify the required accuracy for their purposes, but as this proved very difficult to define in an objective and uniform manner, it was decided that the accuracy of estimation which could be obtained with the existing network would serve as the baseline. The statistical method for evaluating the network comprised several steps and used spatial correlation analysis to determine the decay of correlation with distance between gauges and to estimate interpolation errors to ungauged points or areas. An optimal point interpolation procedure was developed to identify a basic network of uniformly spaced gauges. This 'objectively' defined network was then refined by an iterative approach, adding or replacing gauges by those in the existing network which fulfilled certain criteria, i.e. (a) gauges that would be retained anyway due to some special user requirement, (b) gauges with long historic records and (c) very reliably operated gauges. For assessment purposes maps of interpolation errors were constructed in order to identify areas where the network would be inadequate for user requirements. In using any method based on the correlation between gauges it is of paramount importance to ensure the homogeneity of record of each gauge. Unrecorded changes in siting or exposure may weaken correlations between gauges, resulting in networks that are denser than necessary.

Weather radar and satellites
However great the density of existing networks of raingauges, they can only give an approximation to the actual spatial pattern of precipitation. Since the 1950s extensive development work to investigate the potential of radar for areal rainfall measurement and monitoring has been carried out in a number of countries, in particular the USA, USSR, Japan, Switzerland and the UK. In the USA a great deal of work has centred on the use of radar to detect and give warning of severe storms, including extratropical cyclones and hurricanes, and a national network of over 100 radars is planned (Milner, 1986). In

Europe, where such hazardous systems are much rarer, interest has been concentrated on mapping quantitatively the distribution of rainfall. This aspect is of more general interest to the hydrologist, and is discussed below.

There are several ways of measuring rainfall using radar, including doppler radar and attenuation, but the most widely adopted approach is based on radar echo, or reflectivity. The relation between radar reflectivity and rainfall rate depends on a number of factors including the concentration of drops, their size distribution and the pattern of vertical wind velocity. Marshall and Palmer (1948) analysed extensive experimental results on drop size distributions to yield an empirical relation between rainfall rate (R) and radar reflectivity (Z) of the form: $Z = AR^B$. Values of $A = 200$ and $B = 1.6$ are most commonly used, having been developed for temperate zone conditions. Other workers have, however, found great variations in these parameters with ranges in A from 70 to 500 and in B from 1.0 to 2.0 (see summaries in Borovikov and Kostarev, 1970; Battan, 1973). It is this unpredictable variation in the relation between radar reflectivity and rainfall that has for many years prevented the use of radar for quantitative rainfall measurement, a situation which led Miller (1977) to suggest that 'radar can provide pictures but not numbers'.

According to Collier (1987) the main reasons in practice for the variation between radar reflectivity and rainfall are:

(a) Variations in raindrop size, or the presence of hail, snow or melting snow. Melting snow has a much higher reflectivity than rainfall, giving rise to an enhanced 'bright band' in the vertical reflectivity profile.

(b) Raindrop growth or evaporation between the height at which the radar beam intercepts the precipitation and the ground. For example, on the windward sides of hills, orographic rainfall can result in raindrop growth at levels below the beam height.

Additional less important reasons for variation in the $Z:R$ relation include:

(c) Performance of the radar system.

(d) Attenuation of the signals due to heavy rain along the beam.

(e) Ground echoes due to hills and tall buildings.

Thus the parameters of the $Z:R$ relation can vary widely, not only between rains in different climate zones but even between storms at the same location. A number of studies have found an improvement when the parameters are varied according to the storm rainfall type, e.g. frontal storms, thunderstorms (Battan, 1973).

The major advance enabling radar to be used for quantitative estimates has undoubtedly come from studies that have used measured rainfall rates, not for assessing the accuracy of the radar estimates but as part of the calibration procedure for making the estimates; that is the radar precipitation field is calibrated by point raingauge observations while retaining the areal pattern of precipitation variation observed from the radar. Although this concept has been proposed for many years (Hitschfeld and Bordan, 1954), it has required a great deal of work to overcome many practical problems and implement an operational scheme. Arguing that raingauges should be used for real-time adjustment of the $Z:R$ relation, Wilson (1970) demonstrated that even one raingauge significantly improved estimates of 28 storm totals (r.m.s. error reduced by 39 per cent) for a 3500 km² area in Oklahoma when compared to the 'true' rainfall measured by a dense network of 168 raingauges.

Subsequently, Brandes (1975) showed that radar calibration with *several* gauges would further reduce errors by making allowance for the spatial variation of the calibration factors. Rainfall estimates were compared with totals from a dense network of uniformly spaced gauges and the average error in the radar estimate of storm depth for nine storms was reduced from 52 per cent (using radar alone) to 18 per cent using a single calibration factor based on the average factor for a subset of gauges with a density of one per 900 km² (similar to a national climate network). When the factors for the individual calibration gauges were used to fit a radar calibration field or surface by plane fitting, the error was further reduced to 13 per cent. It is worth noting that both calibrated radar rainfall estimates were better than estimates based on the calibration raingauges alone: fitting a surface to the climate network gauges gave an average estimate error of 21 per cent.

In Britain, Harrold *et al.* (1974) used radar over a 1000 km² hilly area of the River Dee in north Wales. Comparing radar to the 'true' pattern of rainfall from 60 recording raingauges set at ground level, they used a single calibration raingauge with the radar for *hourly* calibration. The calibrated radar echo pattern showed average errors of about 20 per cent for two-hour integrations, which reduced to 13 per cent over six-hour periods for subcatchment rainfall (typically 50 km²). The accuracy of the single calibrated radar was compared with that of a conventional raingauge network and found to be dependent on rainfall conditions. For days with widespread rain, a network of 1 gauge per 100 km² was equivalent in accuracy, while on showery days 1 gauge per 50 km² would be required. Ninomiya and Akiyama (1978) used calibrated radar to study intense falls of rain (up to 100–150 mm/h) which occur over south-west Japan in the pre-summer rain season for short durations over relatively small areas.

In a useful review of a number of studies in which radar data had been calibrated by raingauge data, Wilson and Brandes (1979) reported typical errors of 43–55 per cent for uncalibrated radar data. This reduced to 18–35 per cent using a single calibration raingauge and to 13–27 per cent for spatially variable calibration. These results were for a number of studies, which differed in a number of factors including the summation periods, area and density of calibration gauges.

Current work on radar calibration aims to improve the estimates by a number of means, including making use of more calibration gauges as well as through more sophisticated fitting of calibration fields. The techniques of radar measurement have developed enormously over the last 30 years with hardware advances in radar technology and data processing capacity and with software developments in real-time calibration. An up-to-date review of the UK operational network of weather radar and of the proposed developments for the European weather radar network were given by Collinge and Kirby (1987). The Hameldon Hill radar station in north-west England commenced operation as an unmanned station in 1980 (Hill and Robertson, 1987), and has been used as the basis for the following brief description of a modern weather radar system. The radar aerial rotates at approximately one revolution per minute and in turn scans each of four elevations (0.5°, 1.5°, 2.5° and 4° above the horizontal). Three complete scans are used to produce 15-minute rainfall estimates. While ideally the beam closest to the ground is to be preferred for rainfall estimation, a higher beam may be substituted on bearings where hills and tall structures can interfere with the beam. Five telemetering recording raingauges are used for automatic calibration using an onsite computer, and these data are transmitted from the site to a number of manned centres for use by water authorities, civil aviation authorities and the UK Meteorological Office. The station is designed to run automatically, with only monthly visits for preventive maintenance. Radar output is produced for a 2×2 km grid within 75 km of the installation and on a 5×5 km grid for distances 75–210 km from the radar. Beyond 210 km the radar estimates are not considered to be sufficiently accurate, due to the fact that the return signal becomes too weak and also because the beam becomes too high and diffuse. Based on the experience gained from the Hameldon Hill station, other radars have been installed, and Fig. 2.4 shows the calibrated pattern of hourly rainfall during a storm over southern England, obtained from the Chenies radar station, London.

In areas where raingauges or radar coverage are inadequate, satellite monitoring techniques can be used to study the movements of weather systems and to provide estimates of probable rainfall distributions. Satellites cannot measure rainfall directly and the techniques for estimating rainfall depths are not as accurate as other methods, but may be the only means available in some areas. Barrett and Martin (1981) estimated that as much as 80–90 per cent of the globe has insufficient raingauge coverage to meet the WMO guidelines described earlier. These areas include most of the desert and semi-desert regions, most major mountainous regions, extensive humid regions in the tropics, in addition to the world's oceans. Satellites have some advantages over other types of data. They can provide spatially continuous information (i.e. not just point measurements) and, depending on the orbit, they can provide complete global coverage over a period of time.

Although radar can be used from satellites it has not been used on meteorological satellites because the background reflectivity of the earth's surface is many times greater than that of precipitation in the atmosphere. The reflectivity of land areas is not well understood, but varies with such factors as topography, ground wetness and vegetation, while that of the oceans varies with waves on the surface. In practice, therefore, all operational programmes and most research using satellites for rainfall monitoring have been based on visible and infrared wavelength radiation (Barrett and Martin, 1981). Visible radiation is most strongly related to the albedo of highly reflective surfaces such as ice, snow and desert sands, as well as clouds, which give relatively bright images. High brightness implies a greater cloud thickness and probability of rainfall. Infrared radiation is largely dependent on temperature. Since temperature is often assumed to vary as a function of altitude, this may be interpreted as indicating the cloud top height. Low temperatures imply high cloud tops and large thickness of clouds, with a greater probability of rainfall. In practice, neither bright clouds nor cold clouds necessarily produce rainfall, and the best approach is probably to use both types of information together (Browning, 1987). Rain is more likely in clouds that are both cold and bright.

A number of approaches have been developed to monitor rainfall from satellites, of which the most important techniques include cloud indexing methods, life history methods and bispectral methods. For detailed descriptions and further references the reader should consult Barrett and Martin (1981). Cloud indexing methods have been applied over the widest range of climate conditions, and are based on identifying cloud types and their areal coverage and applying a probability of rainfall to each given cloud type. Of all the available techniques, this method is the least dependent upon sophisticated hardware and software and is perhaps the most promising approach to date. Estimates of rain depths can be made for periods of a day or longer, on the basis of empirical relationships between satellite and conventional weather data (Barrett, 1970,

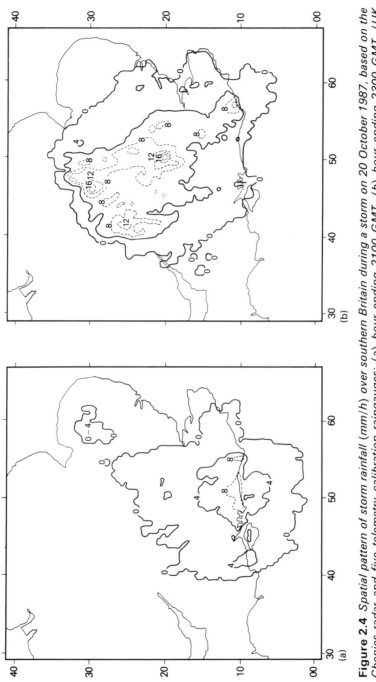

Figure 2.4 *Spatial pattern of storm rainfall (mm/h) over southern Britain during a storm on 20 October 1987, based on the Chenies radar and five telemetry calibration raingauges: (a) hour ending 2100 GMT, (b) hour ending 2300 GMT. (UK Meteorological Office. Copyright reserved. By permission of the Controller of Her Majesty's Stationery Office.)*

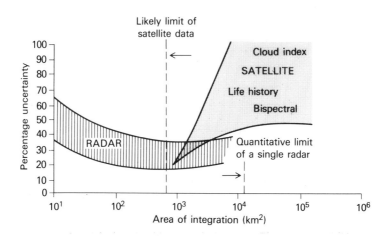

Figure 2.5 *Comparative accuracy of hourly rainfall measurements from a ground-based calibrated radar and from satellite techniques (redrawn after Collier, 1984).*

1973). However, there is still no general agreement on the best method of satellite monitoring of rainfall since the requirements for rainfall data differ so widely between users. The WMO (1967) specified certain objectives within the World Weather Watch for satellite measurement of weather parameters including radiation, humidity and precipitation intensity, and although most of these objectives have been met, that of estimating 12-hourly rainfall to a level of accuracy of ±20 per cent has not been achieved (Barrett and Martin, 1981). As Browning (1987) noted, the relation between cloud properties and surface rainfall is rather variable, and although it provides a useful indication of precipitation, other methods should be used whenever possible. In comparing satellite and weather radar methods for remote sensing of rainfall, Collier (1984) concluded that radar and satellite techniques are complementary. Radar is more appropriate for areas smaller than about 10 000 km² while, due to the limited range of surface radar systems, satellites are better for larger areas (Fig. 2.5). In both cases, however, surface rainfall measurements are necessary for calibration and checking purposes.

2.5 Temporal variations in precipitation records

Point precipitation records exhibit great variation from hour to hour, week to week and even from year to year. This variation is far larger than that of any other component of the hydrological cycle. Evaporation, for example, is strongly related to the radiational output from the sun and the wetness of the ground, while streamflow represents a much moderated pattern of the precipitation inputs. In principle, the pattern of precipitation is deterministic, being related to the synoptic weather conditions and the properties of the air masses. Following the pioneering work of Richardson (1922), considerable advances have been made using numerical prediction methods to produce digital models of weather systems for forecasting purposes (Atkins, 1985). In practice, for hydrological purposes, however, the analysis of rainfall data is often based on the statistical properties of observed rainfall time series.

Variations in precipitation records may incorporate three time series components: stochastic, periodic and secular. Stochastic variations result from the probabilistic or random nature of precipitation occurrence, and may be so great that they effectively dominate the time series. Periodic or cyclic variations are related, for example, to astronomical cycles such as the diurnal and annual cycles. Finally, secular or long-term variations, which are often referred to as 'climatic change', may incorporate both cyclic or trend characteristics.

2.5.1 *Stochastic variations*

The reason for the great variability in rainfall totals can be seen if one looks at the frequency distribution of rainfall, since it is apparent that a small proportion of storms or raindays in the course of a year may provide a disproportionate amount of the total rainfall (Fig. 2.6). The presence or absence of only a small number of storms may therefore have a considerable effect upon the total precipitation. The variability of annual rainfalls is greater for areas with low annual precipitation, where rain may fall only occasionally, than for, say, equatorial regions where rain may fall on nearly every day. Thus, estimates of water resources in arid and semi-arid areas are particularly sensitive to short lengths of precipitation records (French, 1988).

In addition to rainfall amounts, the time intervals between storms are of great interest to the hydrologist, especially in the drier parts of the world. The importance of the time interval depends upon the storage capacity and the depletion characteristics of the particular system of interest, such as a column of soil or a water supply reservoir. Sandy soils, for example, have a low water holding capacity and water will drain out easily. Due to differences in the vapour inflows and the mode of uplift of the air between passing

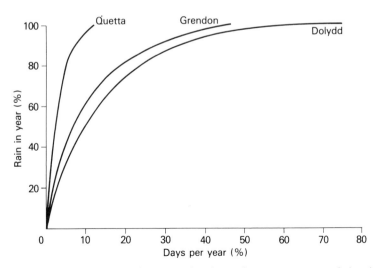

Figure 2.6 *Percentage of annual rainfall occurring in a given percentage of the time for Dolydd (mid-Wales, 1780 mm/yr) and Grendon (southern England, 630 mm/yr) (based on data for 1969–81 from the Institute of Hydrology); and Quetta (northern Pakistan, 205 mm/yr) (data from Packman, 1987).*

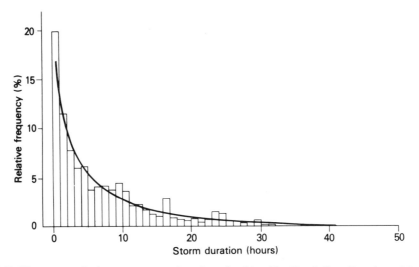

Figure 2.7 *Histogram of rain storm event durations for Ahoskie, North Carolina, from May 1964 to September 1973, with the fitted Weibull probability density function (from an original diagram by W. C. Mills, 1982).*

weather systems there is a tendency for rainy days to cluster in groups. This tendency for 'persistence' or 'serial correlation' has often been described using Markov chain analysis (Essenwanger, 1986) (see also Section 7.7).

The timing and magnitude of individual storms is largely stochastic in nature, and a number of studies have represented parameters such as the time interval between storms, the storm durations and the precipitation depths by statistical frequency distributions (e.g. Grace and Eagleson, 1966; Mills, 1982) (Fig. 2.7). The pattern of rainfall during a storm, however, will be largely deterministic since it depends upon the weather system. Convectional rainfall is usually of higher intensity and shorter duration than rain from frontal systems. In general, convective and frontal type storms tend to have their peak rates near the beginning, while cyclonic events reach their maximum intensity nearer the middle of the storm period (Eagleson, 1970). The variety of storm rainfall profile shapes is almost infinite, depending not only on the rainfall type, and the state of development or decay of the rainfall system as it passes over the rainfall measurement point, but also on the speed of movement of the system (Keers and Westcott, 1977). If enough storms are examined for a particular site then their shapes can be summarized statistically. Extensive research carried out by the UK Meteorological Office on storm profiles found great variety in shapes, and for hydrological design purposes it was decided to define storm shape in terms of the profile 'peakedness' which exceeds that of a given percentage of observed storms (Keers and Westcott, 1977).

2.5.2 *Periodic variations*

These are regular cyclic variations with rainfall minima and maxima recurring after approximately equal time intervals. The best known are the diurnal and annual cycles which give most rain over land at the warmest times of the day and during the warmest seasons, when the water vapour content of the air is high and thermal convection is strongest, and least rain around dawn and in late winter.

Diurnal variations occur in areas where a large proportion of the rainfall derives from purely convective storms generated by local surface heating. This diurnal pattern is most often found in warm tropical climates although daily patterns have been noted in other areas. It is typified by a maximum in the afternoon, with the rainfall often accompanied by heavy thunderstorms. In Quetta in northern Pakistan, for example, 80 per cent of the rain falls between 1400 and 2000 hours (Rudloff, 1981). The importance of such convectional rainfall patterns will vary through the year with changes in the degree of radiant heating and convection, and variations in evaporation and hence the vapour content of the air (Fig. 2.8). The diurnal rainfall pattern may be modified by the interaction of land and sea breezes near the coast in places such as Malaysia and Florida (Ramage, 1964; Burpee and Lahiff, 1984) and the effect of topography. In mountainous areas the terrain enhances thunderstorm activity and can lead to a repetitive daily convective cycle. In Colorado, this phenomenon has been studied using both weather radar data, to monitor rainfall variations (Karr and Wooten, 1976), and satellite imagery using visible and infrared wavelength data (Klitch et al., 1985), to identify the timing and occurrence of probable deep convection. For most parts of the world, however, there is no systematic pattern of rainfall over the course of the day. A far more widespread cycle in rainfall patterns is that associated with the changing seasons of the year.

The annual cycle is the most obvious weather cycle and results from the regular seasonal shifts in the zones of atmospheric circulation in response to the changes in the heating patterns accompanying the migration of the zenith sun between latitudes 23°N and 23°S (Lamb, 1972). Near to the equator these movements usually result in two maxima, while in tropical areas there is often a distinct summer rainfall maximum. Such

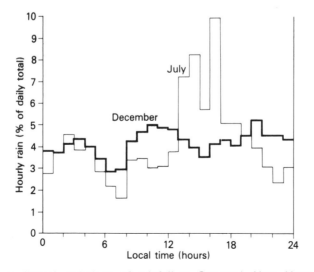

Figure 2.8 *Average diurnal variations of rainfall at Concord, New Hampshire, in July and December, over a 26-year period (from Foster, 1948).*

seasonal changes in rainfall are most pronounced in areas on the fringes of arid zones, such as the Mediterranean region, where cyclonic depressions bring rain in the winter but the summers are dry. The annual rainfall regimes at locations in different climatic zones of the world are adequately described in numerous climatology textbooks, and will not be discussed further here.

One aspect of the weather that may be of particular interest to hydrologists is the fairly regular seasonal weather sequences which have been called 'singularities'. Examination of the long weather records for Europe indicates that certain types of weather tend to occur with greater frequency at the same time each year (Lamb, 1950, 1972). The general circulation of the atmosphere is strongest in winter, when global heat gradients are greatest, and this often results in cyclonic activity and high rainfall. Circulation strength decreases to a minimum in spring (March to May) which for many parts of north-west Europe is the time of lowest rainfall (Lockwood, 1974). In an account of severe flooding in mid-Wales in August 1973, Newson (1975b) noted that both its occurrence and that of the previous most severe flood, 17 years before, coincided with the singularity associated with the period of maximum cyclonic storm activity (Lamb, 1950). Although singularities do not provide definite forecasts, they do provide a reasonable and sound guide, with a physical basis, to periods of unsettled weather and a higher risk of extreme storm rainfalls.

2.5.3 *Secular variations*

There is ample evidence that climate has changed in the past and no reason to suppose that it will not change in the future. It is known, for example, from historical documents that there have been significant variations in the extent of the polar ice caps over the last millennium. A number of studies have indicated warmer periods of 'climatic optimum' for Europe around 1100 and 1900 AD with lower temperatures from 1300 to 1600 AD, and in this century after about 1950 (Lamb, 1977). The question of whether such secular, or long-term, changes in climate are cyclic in pattern remains a controversial topic in hydrology. Numerous attempts have been made to detect regular periodicities in series of annual rainfall totals in the hope that they could be used to predict future variations. In a comprehensive review of the evidence for cycles of different lengths, Lamb concluded that the daily and annual cycles were well documented and there was reason to accept that the well-known astronomical oscillations in the earth's orbit (Milankovitch cycles), of the order of 10 000–100 000 years wavelength, would be likely to affect climate. For periods between these cycles, however, there was little evidence. It is worth quoting his opinion that 'A search for dependable periodicities—inspired by the hope of finding some predictable element in the year to year and decade to decade, or longer variations in climate—has long provoked effort out of all proportion to the results so far attained' (Lamb, 1972, p. 214). It certainly appears doubtful whether anyone has used a cyclic pattern of such wavelengths to predict changes; indeed, great caution would be needed before accepting any cycle as the basis for a forecast without understanding the physical processes involved to explain why it occurs and how it operates. Analysis of thousands of annual precipitation series for the last 100 to 150 years by Yevjevich (1963) failed to identify any significant periodicities or trends. Nevertheless, it has recently been suggested that since rainfall is so spatially variable the search for cycles should be based not on individual gauges but on groups of raingauges in a region. This also has the advantage of enabling the records from a gauge to be checked for homogeneity since unrecorded changes to a gauge could obscure or create apparent changes in catch over time. Tabony (1981) assembled about 100 years of data from some 180 gauges in Western Europe, and the conclusions of this and other regional studies (Vines, 1985) indicated that much further work will be necessary before the debate regarding the existence, or otherwise, of rainfall cycles is finally resolved.

Although attempts to find generally applicable cycles of precipitation have been largely unsuccessful, investigations have demonstrated numerous examples of non-cyclic secular variations of precipitation, and some studies have considered the consequent effects on runoff. Howe *et al.* (1967) demonstrated a marked increase in the frequency of storm rainfall (>about 65 mm/day) in mid Wales between the period 1911–40 and 1940–64, and discussed this in relation to the observed increase in river flooding. Leopold (1951) analysed long rainfall records for New Mexico and found an increase in the proportion of the annual rainfall that occurred in heavy daily amounts (>25 mm) in the latter half of the nineteenth century. This broadly coincided with a period of intensive erosion.

The present growing awareness and acceptance that climate change is a reality is in great contrast with earlier opinions. In the late nineteenth century Binnie (1892) believed that a continuous precipitation record of 35 years would have a mean value within 2 per cent of the 'true' mean. Morris and Marsh (1985) used a 219-year rainfall series for England and Wales to study the effect of using different base periods to estimate average annual precipitation. Differences between 30-year periods varied by up to 6 per cent of the long-term average value and were much greater for shorter periods.

As a consequence of this variation, there is international agreement on the use of a standard period of record for the calculation of climatological normals, including precipitation. This is, at present, the 30-year span 1931–60 inclusive. It is not always practicable to use this period and so, whenever comparisons are made between areas, efforts should be made to ensure that a common time period is used. In any case average rainfall values should specify the period of record used.

Investigations have emphasized that many non-cyclic variations in precipitation are caused directly by a combination of geographical and climatological factors. Observed changes in rainfall amounts have been attributed to shifts in the global wind circulation (Kraus, 1958; Winstanley, 1973), and may be partly due to changes in the paths of rain-bearing winds and partly due to changes in orographic precipitation enhancement as a result of different frequencies of wind directions. Changes in global wind and moisture fluxes related to sea surface temperature anomalies (between northern and southern hemisphere oceans) have been suggested as a possible cause of the failure of rainfall in recent years over the Sahel region of Africa (Folland *et al.*, 1986).

2.6 Analysis of precipitation data

There are a number of aspects concerned with the use and interpretation of precipitation data which are of direct concern to the hydrologist. A basic requirement is to estimate the average rainfall over an area from a number of point measurements, or perhaps to determine the spatial pattern and movement of an individual storm, often from comparatively widely separated gauges. Hydrologists are also interested in the frequency of occurrence of rainfalls of different magnitudes, and so study the statistical properties of rainfall data. Finally, there is the special case of trying to estimate the largest rainfall that is physically possible over a given area, i.e. what is the Probable Maximum Precipitation?

2.6.1 *Catchment mean rainfall*

There is often a need to estimate the rainfall input to a catchment, as in the case of water balance studies. The usual spatial variability of rainfall necessitates combining point raingauge measurements at many places to produce estimates of areal averages. The simplest method is to take the arithmetic mean of all the raingauge totals. This may be satisfactory for areas of flat topography with little variation in rainfall and a uniform distribution of gauges. Such conditions are not generally found in practice and there is often a tendency for gauges to be most widely spaced in mountainous areas where rainfall depths and spatial variability are greatest. The Thiessen polygon method (Thiessen, 1911) has been widely adopted as a better method for calculating areal rainfall since it 'weights' the catches at each gauge by the proportion of the catchment area that is nearest to that gauge. This geometrical method gives some allowance for an uneven distribution of rainfall (and gauges) over the area on the simple basis of the distance between gauges. The method can be programmed for computer application (Diskin, 1969, 1970) but is difficult to apply in situations where there are frequent changes to the raingauge network. A number of other geometrical methods can be used to define areal weights to gauges, although they have been less popular than Thiessen polygons. Several authors have suggested a triangular weighting method in which the locations of raingauges are connected on a map to form a number of triangular subareas. The average depth of rain on each subarea is then calculated assuming a linear variation between the three corner gauge points (Akin, 1971; Goel and Aldabagh, 1979).

The isohyetal method is the most reliable, but subjective, of the standard methods of areal precipitation calculation (Shaw, 1988). It involves drawing isohyets between gauge points by hand, making allowance for factors such as topography, distance from the sea and rainshadow effects. Areal precipitation is then computed by planimetering the areas between the isohyets. The isohyetal method thus uses all the data and knowledge about rainfall patterns in an area, but can involve a considerable amount of time to construct the maps (Wiesner, 1970).

In recent years a number of mathematical techniques have been applied to describe spatial variations in rainfall over areas with fairly dense networks of gauges. Several investigations have used trend surface analysis to fit a mathematical function to the rainfall depths at the gauges. Chidley and Keys (1970) and Mandeville and Rodda (1970) fitted polynomial surfaces to rainfall totals. These were then integrated to give the total rainfall over the area. Shaw and Lynn (1972), however, noted that polynomial surfaces can become unstable and produce wildly exaggerated isohyetal surfaces in situations where rainfall is strongly spatially variable, as, for example, over mountainous areas. They found that a multiquadric surface largely overcame the problem of instability. Kriging is a statistical method that uses the variograph of the rainfall field to optimize the gauge weightings to minimize the estimation error. Bastin *et al.* (1984) used it to estimate areal rainfall as well as to indicate the degree of redundancy in a gauge network and to identify the locations where additional gauges might be most useful. Other methods of estimating areal rainfall represent the area by a regular grid. Rainfall is then estimated for each grid and these are averaged to give the areal mean. A common method for computing depths at grid points is to weight the values at the nearby gauges by the inverse square of their distance to the grid point (Shearman, 1975; Essenwanger, 1986).

A variety of techniques have been developed to estimate areal rainfall from point measurements, and the selection of the most appropriate method for a particular problem will depend upon a number of factors, including the time and expertise available, the density of the gauge network and the known spatial variability of the rainfall field. In general, the accuracy of these methods in estimating areal rainfall will increase with (a) the density of gauges, (b) the length of period and (c) the size of area.

The foregoing discussion is applicable to situations for which radar data are either not available or inappropriate. These include locations that are distant from radar installations or are very hilly (leading to obstruction of the beam), and to cases where information is required over very small areas (current weather radar systems work on grid sizes of several kilometres). May (1986, 1988) described a method developed in the United Kingdom to use the network of weather radars with the national set of daily raingauges to provide the best estimate of daily areal rainfall for the country on a 5 × 5 km grid. First, the radar and gauge estimates are compared to given an adjustment factor for each gauge. Then a least-square surface is fitted to these values to provide a means to estimate the rainfall at ungauged locations. Depending on a number of factors, both static (e.g. distance from radar installation or raingauges) and variable (rainfall type and duration), in some areas the raingauge estimates may be the best guide to the true rainfall, while in other areas the radar estimate may be preferable. An automatic algorithm has been developed to select between the raingauge and radar information, to provide a best estimate of daily rainfall for each grid square over the country.

2.6.2 *Storm precipitation patterns*

Calibrating raingauges can be used in real-time to apply a correction or 'assessment' factor to periods as short as hourly radar estimates. Collier *et al.* (1983) demonstrated that this factor could vary widely (but generally between 0.1 and 10) due to changes in the drop size distribution, 'bright band' enhancement due to melting snow and evaporation losses or raindrop growth below the level of the radar beam (Fig. 2.9). The calibration factor can be determined in 'real-time' using telemetry gauge information, and the nature of the temporal changes in the factor can be used to automatically identify the rainfall type. For each rainfall type Collier *et al.* (1983) identified a number of rainfall calibration domains. Correction factors based on different subsets of calibration gauges are then applied to the radar data for each domain. By this means calibrated radar can give a detailed quantitative record of the movement of storm systems over large areas. Prior to the availability of reliable radar estimates, storm cell movement had been studied for small areas (about 25 km²) using dense networks of recording gauges. This work was mainly for purposes of urban storm drainage design (Shearman, 1977; Marshall, 1980), but it was hampered by the problems of digitizing and synchronizing large amounts of data. There is now an operational network of weather radars over a large proportion of England and Wales, providing detailed information on storm precipitation intensities and distributions (Fig. 2.10). It is planned that eventually there will be an integrated network of weather radars for western Europe.

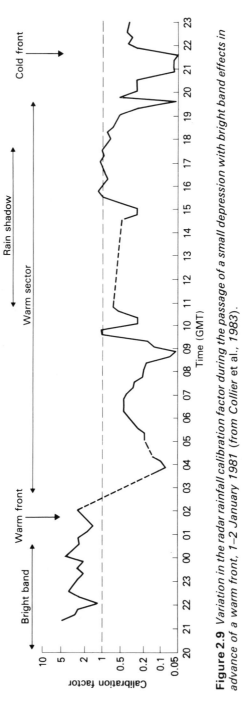

Figure 2.9 *Variation in the radar rainfall calibration factor during the passage of a small depression with bright band effects in advance of a warm front, 1–2 January 1981 (from Collier et al., 1983).*

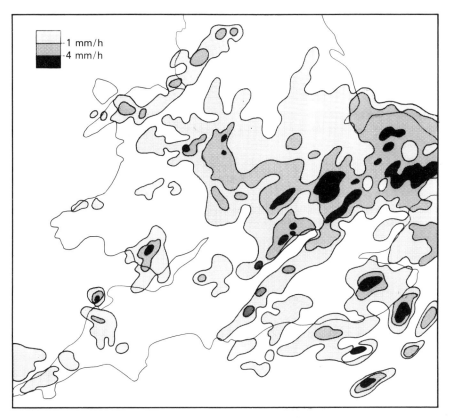

Figure 2.10 *Composite rainfall map from a network of five radars, showing the distribution of rain associated with a frontal system. Rain intensity (mm/h) at 2030 GMT on 23 November 1984. (UK Meteorological Office. Copyright reserved. By permission of the Controller of Her Majesty's Stationery Office.)*

2.6.3 *Rainfall statistics*

The frequency of heavy rainfall is of interest to the hydrologist for a number of reasons including the susceptibility of a catchment to flooding and for the design of engineering structures such as bridges and culverts. In Britain, designs are commonly based on rainfall depths with a frequency (also called the return period) of between once in 2 years and once in 100 years. The duration of rainfall that is considered may vary from several days, for very slowly responding river catchments, to only a few hours, for medium and small catchments, while in the extreme case of fast-responding impermeable urban catchments, rainfall input data at one-minute intervals may be appropriate.

The choice and fitting of alternative frequency distributions to rainfall data are discussed in various texts (cf. Sevruk and Geiger, 1981), to which the reader may refer for further details. Geomorphologists are also interested in the impact on different elements in the landscape of storms of various magnitudes and frequencies, and a number of studies have been made of the erosional features resulting from extreme meteorological events (Newson, 1975b; Anderson and Calver, 1977).

Point rainfall frequencies

Daily read raingauge records are the most commonly analysed rainfall data, largely because of their greater availability both in terms of the number of measurement points and the length of records, compared with shorter time interval information (Keers and Westcott, 1977; Sevruk and Geiger, 1981). Daily maximum values may be broadly related to the average annual rainfall and, in a study of records from 121 sites in the United Kingdom, Rodda (1967) calculated the daily falls of given return periods at each site and found a good correlation with the average annual rainfall for short return periods, although there was a poorer correlation with longer return interval daily falls. A problem with the use of daily data is that the time interval must correspond to a set 'rain day'. In the case of the United Kindom, raingauges are traditionally read and emptied at 0900 GMT each day, and consequently the precipitation from storms which span this interval is split between two rain days. According to Sevruk and Geiger (1981) the maximum falls in rain day periods are on average 14 per cent smaller than the maximum falls over a 24-hour duration. The use of daily totalled rainfall can be even more misleading for short-duration storm intensities since rain may only fall for a small part of the day. In general, the relation between daily falls and short-duration, high-intensity storms is poorer than for annual totals.

Short-duration rainfall statistics have been studied by a number of investigators using data from recording raingauges. In a pioneering study in the British Isles, Bilham (1936) used 10 years of data from 12 lowland gauges to estimate the depth of falls of 5–120 minutes' duration at different return periods. His work was widely used and was subsequently updated with additional data, and was extended to storm durations up to 25

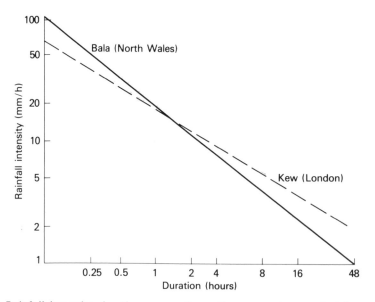

Figure 2.11 *Rainfall intensity duration curves for a five-year return period for two areas with contrasting rainfall regimes.* (*From Keers and Westcott, 1977. By permission of the Controller of Her Majesty's Stationery Office.*)

hours by Holland (1967). In a major national study of rainfall statistics the UK Meteorological Office analysed data from approximately 200 recording gauges, half of which had over 20 years of record, in addition to records from over 6000 daily read gauges (NERC, 1975, vol. II). The analyses were based on durations of one-hour and two-day rain depths (the latter to avoid problems of storms which straddled the division between 'rain days') in addition to some very short time interval records for a subset of gauges. The method that was developed in the study enables the return period to be estimated for rainfall depths over any duration from one minute up to 48 hours. Depth–duration frequency analyses indicate that very high intensities occur only rarely and also that they contribute less overall to high totals of rainfall than smaller but more frequent falls. This explains why, as noted earlier, heavy short-duration rainfall is less easily correlated with areas of high annual rainfall, since storms of this type frequently result from intense convectional activity, often associated with the distribution of severe thunderstorms (Holland, 1964). Depth–duration frequency curves vary from place to place, being steeper for areas with convectional rain than for those with predominantly frontal storms characterized by longer, less intense rainfall (Keers and Westcott, 1977) (Fig. 2.11). The frequency assigned to a given storm may also be very sensitive to the duration within the storm which is considered (Jack, 1981) (Fig. 2.12).

For countries that do not have sufficient rainfall data to construct these curves, Bell (1969) derived generalized depth–duration frequency relations using data from a range of countries and climatic conditions. He demonstrated a general uniformity of results for

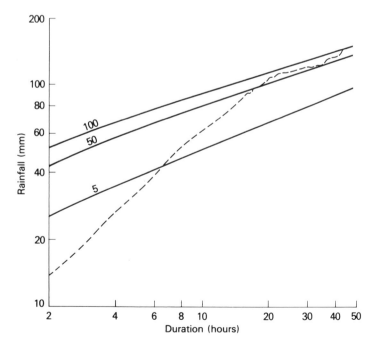

Figure 2.12 *Depth–duration frequency curves of 5, 50 and 100 year return periods based on NERC (1975) for a site in South Wales, together with the profile of a large storm on 26–27 December 1979 in South Wales (from Jack, 1981).*

short-duration storms of less than two hours, and attributed this to the fact that these falls were mostly caused by short-duration high-intensity convective cells which have similar physical properties in many parts of the world.

Areal rainfall frequencies
For many purposes the hydrologist is interested not just in the frequency of point rainfall but also in that of the rainfall over an area. The areal rainfall can be determined from the isohyetal method by analysing a number of storms to give depth–area relationships for different durations. Several storms of similar durations should be analysed and an upper envelope curve drawn (Shaw, 1988). Similar curves can be drawn for different duration storms. Such depth–area–duration curves show the *maximum* observed falls for different durations over a range of areas (Fig. 2.13). The construction of these curves is quite time consuming and requires detailed rainfall data. A number of depth—area relations for areas in the USA and Europe were analysed by Court (1961) who proposed a formula for convectional storms, assuming an elliptical pattern of isohyets and a Gaussian bell-shaped cross-section. This work was subsequently supported and extended in numerous studies (Fogel and Duckstein, 1969; Huff, 1970). It must be remembered, however, that since short-duration storms tend to have steep rainfall gradients and cover smaller areas than longer-duration storms, there can be no general formula that will be applicable for all areas and for all durations (Linsley *et al.*, 1975; Shaw, 1988).

A method of estimating the areal rainfall of a given return period from the frequency of point values of given durations was described in the *Flood Studies Report* (NERC, 1975). It involves multiplying the point depth by an Areal Reduction Factor (ARF). The ARF is the ratio of the rainfall over the area and the mean of the point rainfall depths in the area, for the same duration and return period (Bell, 1976) (Fig. 2.14). According to NERC

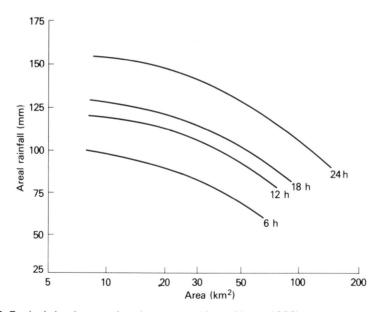

Figure 2.13 *Typical depth–area–duration curves (from Shaw, 1988).*

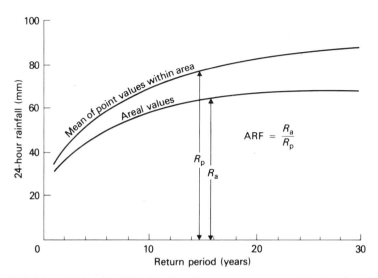

Figure 2.14 *Definition of the Areal Reduction Factor to convert point depths to average areal depths having the same return period (from an original diagram by Bell, 1976).*

(1975, vol. II) and Bell (1976), for a given storm duration this ratio appears to be constant for different return periods and for different regions of the United Kingdom (although not necessarily for other parts of the world). The *Flood Studies Report* contained a table of ARF values that approach unity for longer storm durations and for smaller areas.

As well as being concerned with the distribution and amount of above-average rainfall, the hydrologist is also concerned with periods of *drought*. This is a sustained and regionally extensive occurrence of below-average precipitation and is not to be confused with *aridity*, which is a perennial state of water shortage (Beran, 1987). There are many different ways of defining a drought, depending on the particular purpose of study, and over 150 methods are described in the literature (Barry and Chorley, 1987). Farmers, for example, will be more concerned with a shortfall in precipitation during the growing season, while a reservoir manager might be more concerned with the total rainfall depth over one or more years. Statistical relationships between drought severity and return period can be derived from rainfall records in a similar way as for large rainfall events. Maps can then be produced showing the minimum rainfall of a given return period over a specified duration.

Determining the Probable Maximum Precipitation (PMP)
Previous sections have discussed the relation between storm magnitude and return period, but it can be easily recognized that there must be a physical upper limit to the amount of precipitation that can fall on a given area in a given time. An accurate estimate is both desirable from the academic point of view and virtually essential for a range of engineering design purposes, yet it has proved very difficult to estimate such a value accurately. The upper limit to precipitation has become known as the Probable Maximum Precipitation (PMP) and has been defined as 'The theoretically greatest depth of

precipitation for a given duration that is physically possible over a particular drainage area at a certain time of year' (American Meteorological Society, 1959). The world 'probable' is intended to emphasize that, due to inadequate understanding of the physics of atmospheric processes and with imperfect meteorological data, it is impossible to define with certainty an absolute maximum precipitation. It is not intended to indicate a particular level of statistical probability or return period. In recognition of the uncertainties in estimating PMP, yet the great usefulness of such a concept, Miller (1977) called PMP a 'convenient fiction'.

There are various methods for estimating PMP (WMO, 1986) and Wiesner (1970) reviewed the basic literature and discussed the more important methods in some detail. In brief, there are two main approaches: first, the maximization and transposition of real or model storms and, second, the statistical analysis of extreme rainfalls. The maximization technique involves the estimation of the maximum limit on the humidity concentration in the air that flows into the space above a basin, the maximum limit to the rate at which wind may carry the humid air into the basin and the maximum limit on the fraction of the inflowing water vapour that can be precipitated. PMP estimates in areas of limited orographic control are normally prepared by the maximization and transposition of real, observed storms while in areas in which there are strong orographic controls on the amount and distribution of precipitation, storm models have been used for the maximization procedure for long-duration storms over large basins (Wiesner, 1970). For some large regions PMP estimates have been made encompassing numerous catchments of various sizes and generalized maps produced showing the regional variation of PMP for various basin sizes and storm durations (WMO, 1986).

The maximization/transposition techniques require a large amount of data, particularly volumetric rainfall data. In the absence of suitable data it may be necessary to transpose storms over very large distances despite the considerable uncertainties involved. In this case reference to published values of maximum observed point rainfalls will normally be helpful. Foster (1948) published worldwide maximum falls for various durations. These data were subsequently updated in 1967 by UNECAFE (reported in Todd, 1970). The world envelope curves for data recorded prior to 1948 and 1967 are shown in Fig. 2.15, together for comparison with maximum recorded falls in the United Kingdom (from Rodda, 1970, and Reynolds, 1978). By comparison with the world maxima, the British falls are rather small, as it is to be expected that a temperate climate area would experience less intense falls than tropical zones subject to hurricanes or the monsoons of southern Asia. The maxima shown for La Réunion, a rugged mountainous island (up to 3000 m elevation) in the Indian Ocean, resulted from just two intense tropical storms—in 1952 and 1964. Thus for areas of less rugged topography and cooler climate lower values of PMP might be expected.

Since the plotted points represent the largest rainfalls in the period of record the envelope curves will be subject to revision upwards as further storms are observed. This is demonstrated by the two world curves which may be approximated as:

Prior to 1948: $$P = 255 \, D^{0.5} \tag{2.1}$$

Prior to 1967: $$P = 420 \, D^{0.475} \tag{2.2}$$

where P is the precipitation in millimetres over the duration D, in hours. The two world

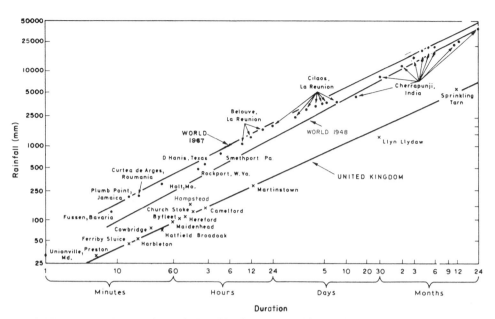

Figure 2.15 *Magnitude–duration relationship for the world and the United Kingdom extreme rainfalls (redrawn after a diagram by Rodda, 1970), based on data from various sources and showing envelope curves for the world maxima recorded prior to 1948 and prior to 1967.*

curves show that the effect of a longer data period has been, on average, to increase the maximum observed fall for a given duration by about 50 per cent.

From an analysis of the maximum rainfall in each year at several thousand gauges, Hershfield (1961) used a general formula for the analysis of extreme value data developed by Chow (1951) to relate the PMP for a selected duration to the mean (X) and standard deviation (σ) of the largest falls in each year:

$$PMP = X + K\sigma \qquad (2.3)$$

The parameter K was originally set to 15, but this was found to vary widely between sites (e.g. Rodda, 1970), and Hershfield (1965) subsequently modified the method to allow K to vary with the mean annual maxima, X, and the duration. The method has the advantage that it is easy to use and is based on observed data. Furthermore, since the processes in short, intense, thunderstorms are similar for different parts of the world it could be widely applicable for such rainfall conditions. The disadvantages are that, like all statistical methods, its success depends upon the length and nature of the available record, and parameter K may well depend on factors other than rainfall duration and the mean of the annual maxima. The computed value of K may be too high in some regions and too low in others, and this 'quick' approach should be used in conjunction with other methods (Wiesner, 1970). In the final analysis, there is no objective way of assessing the level of a PMP estimate, and judgement based on an understanding of the meteorological processes is most important. If an estimate is too low it will probably be exceeded fairly quickly, but if it is too high then this may not become apparent until our knowledge has improved sufficiently to allow for more refined and accurate methods of estimating PMP (Miller, 1969).

2.7 Hydrological aspects of snow

Snow and ice account for about 77 per cent of the earth's fresh water, but most of this is held as ice in Antarctica and Greenland, with a residence time of the order of 10 000 years. Of more relevance to the hydrologist is the fact that about 5 per cent of the global precipitation falls as snow (Hoinkes, 1967). For the United States it has been estimated that about 13 per cent or 100 mm out of the mean annual precipitation of 760 mm occurs in this form (Wilson, 1959). Its hydrological importance, however, is quite out of proportion to its magnitude because of the time lag it creates between the occurrence of precipitation in the form of snow and the active participation of that snow, after melting, in the hydrological cycle, including soil moisture recharge and runoff processes. For this reason, the hydrologist is normally less interested in the timing of snowfall than *where* it falls, *how much* has fallen, and when and how rapidly *melting* of the accumulated snow occurs.

2.7.1 *Distribution of snow*

The majority of snow and ice is in the permanent ice sheets of Antarctica and Greenland, but seasonal snow cover extends over large areas in Asia, Europe and North America. Snowfall is the predominant form of precipitation when the temperature in the lower atmosphere is below 0°C, and ground temperatures below freezing are necessary for the deposited snow to remain unmelted. In addition to providing a store of water, snow cover can also serve as a protective, insulating cover for the soil and crops through the winter and, on a larger scale, seasonal snow cover may have an important role affecting global atmospheric circulation and climatic change. In a study of the duration of snow cover in the British Isles, Jackson (1978) used data from a network of stations to map the median number of days per year that snow was lying (Fig. 2.16). This map shows that the duration increases northwards and with increasing altitude and, less expectedly, also eastwards. The latter is because snow is unlikely from westerly or south-westerly winds as the temperature of the Atlantic Ocean keeps the freezing level in the atmosphere at a sufficiently high altitude for solid precipitation to be melted before it reaches the ground. Those areas exposed to northerly or easterly winds suffer most from snow showers, especially where high ground lies close to the coast as, for instance, the North York Moors and even north Norfolk (Atkinson and Smithson, 1976). The dependence of snowfall on synoptic conditions results in a great variation in the extent and duration of snow from one year to the next. At Kew Observatory, London, there were 45 days of snow lying in the severe winter of 1962–3 (with a return period of approximately 300 years) while the next longest period in the 25 years 1948–73 was only 13 days (Jackson, 1978). Just as with liquid precipitation, there is evidence of long-term variation in snowfall over time. The period 1900–40, for example, was relatively mild in comparison with the long-term British climate and so earlier work by Manley (1940, 1947) based on snow data from this period may understimate present snowfall distributions. Since 1940 the number of days per year with snow has increased, and is now more in accord with values in the middle of the last century (Jackson, 1978).

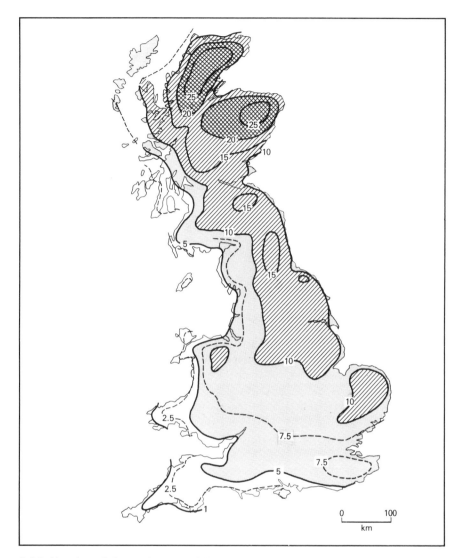

Figure 2.16 *Number of days with snow lying in a median winter in Britain, 1941–70, reduced to sea level (redrawn after Jackson, 1978).*

Information on the spatial distribution of snow was traditionally based on reports from local observers. This method is somewhat subjective in areas of patchy snow cover, and Manley (1969) noted the apparent reluctance of some observers in health resort towns in Britain to record snow! Another problem is that in the milder areas light snow present in the morning may have melted later in the day (Atkinson and Smithson, 1976). In addition to such problems in the interpretation and recording of site conditions there is the need to interpolate between these point observations to make estimates of the areal extent of snow cover.

Since the 1960s remote sensing, from aircraft or satellites, has revolutionized the assessment of snow cover. This provides the means to map rapidly the extent of snow cover over large areas, and, by repeated observation, the seasonal expansion and contraction of snow cover can be monitored (Wiesnet et al., 1987). Due to its high albedo, snow cover can be readily identified from bare ground using visible radiation reflectance, although it is often difficult to distinguish it from cloud cover. This problem may be reduced by repeat photography over time in order to filter out the variable cloud pattern, or else by using the very high near-infrared reflectance of snow to distinguish it from cloud cover.

Even greater than the difficulties of defining the presence or absence of snow are those encountered when trying to *quantify* the amount of snow lying on the ground.

2.7.2 *Amount of snowfall*

It is often not sufficient to be aware of the presence of snow, and the hydrologist may require some measure of the quantity of snow. Rather than the depth of lying snow it is the *water equivalent* of the snow which is important, i.e. the equivalent water depth of the melted snow. This is the quantity that is potentially available for runoff and for soil moisture replenishment.

The difficulties in collecting and measuring the amount of snowfall in gauges are even greater than those of rainfall. Snowflakes are even more prone than raindrops to turbulence around gauges, resulting in severe undercatches, and although wind effects can be greatly reduced by using wind shields around the gauges, the errors due to undercatches are often still too great to be acceptable (Weiss and Wilson, 1958; Larson and Peck, 1974; Sevruk, 1982). There are also errors resulting from fallen snow subsequently being blown into, or out from, the funnel of the gauge. The amount of unmelted snow accumulated in the funnel may be melted by the addition of a known quantity of water and then measured in the normal way (Meteorological Office, 1982), although the value of such measurements is inevitably limited.

A better method is to measure the depth of snow at particular locations and then convert this to the water equivalent using the density of the snow (Martinec, 1976). The density may be assumed to have a constant value, often 10 per cent, i.e. 10 cm snow = 1 cm water, or more commonly it is measured by taking a core through the snow and weighing it. The density of freshly fallen snow varies from 5 to 20 per cent, depending on the temperature during the storm. The density increases over time due to settling and compaction under gravity as well as to any light melting and refreezing of the snow pack, and may reach values of 50 or 60 per cent. As well as variations in snow density, the depth of snow in an area may vary greatly with topography and with drifting, resulting in spatial sampling problems that are even more severe than those already discussed for liquid precipitation. For this reason depth and density measurements may be made along predetermined snow courses selected to be representative of conditions over a wide area (US Army, 1956).

These techniques disturb the snowpack, making it difficult to repeat measurements at the same point to study changes over time. One solution is to measure the water equivalent directly by automatically weighing the snow that falls onto a snow pillow (Beaumont, 1965). This is a rubber pillow, up to 3.7 m in diameter, which is filled with a liquid which will not freeze during winter exposure. The overlying weight of snow depresses the surface of the pillow, and this is recorded using a manometer or a pressure transducer. Results of snow pillow measurements were discussed by Warnick and Penton (1971).

Other methods of directly measuring the water equivalent at a site include the use of radioactive isotopes. A gamma source and detector may be positioned in separate vertical tubes to provide information on density changes through the snowpack by measuring the attenuation of the gamma emissions; alternatively, the detector may be positioned on the ground, beneath the snow, and the source positioned vertically above the top of the snow (Warnick and Penton, 1971; Martinec, 1976). Another method to measure the water equivalent is to move a neutron probe source and counter up and down a vertical access tube to record the amount of backscatter in a manner similar to soil moisture measurements (Cooper, 1966; Harding, 1986) (see also Section 5.3.6 under 'Measurement of soil water content').

The above methods of making point measurements can provide only limited information on the uneven areal distribution of snow cover, unless a large amount of replication is possible. The use of remote sensing in combination with conventional snow surveying methods offers the opportunity to obtain quantitative information on the areal distribution of snow cover. Collier and Larke (1978) reported promising results when they calibrated radar measurements of snowfall against a network of twice-daily ground measurements of snow depth. The natural radioactivity of the earth also provides a means of measuring the snow water equivalent over large areas. Natural gamma radiation from the upper layer of the soil will be attenuated by the overlying snow. Airborne measurements carried out when snow is lying on the ground can be compared with results when the ground is snow free to provide information on the snow water equivalent (Peck et al., 1971). The method cannot be used for wetlands or swamps and allowance must be made for differences in the moisture content of the soil, but nevertheless this is by far the most widely used technique for estimating the snow water equivalent, and is used operationally in a number of countries, including the USA, USSR, Norway and Finland (Kuittinen, 1986; Carroll, 1987). The use of microwave sensors to measure both snow depth and density appears to offer great potential, although further work is needed to develop the techniques.

These techniques provide the hydrologist with information on the quantity of water stored in the snowpack and which is available on melting to recharge soil mosture and to generate runoff in streams.

2.7.3 *Snowmelt*

The prediction of snowmelt is of great importance in areas of seasonal snow cover where snow comprises much of the annual precipitation. It is needed to estimate seasonal flood risk and, in arid areas bordering high mountains, information is needed on the amount and timing of water that will be available for irrigation. Various approaches have been adopted to estimate snowmelt rates and may be grouped broadly into empirical regression models, using linear equations between weather variables and snowmelt, and conceptual models, in which the various hydrological processes are represented by separate equations (Morris, 1985).

Before snowmelt can occur, a metamorphosis of the snowpack must take place. Over time the properties of a snowpack change; density increases, snow crystals become large grained and the albedo becomes smaller (Miller, 1977). Melting and recrystallization are important for preparing the snowpack for melting (Dunne and Leopold, 1978). With the onset of warmer temperatures, meltwater from the surface layer of the snow percolates down to lower, colder layers and refreezes. This freezing releases latent heat, warming the lower layers of the snowpack and, over time, tending to equalize temperatures at 0°C over the whole snow depth. The snow can hold a certain amount of water in pores against gravity (usually 2–8 per cent by volume; Dunne and Leopold, 1978) and when this takes place the snowpack is said to be *ripe*: any further addition of energy will result in meltwater runoff.

The rate of snowmelt depends upon the energy exchange between the snowpack and its environment. The energy balance of an isothermal snowpack at 0°C can be written as

$$H_m = H_r + H_a + H_e + H_p + H_g \qquad (2.4)$$

where H_m is the energy available for melting snow, H_r is the net short- and long-wave radiation, H_a is the sensible heat flux, H_e is the latent heat flux, H_p is the heat gain from precipitation and H_g is the heat exchange with the underlying ground.

The last two terms are often ignored since the energy exchange with precipitation or with the ground is usually very small (Foster, 1948; Martinec, 1976).

Net radiation (H_r) can be measured directly using a net radiometer, although such data are rarely measured (Price and Dunne, 1976), and the individual fluxes of different wavelengths usually have to be measured or estimated separately. Sensible heat and latent heat fluxes are closely related and can be considered together as a single process. Heat and moisture transfer above the snow surface both occur as a result of turbulent mixing, and depend upon the vapour pressure gradient between the air and the snow surface and on the windspeed. (See also the discussion of evaporation processes in Chapter 4.) The complexity of the snowpack energy balance was discussed in detail in the standard work *Snow Hydrology* (US Army, 1956). The relative importance of the various components of the energy balance is difficult to determine with accuracy and has been found to vary with time, both seasonally and diurnally, and also between days with different weather conditions. Kuusisto (1986) summarized the findings of 20 studies of energy fluxes of melting snowpacks. Net radiation will be negative at night, but on sunny days with little wind it may be the dominant source of energy gain. Its importance increases as the spring season advances due to increasing solar radiation as well as the decline in albedo with the age of the snowpack. Turbulent heat exchange will dominate during the night and on cloudy days. It is greatest when there are strong, moist winds. Although dry air can cause a certain amount of evaporation (sublimation) of the snow, humid air can have a much greater effect on the snowpack since the condensation of water vapour on snow releases sufficient latent heat to melt a much larger quantity of ice. There are differences between snowmelt for forested and open sites. Forests dampen turbulent fluxes and shade direct solar radiation, although long-wave radiation is higher from the warmed tree branches. In general, rates of snowmelt are lower from forested catchments than from open ground. The role of tree canopy storage of snow is discussed in Chapter 3.

In many situations insufficient meteorological data are available to compute the energy budget of the snowpack, and empirical methods have been used to predict the magnitude and timing of snowmelt. Generally these correlations are made between aspects of air temperature and snowmelt (US Army, 1956), although it is clear from the preceding discussion that due to the variation in the relative importance of the various heat transfer processes, no single index or method of estimating snowmelt will be applicable to all areas and for all weather conditions. Zuzel and Cox (1975), in a study in Idaho, found that air temperature was the most important *single* meteorological variable in regression models of snowmelt, but that better estimates could be obtained with a *combination* of other meteorological variables, namely vapour pressure, net radiation and wind. Harding (1986) obtained a full energy balance for a level site in southern Norway where air turbulence was much reduced. He compared measured snowmelt rates with predictions using a degree-day model based on daily air temperature with empirical parameter values from the literature, including US Army (1956). The model failed to simulate either the total melt or its day-by-day variations, and this was attributed to the large proportion of input energy from net radiation at the site and the control on air temperature by the melting snow surface.

2.8 Human modification of precipitation patterns

The preceding sections have treated precipitation as a process independent of man, who must passively accept what nature does, or does not, provide. In fact it is now becoming apparent that man has an increasing ability to modify rainfall at certain scales, intentionally or otherwise.

Over the last 30 years rain-seeding experiments have been carried out, adding artificial nuclei to clouds to encourage precipitation. Although greatly exaggerated claims were made by the early cloud seeding experimenters, these have now largely been discredited and current claims are much more modest. While the mechanism can be observed at the local scale of individual clouds, its wider effect is less clear. There is little convincing evidence that large increases in rainfall can be produced consistently over large areas (Mason, 1975; Essenwanger, 1986).

At the regional scale there has long been controversy as to whether vegetation cover could influence rainfall via its effect on evaporation losses. Early notions that vegetation could have such an effect were largely discredited by later work which emphasized the large-scale nature of water vapour transfer, with often great distances between the evaporation of water and its subsequent precipitation (Penman, 1963). This view may need to be modified, however, as studies in the extensive tropical basin of the Amazon forest indicated that about half of the rainfall originates from forest evaporation (Salati and Vose, 1984; Shuttleworth, 1988). Continued large-scale deforestation is likely to lead to reduced evaporation, to increased runoff and, ultimately, to reduced precipitation in that region. Further, it was suggested that vegetation changes could be the cause of reduced rainfall over the Sahelian region of north Africa (Charney, 1975). Overgrazing would reduce the vegetation cover and expose more sandy soil, increasing the albedo of the ground surface. This in turn would lower the ground surface temperatures, reducing the likelihood of convective precipitation. Subsequent work by Sud and Fennessy (1982) gave support to this hypothesis.

At the global scale it is known that the burning of fossil fuels has led to an increase in the carbon dioxide content of the atmosphere. This is one of a number of gases that absorb outgoing long-wave radiation from the earth, which would otherwise be lost into space. There is evidence that this is causing a general warming up of the atmosphere (Barry and Chorley, 1987). This could lead to changes in circulation patterns, with a poleward migration of climatic zones and spatially variable effects on precipitation (Manabe and Wetherald, 1980).

References

Akin, J. E. (1971) Calculation of mean areal depth of precipitation, *J. Hydrol.*, **12**: 363–76.

American Meteorological Society (1959) *Glossary of meteorology*, R. E. Huschke (ed.), Boston, Mass., 638 pp.

Anderson, M. G. and A. Calver (1977) On the persistence of landscape features formed by a large flood, *Trans. IBG*, **2**: 243–54.

Atkins, M. J. (1985) Using computers to forecast the weather, in *National electronics review*, National Electronics Council, London, pp. 51–5.

Atkinson, B. W. and P. A. Smithson (1976) Precipitation, Ch. 6 in *The climate of the British Isles*, T. J. Chandler and S. Gregory (eds), Longman Group Ltd, London, pp. 129–82.

Barrett, E. C. (1970) The estimation of monthly rainfall from satellite data, *Mon. Wea. Rev.*, **98**: 322–7.

Barrett, E. C. (1973) Forecasting daily rainfall from satellite data, *Mon. Wea. Rev.*, **101**: 215–22.

Barrett, E. C. and D. W. Martin (1981) *The use of satellite data in rainfall monitoring*, Academic Press, London, 340 pp.

Barry, R. G. and R. J. Chorley (1987) *Atmosphere, weather and climate*, 5th edn, Methuen, London, 460 pp.

Bastin, G., B. Lovert, C. Duque and M. Gevers (1984) Optimum estimation of the average areal rainfall and optimal selection of rain gauge locations, *WRR*, **20**, 463–70.

Battan, L. J. (1973) *Radar observation of the atmosphere*, University of Chicago Press, Chicago, 324 pp.

Beaumont, R. T. (1965) Mt. Hood pressure pillow snow gauge, *Proc. 33rd Annual Meeting Western Snow Conf.*, Colorado State University, Fort Collins, Colo, pp. 29–35.

Bell, F. C. (1969) Generalized rainfall-duration–frequency relationships, *Proc. ASCE, J. Hydraul. Div.*, **95** (HY1): 311–27.

Bell, F. C. (1976) The areal reduction factor in rainfall frequency estimation, *Report 35*, Institute of Hydrology, Wallingford, 58 pp.

Beran, M. (1987) Data collection and use for modelling, simulation and forecasting of drought, *Water Supply*, **5**: 11–21.

Bergeron, T. (1960) Problems and methods of rainfall investigation, in *Physics of precipitation*, Geophysical Monograph 5, Washington, pp. 5–30.

Bilham, E. G. (1936) Classification of heavy falls of rain in short periods, in *British Rainfall, 1935*, pp. 262–80.

Binnie, A. R. (1892) On mean or average rainfall and the fluctuations to which it is subject, *Proc. ICE*, **109**: 89–172.

Binnie, G. M. (1981) *Early Victorian water engineers*, Thomas Telford Ltd, London, 310 pp.

Biswas, A. K. (1969) A short history of hydrology, in *The progress of hydrology*, University of Illinois, Urbana, pp. 914–36.

Biswas, A. K. (1970) *History of hydrology*, North-Holland Publishing Co., Amsterdam, 336 pp.

Borovikov, A. M. and V. V. Kostarev (eds) (1970) *Radar measurement of precipitation rate*, Israel Program for Scientific Translations, Jerusalem, 112 pp. Translated from Russian by A. Moscona.

Brandes, E. A. (1975) Optimising rainfall estimates with the aid of radar, *J. Appl. Met.*, **14**: 1339–45.

British Rainfall. Published annually by the UK Meteorological Office, Bracknell. 1860 to Present.

Browning K. A. (1987) Towards the more effective use of radar and satellite imagery in weather forecasting, Ch. 16 in *Weather radar and flood forecasting*, V. K. Collinge and C. Kirby (eds), John Wiley and Sons, Chichester, pp. 239–69.

Browning, K. A. and F. F. Hill (1981) Orographic rain, *Weather*, **36**: 326–9.

Browning, K. A., R. B. Bussell and J. A. Cole (1977) Radar for rain forecasting and river management, in *Water power and dam construction* (December).

Burpee, R. W. and L. N. Lahiff (1984) Area average rainfall variations on sea breeze days in S. Florida, *Mon. Wea. Rev.*, **112**: 520–34.

Caffey, J. E. (1965) Inter-station correlations in annual precipitation and in annual effective precipitation, *Hydrology Papers 6*, Colorado State University, Fort Collins, Colorado.

Carroll, T. R. (1987) Operational airborne measurement of snow water equivalent and soil moisture using terrestrial gamma radiation in the United States, *IAHS Publ.*, **166**: 213–23.

Catterall, J. W. (1972) An *a priori model* to suggest raingauge domains, *Area*, **4**: 158–63.

Chandler, T. J. and S. Gregory (eds) (1976) *The climate of the British Isles*, Longman Group Ltd, London, 390 pp.

Charney, J. G. (1975) Dynamics of deserts and drought in the Sahel, *QJRMS*, **101**: 193–202.

Chidley, T. R. E. and K. M. Keys (1970) A rapid method of computing areal rainfall, *J. Hydrol.*, **12**: 15–24.

Chow, V. T. (1951) A general formula for hydrologic frequency analysis, *Trans. AGU*, **32**: 231–7.

Clarke, R. T., M. N. Leese and A. J. Newson (1973) Analysis of data from Plynlimon raingauge networks, April 1971–March 1973, *Report 27*, Institute of Hydrology, Wallingford, 75 pp.

Collier, C. G. (1984) Remote sensing for hydrological forecasting, Ch. 1 in *Facets in hydrology*, J. C. Rodda (ed.), vol. II, John Wiley, New York, pp. 1–24.

Collier. C. G. (1987) Accuracy of real-time radar measurements, Ch. 6 in *Weather radar and flood forecasting*, V. K. Collinge and C. Kirby (eds), John Wiley and Sons, Chichester, pp. 71–95.

Collier, C. G. and P. R. Larke (1978) A case study of the measurement of snowfall by radar: an assessment of accuracy, *QJRMS*, **104**: 615–21.

Collier, C. G., P. R. Larke and B. R. May (1983) A weather radar correction procedure for real-time estimation of surface rainfall, *QJRMS*, **109**: 589–608.

Collinge, V. K. and C. Kirby (eds) (1987) *Weather radar and flood forecasting*, John Wiley and Sons, Chichester, 296 pp.

Cooper, C. F. (1966) Sampling characteristics of neutron probe measurements in a mountain snow pack, *Journal of Glaciology*, **6**: 289–98.

Court, A. (1961) Area–depth rainfall formulas, *JGR*, **66**: 1823–31.

Craddock, J. M. (1976) Annual rainfall in England since 1725, *QJRMS*, **102**: 823–40.

Diskin, M. H. (1969) Thiessen coefficients by a Monte Carlo procedure, *J. Hydrol.*, **8**: 323–35.

Diskin, M. H. (1970) On the computer evaluation of Thiessen weights, *J. Hydrol.*, **11**: 69–78.

Dunne, T. and L. B. Leopold (1978) *Water in environmental planning*, Freeman, San Francisco, 818 pp.

Eagleson, P. S. (1970) *Dynamic hydrology*, McGraw-Hill, New York, 462 pp.

Essenwanger, O. M. (1986) *Elements of statistical analysis*, World Survey of Climatology, vol. 1B, Elsevier, Amsterdam, 424 pp.

Flohn, H. (1969) *Climate and weather*, World University Library, London, 253 pp.

Fogel, M. M. and L. Duckstein (1969) Point rainfall frequencies in convective storms, *WRR*, **5**: 1229–37.

Folland, C. K. and M. G. Colgate (1978) Recent and planned rainfall studies in the Meteorological Office with an application to urban drainage design, in *Urban storm drainage*, P. R. Helliwell (ed.), Pentech Press, London, pp. 51–70.

Folland, C. K., T. N. Palmer and D. E. Parker (1986) Sahel rainfall and worldwide sea temperatures, *Nature*, **320**: 602–7.

Foster, E. E. (1948) *Rainfall and runoff*, Macmillan, New York, 487 pp.

French, R. H. (1988) Effects of the length of record on estimates of annual and seasonal precipitation, in *Computational methods and water resources*, D. Ouzar, C. A. Brebia and V. de Kosinsky (eds), vol. 3, Computational Mechanics Publications, Southampton, pp. 3–13.

Goel, S. M. and A. S. Aldabagh (1979) A distance weighted method for computing average precipitation, *JIWES*, **33**: 451–4.

Grace, R. A. and P. S. Eagleson (1966) The synthesis of short time increment rainfall sequences, *Report 91*, Department of Civil Engineering Hydrodynamics Laboratory, Massachusetts Institute of Technology, 105 pp.

Green, M. J. and P. R. Helliwell (1972) The effect of wind on the rainfall catch, *Proc. Symp. on the Distribution of Precipitation in Mountainous Areas*, vol. 2, World Meteorological Organization, pp. 27–46.

Harding, R. J. (1986) Exchanges of energy and mass associated with a melting snowpack, *IAHS Publ.*, **155**: 3–15.

Harrold, T. W., E. J. English and C. A. Nicholass (1974) The accuracy of radar-derived rainfall measurements in hilly terrain, *QJRMS*, **100**: 331–50.

Hendrick, R. L. and G. H. Comer (1970) Space variations of precipitation and implications for raingauge network design, *J. Hydrol.*, **10**: 151–63.

Hershfield, D. M. (1961) Estimating the probable maximum precipitation, *Proc. ASCE, J. Hydraul. Div.*, **87** (HY 5): 99–116.

Hershfield, D. M. (1965) Method for estimating probable maximum precipitation, *JAWA*, **57**: 965–72.

Hill, G. and R. B. Robertson (1987) The establishment and operation of an unmanned weather radar, Ch. 5 in *Weather radar and flood forecasting*, V. K. Collinge and C. Kirby (eds), John Wiley and Sons, Chichester, pp. 55–69.

Hitschfeld, W. and J. Bordan (1954) Errors inherent in the radar measurement of rainfall at attenuating wavelengths, *J. Meteorol.*, **2**: 58–67.

Hoinkes, H. (1967) Glaciology in the International Hydrological Decade, *IASH Publ.*, **97**: 7–16.

Holland, D. J. (1964) Rain intensity frequency relationships in Britain, *Memoranda 33*, Meteorological Office of Hydrology, 28 pp.

Holland, D. J. (1967) Rainfall intensity frequency relationships in Britain, in *British Rainfall 1961*, pt III, pp. 43–51.

Howe, G. M., H. O. Slaymaker and D. M. Harding (1967) Some aspects of the flood hydrology of the upper catchments of the Severn and the Wye, *Trans. IBG*, **41**: 35–58.

Hudleston, F. (1934) A summary of seven years experiments with rain gauge shields in exposed positions, 1926–32 at Hutton John, Penrith, *British Rainfall 1933*, pp. 274–93.

Huff, F. A. (1970) Spatial distribution of rainfall rates, *WRR*, **6**: 254–60.

Hutchinson, P. (1970) A contribution to the problem of spacing raingauges in rugged terrain, *J. Hydrol.*, **12**: 1–14.

Jack, W. L. (1981) Rainfall return periods for December 1979, *Weather*, **36**: 274–6.

Jackson, M. C. (1978) Snow cover in Great Britain, *Weather*, **33**: 298–309.

Jaeger, L. (1983) Monthly and areal patterns of mean global precipitation, in *Variations in the global water budget*, A. Street-Perrot, M. Beran and R. Ratcliffe (eds), D. Reidel Publishing Co., Lancaster, pp. 129–40.

Karr, T. W. and R. L. Wooten (1976) Summer radar echo distribution around Limon, Colorado, *Mon. Wea. Rev.*, **104**: 728–34.

Keers, J. F. and P. Westcott (1977) A computer based model for design rainfall in the United Kingdom, *Meteorological Office Scientific Paper 36*, HMSO, London, 14 pp.

Kelway, P. S. (1975) The rainfall recorder problem, *J. Hydrol.*, **26**: 55–77.

Klitch, M. A., J. F. Weaver, R. P. Kelly and T. H. Van der Haar (1985) Convective cloud climatologies constructed from satellite imagery, *Mon. Wea. Rev.*, **113**: 326–37.

Koschmieder, H. (1934) Methods and results of definite rain measurements, *Mon. Wea. Rev.*, **62**: 5–7.

Kraus, E. B. (1958) Recent climatic changes, *Nature*, **181**: 666–8.

Kuittinen, R. (1986) Determination of areal snow-water equivalent values using satellite imagery and aircraft gamma-ray spectrometry, *IAHS Publ.*, **160**: 181–9.

Kurtyka, J. C. (1953) *Precipitation measurement study*, Report of Investigation 20, State Water Survey Division, Urbana, Illinois, 178 pp.

Kuusisto, E. (1986) The energy balance of a melting snow cover in different environments, *IAHS Publ.*, **155**: 37–45.

Lamb, H. H. (1950) Types and spells of weather around the year in the British Isles: annual trends, seasonal structure of the year, singularities, *QJRMS*, **76**: 393–438.

Lamb, H. H. (1972) *Climate: past, present and future. I. Fundamentals and climate now*, Methuen and Co. Ltd, London, 613 pp.

Lamb, H. H. (1977) *Climate: past, present and future. II. Climatic history and the future*, Methuen and Co. Ltd, London, 835 pp.

Larson, L. W. and E. L. Peck (1974) Accuracy of precipitation measurements for hydrologic modelling, *WRR*, **10**: 857–63.

Leopold, L. B. (1951) Rainfall intensity: an aspect of climatic variation, *EOS, Trans. AGU*, **32**: 347–57.

Linsley, R. K., M. A. Kohler and J. L. Paulhus (1975) *Applied hydrology*, McGraw-Hill, New York.

Lockwood, J. G. (1974) *World climatology, an environmental approach*, Edward Arnold Ltd, London, 330 pp.

McKay, G. A. (1965) Meteorological measurements for watershed research, *Research Watersheds: Proc. Hydrol. Symp. 4*, University of Guelph, Ontario, pp. 185–209.

Manabe, S. and R. T. Wetherald (1980) On the distribution of climatic change resulting from an increase in CO_2 content of the atmosphere, *Journal of Atmospheric Science*, **37**: 99–118.

Mandeville, A. N. and J. C. Rodda (1970) A contribution to the objective assessment of areal rainfall amounts, *J. Hydrol. (N.Z.)*, **9**: 281–91.

Manley, G. (1940) Snowfall in the British Isles, *Met. Mag.*, **75**: 41–8.

Manley, G. (1947) Snow-cover in the British Isles, *Met. Mag.*, **76**: 28–36.

Manley, G. (1969) Snowfall in Britain over the last 300 years, *Weather*, **24**: 428–37.

Marshall, J. S. and W. M. Palmer (1948) The distribution of raindrops with size, *J. Meteorol.*, **5**: 165–6.

Marshall, R. J. (1980) The estimation and distribution of storm movement and storm structure using a correlation analysis technique and raingauges data, *J. Hydrol.*, **48**: 19–39.

Martinec, J. (1976) Snow and ice, Ch. 4 in *Facets of hydrology*, J. C. Rodda (ed.), vol. I, John Wiley and Sons, London, pp. 85–118.

Mason, B. J. (1971) *The physics of clouds*, 2nd edn, Oxford University Press, London, 671 pp.

Mason, B. J. (1975) *Clouds, rain and rainmaking*, 2nd edn, Cambridge University Press, Cambridge, 189 pp.

May, B. R. (1986) Discrimination in the use of radar data adjusted by sparse gauge observations for determining surface rainfall, *Met. Mag.*, **115**: 101–15.

May, B. R. (1988) Progress in the development of Paragon, *Met. Mag.*, **117**: 79–86.

Meteorological Office (1982) *Observer's handbook*, 4th edn, HMSO, London, 220 pp.

Miller, D. H. (1977) *Water at the surface of the earth*, International Geophysics Series Vol. 21, Academic Press Inc., New York, 557 pp.

Miller, J. F. (1969) Hydrometeorological studies, in *The progress of hydrology*, University of Illinois, Urbana, pp. 521–62.

Mills, W. C. (1982) Stochastic modelling of rainfall for deriving distributions of watershed input, in *Statistical analysis of rainfall and runoff*, V. P. Singh (ed.), Water Resources Publications, Littleton, Colo., pp. 103–18.

Milner, S. (1986) NEXRAD: the coming revolution in radar storm detection and warning, *Weatherwise*, April: 72–85.

Moore R. J. (1987) Towards more effective use of radar data for flood forecasting, Ch. 15 in *Weather radar and flood forecasting*, V. K. Collinge and C. Kirby (eds), John Wiley and Sons, Chichester, pp. 223–38.

Morris, E. M. (1985) Snow and ice, Ch. 7 in *Hydrological forecasting*, M. G. Anderson and T. P. Burt (eds), John Wiley and Sons, London, pp. 153–82.

Morris, S. E. and T. J. Marsh (1985) United Kingdom rainfall 1975–84: evidence of climatic instability?, *J. Meteorol.*, **10**: 324–32.

NERC (1975) *Flood studies report, II. Meteorological studies*, Natural Environment Research Council, London, 81 pp.

Newson, A. J. and R. T. Clarke (1976) Comparison of the catch of ground-level and canopy-level raingauges in the upper Severn experimental catchment, *Met. Mag.*, **105**: 2–7.

Newson, M. D. (1975a) *Flooding and flood hazard in the United Kingdom*, Oxford University Press, London, 60 pp.

Newson, M. D. (1975b) The Plynlimon floods of August 5th/6th, 1973, *Report 26, Institute of Hydrology*, Wallingford, 58 pp.

Nicholass, C. A., P. E. O'Connell and M. R. Senior (1981) Raingauge network rationalization and its advantages, *Met. Mag.*, **110**: 92–102.

Ninomiya, K. and T. Akiyama (1978) Objective analysis of heavy rainfalls based on radar and gauge measurements, *JMSJ*, **56**: 206–10.

O'Connell, P. E., R. J. Gurney, D. A. Jones, J. B. Miller, C. A. Nicholass and M. R. Senior (1978) Rationalization of the Wessex Water Authority raingauge network, *Report 51*, Institute of Hydrology, Wallingford. 179 pp.

Packman, J. C. (1987) Baluchistan daily rainfall, Report to UK Overseas Development Administration, Institute of Hydrology, Wallingford.

Peck, E. L., V. C. Bissell, E. B. Jones and D. L. Burge (1971) Evaluation of snow water equivalent by airborne measurements of passive terrestrial gamma radiation, *WRR*, **7**: 1151–9.

Peixoto, J. P. and A. H. Oort (1983) The atmospheric branch of the hydrological cycle and climate. In *Variations in the global water budget*, A. Street-Perrot, M. Beran and R. Ratcliffe (eds), D. Reidel Publishing Co., Lancaster, pp. 5–65.

Penman, H. L. (1963) *Vegetation and hydrology, Technical Communication 53*, Commonwealth Agricultural Bureau, Farnham, 124 pp.

Price, A. G. and T. Dunne (1976) Energy balance computations of snowmelt in a subarctic area, *WRR*, **12**: 686–94.

Ramage, C. S. (1964) Diurnal variation of summer rainfall of Malaya, *Journal of Tropical Geography*, **19**: 62–8.

Reynolds, G. (1978) Maximum precipitation in Great Britain, *Weather*, **33**: 162–6.

Richardson, L. F. (1922) *Weather prediction by numerical process*, Cambridge University Press, London.

Robinson, A. C. and J. C. Rodda (1969) Rain, wind and the arerodynamic characteristics of raingauges, *Met. Mag.*, **98**: 113–20.

Rodda, J. C. (1967) A country-wide study of intense rainfall for the United Kingdom, *J. Hydrol.* **5**: 58–69.

Rodda, J. C. (1968) The rainfall measurement problem, *IASH Publ.*, **78**: 215–31.

Rodda, J. C. (1969) The assessment of precipitation, in *Water, earth and man*, R. J. Chorley (ed.), pp. 130–4, Methuen, London.

Rodda, J. C. (1970) Rainfall excesses in the United Kingdom, *Trans. IBG*, **49** (March): 49–60.

Rodda, J. C. (1973) Annotated bibliography on precipitation measurement instruments. *Projects Report 17*, WMO/IHD, Geneva, 278 pp.

Rodda, J. C. and S. W. Smith (1986) The significance of the systematic error in rainfall measurement for assessing atmospheric deposition, *Atmos. Environ.*, **20**: 1059–64.

Rudloff, W. (1981) *World climates with tables of climatic data and practical suggestions*, Wissenschaftliche Verlagsgesellschaft mbH, Stuttgart, 632 pp.

Salati, E. and P. B. Vose (1984) Amazon basin: a system in equilibrium, *Science*, **225**: 129–225.

Sevruk, B. (1982) Methods of correction for systematic error in point precipitation measurement for operational use, *Operational Hydrology Report 21*, World Meteorological Organization, Geneva, 91 pp.

Sevruk, B. and H. Geiger (1981) Selection of distribution types for extremes of precipitation, *Operational Hydrology Report 15*, World Meteorological Organization, Geneva, 64 pp.

Shaw, E. M. (1988) *Hydrology in Practice*, 2nd edn, Van Nostrand Reinhold, Wokingham, 539 pp.

Shaw, E. M. and P. P. Lynn (1972) Areal rainfall evaluation using two surface fitting techniques, *Bull. IAHS* **17**: 419–33.

Shearman, R. J. (1975) Computer quality control of daily and monthly rainfall data, *Met. Mag.*, **104**: 102–8.

Shearman, R. J. (1977) The speed and direction of storm rainfall patterns with reference to urban storm sewer design, *Hydrol. Sci. Bull.*, **22**, 421–31.

Shuttleworth, W. J. (1988) Evaporation from Amazonian rainforest, *Proc. Roy. Soc. London, B*, **233**: 321–46.

Stol, Ph. Th. (1972) The relative efficiency of the density of raingauge networks, *J. Hydrol.*, **15**: 193–208.

Sud, Y. C. and M. Fennessy (1982) A study of the influence of surface albedo on July circulation in semi arid region using the GLAS General Circulation Model, *J. Climatol.*, **2**, 105–25.

Tabony, R. C. (1981) A principal component and spectral analysis of European rainfall, *J. Climatol.*, **1**: 283–94.

Thiessen, A. H. (1911) Precipitation averages for large areas, *Mon. Wea. Rev.*, **39**: 1082–4.

Todd, D. K. (ed.) (1970) *The water encyclopaedia*, Water Information Centre, New York, 559 pp.

US Army (1956) *Snow hydrology*, US Army Corps of Engineers, Portland, 437 pp.

US Weather Bureau (1947) Thunderstorm rainfall, *Hydrometeorological Report 5*, US Weather Bureau.

Vines, R. G. (1985) European rainfall patterns, *J. Climatol.*, **5**: 607–16.

Warnick, C. C. and V. E. Penton (1971) New methods of measuring water equivalent of snow pack for automatic recording at remote mountain locations, *J. Hydrol.*, **13**: 201–15.

Weiss, L. L. and W. T. Wilson (1958) Precipitation gauge shields, *IASH Publ.*, **43**(1): 462–84.

Wiesner, C. J. (1970) *Hydrometeorology*, Chapman and Hall Ltd, London, 232 pp.

Wiesnet, D. R., C. F. Ropelewski, G. J. Kukla and D. A. Robinson (1987) A discussion of the accuracy of NOAA satellite-derived global seasonal snow cover estimates, *IAHS Publ.*, **166**: 291–304.

Wilson, J. W. (1970) Integration of radar and raingauge data for improved rainfall measurement, *J. Appl. Met.*, **9**: 489–97.

Wilson, J. W. and E. A. Brandes (1979) Radar measurement of rainfall—a summary, *Bull. Amer. Met. Soc.*, **60**: 1048–58.

Wilson, W. T. (1959) Snow, Ch. 10 in *Hydrology*, C. O. Wisler and E. F. Brater (eds), 2nd edn, John Wiley and Sons, New York, pp. 301–18.

Winstanley, D. (1973) Rainfall patterns and general circulation, *Science*, **245**: 190–4.

WMO (1967) The role of meteorological satellites in the World Weather Watch, *World Weather Watch Planning Report 18*, World Meteorological Organization, Geneva, 38 pp.

WMO (1981) *Guide to Hydrological Practices, No. 168*, 4th edn, 2 vols, World Meteorological Organization, Geneva.

WMO (1986) Manual for estimation of probable maximum precipitation, *Operational Hydrology Report 1*, 2nd edn, World Meteorological Organization, Geneva, 269 pp.

Yevjevich, V. (1963) Fluctuations of wet and dry years, Part 1, Research data assembly and mathematical models, *Hydrology Papers 1*, Colorado State University, Fort Collins, Colo.

Zuzel, J. F. and L. M. Cox (1975) Relative importance of meteorological variables in snowmelt, *WRR*, **11**: 174–6.

3. Interception

3.1 Introduction and definitions

The amount of precipitation actually reaching the ground surface is largely dependent upon the nature and density of the vegetation cover, if this exists, or upon the existence of an artificial cover of buildings, roads and pavements. This cover, whether natural or artificial, intercepts part of the following precipitation and temporarily stores it on its surfaces, from where the water is either evaporated back into the atmosphere or falls to the ground.

The three main components are *interception loss*, i.e. water which is retained by plant surfaces and which is later evaporated away or absorbed by the plant; *throughfall*, i.e. water which either falls through spaces in the vegetation canopy or which drips from leaves, twigs and stems to the ground surface; and *stemflow*, i.e. water which trickles along twigs and branches and finally down the main stem or trunk to the ground surface. Only in the case of interception loss is the water prevented from reaching the ground surface and so taking part in the land-bound portion of the hydrological cycle. In this sense, therefore, interception loss may be regarded as a *primary* water loss and it is evidently this component of interception that is of most concern to the hydrologist. In exceptional circumstances, however, vegetation may intercept moisture in the air which would not otherwise have fallen as precipitation and in this case the main hydrological interest lies in the amount of water that is transmitted to the ground as throughfall and stemflow.

Although it is now recognized that part, if not the whole, of interception loss as defined above represents a significant net addition to catchment evaporative losses, this has not always been the case. For many years interception losses were seen as a substitute for other types of evaporative loss from a vegetation-covered surface which could be safely ignored by the practising hydrologist. Remarkably, interception received only a passing reference in two recent British engineering hydrology textbooks (cf. Shaw, 1988; Wilson, 1983). It will be shown in this chapter, however, that interception is hydrologically significant and that hydrologists now have a much clearer, if as yet incomplete, understanding of the energy and water-balance processes involved.

3.2 Factors affecting interception loss from vegetation

The fact that following a dry spell interception loss is usually greatest at the beginning of a storm and reduces with time reflects the interaction of the main factors that affect it. Of these, undoubtedly the most important is the *interception storage* of the vegetation cover, i.e. the ability of the vegetation surfaces to collect and retain falling precipitation. At first, when all the leaves and twigs or stems are dry, this is at a maximum (i.e. the *interception storage capacity*) and a very large percentage of precipitation is prevented from reaching the ground. As the leaves become wetter the weight of water on them eventually overcomes the surface tension by which it is held and thereafter further additions from rainfall are almost entirely offset by the water droplets falling from the lower edges of the leaves.

Even during rainfall, however, a considerable amount of water may be lost by *evaporation* from the leaf surfaces, so that even when the initial interception storage capacity has been filled, there is some further fairly constant retention of falling precipitation to make good this evaporation loss, which is significantly higher from wetted than from non-wetted vegetation surfaces. Indeed, during long continued rains, the interception loss may be closely related to the rate of evaporation, so that *meteorological factors* affecting the latter are also relevant to this discussion. While rain is actually falling *windspeed* is a factor of real significance. Other conditions remaining constant, evaporation tends to increase with increasing windspeed, so that during prolonged periods of rainfall the interception loss is greater in windy than in calm conditions. This observation may, however, be less applicable to rainfalls of short duration during which the effect of high windspeeds in reducing interception storage capacity, by prematurely dislodging water collected on vegetation surfaces, partially outweighs the greater evaporative losses.

The *duration of rainfall* is thus a secondary factor in that it influences interception by determining the balance between the reduced storage of water on the vegetation surfaces, on the one hand, and increased evaporative losses, on the other. Data collected during classic work by Horton (1919) and in numerous subsequent investigations showed that interception loss increases with the duration of rainfall, but only gradually, so that the relative importance of interception decreases with time (cf. Fig. 3.1). Since *rainfall amount* and rainfall duration are often closely related, many investigators have been tempted to relate interception losses and rainfall amount. Evidently, however, provided that the initial interception storage capacity is filled and provided that the subsequent rate of rainfall at

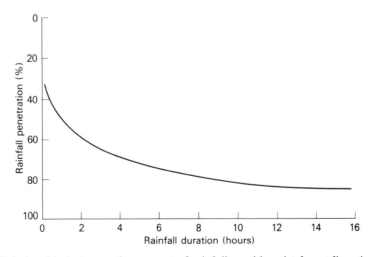

Figure 3.1 *Relationship between the amount of rainfall reaching the forest floor (an approximate reciprocal of interception loss) and rainfall duration. Data for a mixed deciduous forest in Poland (based on an original diagram by Olszewski, 1976).*

least equals the rate of evaporation (the expected situation), rainfall amounts can have no influence upon the magnitude of interception losses. Nevertheless, the relative importance of interception losses will tend to decrease as the amount of rainfall increases. This is illustrated in Fig. 3.2a in which the interception ratio (i.e. interception loss/precipitation) is plotted against storm precipitation amounts in an area of tropical forest in Puerto Rico. The relationship also holds good for annual conditions, as is illustrated in Fig. 3.2b by the graph of annual mean interception ratio against annual precipitation for a number of forest sites in the maritime climate of Great Britain.

Since the greatest interception loss occurs at the beginning of a storm, when the vegetation surfaces are dry and the interception storage capacity is large, it will be apparent that *rainfall frequency*, i.e. the frequency of re-wetting, is likely to be of considerably greater significance than either the duration or amount of rainfall.

Interception loss will also be affected by the *type of precipitation* and particularly by the

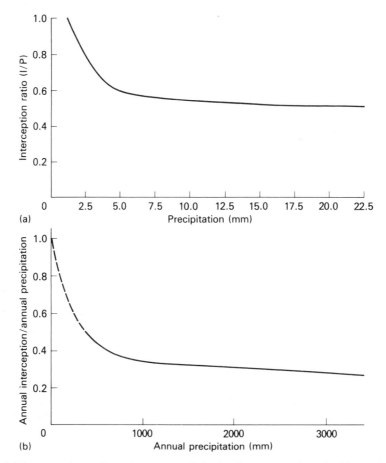

Figure 3.2 (*a*) *Interception ratio and storm precipitation in an area of tropical forest in Puerto Rico* (*based on data in Clegg, 1963*). (*b*) *Annual mean interception ratio and annual precipitation for a number of British sites* (*based on an original diagram in IH, 1982*).

contrast between rain and snow, which will be discussed more fully at a later stage. Another important factor, which also merits a separate discussion below, is the variation of interception loss with the *type and morphology of the vegetation cover.*

3.3 Interception and the water balance

The role of interception in the water balance of catchment areas is now comparatively well understood. This position has, however, been reached largely as a result of quite recent progress in our understanding of the mechanisms that control evaporation and transpiration from vegetation.

In the past three partially conflicting and partially complementary views have coexisted, each having some empirical support but little basis in theory, which maintained respectively that interception losses had a neutral, a negative and a positive effect on the catchment water balance.

The neutral hypothesis emphasized that interception losses are essentially evaporative and that, since only a certain amount of energy is available in any period of time, this will be used either to evaporate water from within the leaf, i.e. in transpiration, or to evaporate water from the surface of the leaf, i.e. interception loss. In their experiments on grasses, Burgy and Pomeroy (1958) and McMillan and Burgy (1960) concluded that wet foliage evaporation was, in fact, equally balanced by a reduction in transpiration loss, so that with this type of vegetation the net interception loss was zero. Similar conclusions were reached by Rakhmanov (1958) in his discussion of forest interception losses. In such cases, interception loss would clearly be an alternative to and not an addition to transpiration and would therefore have little, if any, effect upon the water balance of a catchment area. This hypothesis was undoubtedly given additional credence and a prolonged existence through its apparent espousal by Penman (1963) who affirmed that '. . . the same energy cannot be used twice, and while the intercepted water is being evaporated the drain on soil water is checked'.

Certainly scattered experimental evidence had been available for many years (cf. Hirata, 1929; Kittredge, 1948) to support the negative hypothesis which regarded interception unequivocally as a loss of precipitation that would otherwise have been available at the ground surface for direct evaporation, for infiltration through the surface or for overland flow. This view was implied by Whitmore (1961) who suggested that in South Africa interception probably amounts to about 5–15 per cent of the annual rainfall and it was clearly the view of Law (1956), whose experiments in a small spruce plantation in the Pennines showed that water losses from a forested area exceeded those from adjacent grassland by some 280 mm in a year.

Fundamental questions to be answered were whether evaporation from the wetted surfaces of vegetation could take place at a significantly higher rate than the evaporation and transpiration from unwetted vegetation and whether significant evaporation occurred from wetted dead and dormant vegetation and during the winter and at night when transpiration rates would otherwise be negligibly small. Affirmative answers to these questions then required an explanation of the sources of energy for the additional evaporative losses.

Evidence accumulated rapidly during the 1960s, mostly from forested areas, to support the conclusion that intercepted water evaporates much faster than transpired water, and therefore much of the interception loss represents an additional loss in the catchment water balance. Thorud (1967), for example, applied water artificially to the foliage of potted ponderosa pine and found that 91 per cent of the applied water was a net loss to soil moisture and Rutter (1963) found that evaporative losses from cut wetted branches exceeded those from unwetted branches losing water only by transpiration. More realistically, field experiments by Rutter (1963, 1967), Patric (1966), Helvey (1967), and Leyton et al. (1967) indicated that during the winter period the loss of intercepted water considerably exceeded the transpiration rate in the same environmental conditions. Results from small catchment studies showed that substantial increases in water yield resulted from the removal of forest vegetation (Hewlett and Hibbert, 1961; Hibbert, 1967) and that decreased yields resulted from the conversion of hardwood forest to pine forest (Swank and Miner, 1968), largely as a result of interception effects.

Subsequently a combination of theoretical analysis and field data collection have confirmed that precipitation intercepted by vegetation evaporates at a greater rate than transpiration from the same type of vegetation in the same environment (Murphy and Knoerr, 1975) and that the difference may be of the order of 2–3 times (Singh and Szeicz, 1979) or as much as 5 times the transpiration rate (Stewart and Thom, 1973). If account is taken of the additional water losses resulting from the existence of high night-time rates of evaporation of intercepted water, which Pearce et al. (1980) found for an evergreen mixed forest in New Zealand to be very similar to mean daytime wet canopy evaporation rates, it seems clear that net interception loss may be as high as 84 per cent of gross interception or 52 per cent of total evaporation. Furthermore, as Pearce et al. (1980) observed, net interception loss will increase as the proportion of night-time rainfall duration and amount increases. This means that, in many high-rainfall areas, especially in maritime climates, where at least one-half of rainfall may occur at night, the importance of interception as an evaporative loss and the magnitude of the net loss may be greatly enhanced in relation to areas where rainfall is dominated by daytime convective activity.

In specific conditions other factors may result in additional net interception losses. For example, in winter conditions in cold climates transpiration may be limited more by the availability of water than of energy (Goodell, 1963). Then by increasing the amount of available water, interception would increase the total loss of water from a catchment area. The evaporation of water intercepted by dormant or dead, and therefore non-transpiring, vegetation and by a litter layer would certainly represent a net interception loss (Burgy and Pomeroy, 1958; McMillan and Burgy, 1960), the only factor involved in this case being the interception storage capacity and its depletion by evaporation. It was the storage aspect of interception loss that Zinke (1967) considered might play the greatest part in affecting the catchment water balance.

The primary explanation of the higher evaporation rate from wetted vegetation surfaces, and especially from wetted forest canopies, relates to the relative importance of the two main resistances imposed at the vegetation canopy on the flux of water vapour into the overlying atmosphere. This will be discussed in more detail in Chapter 4, in relation to the Penman–Monteith equation for calculating evaporation. At this stage it will be sufficient to note that the *surface resistance* is a physiological resistance, imposed by the vegetation canopy itself on the movement of water by transpiration, and the *aerodynamic resistance* is a measure of the resistance encountered by water vapour moving from the vegetation surface as wet surface evaporation into the surrounding atmosphere. In dry conditions forest canopies probably have a slightly higher surface resistance than grass and other lower-order vegetation, but when the vegetation surfaces are wet this resistance is effectively 'short-circuited' and reduces to zero for all vegetation types (Calder, 1979). The aerodynamic resistance depends essentially on the roughness of the vegetation surface, which tends to be significantly greater for trees than for grass. Accordingly, the resistance to vapour flux is lower for all wetted vegetation surfaces, compared with dry vegetation, and relatively lower still for forest compared with grasses and other lower-order covers.

The additional energy required to maintain the higher rates of evaporative loss permitted by the dominating role of the aerodynamic resistance for wetted vegetation appears to be attributable to a combination of advected energy and radiation balance modifications. The analysis by Rutter (1967) indicated that in wet canopy conditions evaporation losses may not be controlled predominantly by the radiation balance but rather that the wet canopy acts as a sink for advected energy from the air. Certainly he found that when intercepted water was being evaporated the foliage was measurably cooler than the surrounding air and that the resulting temperature gradient was sufficient to yield a heat flux to supply the energy deficiency. This hypothesis was subsequently confirmed in a number of investigations, mainly of forested areas (cf. Stewart and Thom, 1973; Thom and Oliver, 1977), and elaborated to the extent that it is recognized now that the advected energy may be derived either from the heat content of the air passing over the vegetation canopy (Stewart, 1977; Singh and Szeicz, 1979) or from heat stored in the canopy space and the vegetation itself (Moore, 1976).

In this connection, Singh and Szeicz cautioned that their own forest experiment, and also that of Stewart, was carried out in a comparatively small forested area, surrounded by farmland, where large-scale advection of 'surplus' energy was to be expected. It was possible, though, that in the case of very large tracts of forest as there are on the Canadian shield, where trees extend for many hundreds of kilometres, less surplus energy may be available when large areas are wetted, although localized, spatially random, thunderstorm-type wetting would still permit sensible heat to be released from the dry areas to boost evaporation in the wetted areas.

Additional proof of the role of advection and stored energy in promoting wet canopy evaporation was provided by Pearce *et al.* (1980) who confirmed the evidence of high evaporation rates during the night when no other energy source is available.

Simulation of the energy exchange between the atmosphere and a vegetation surface by Murphy and Knoerr (1975), however, indicated that, in appropriate conditions, radiation balance modifications may also play a significant role. They found that the integrated effect of interception on the energy balance of a forest stand was an increase in the latent heat exchange, at the expense of the long-wave radiation and sensible heat exchange, which varied according to relative humidity and windspeed conditions. As a result they concluded that enhanced evaporation of intercepted water can occur for forests of large areal extent where horizontal advection may be negligible.

A third view of the role of interception in the water balance relates to the fact that in certain circumstances the interaction of water loss and gain in vertical and horizontal interception respectively may result in a *net gain* of water in a catchment area. It will be apparent from a later discussion of horizontal interception in Section 3.9 that this is most likely to be true in forested areas of high relief, where fogs or low cloud are prevalent.

Finally, the interception process may exert microscale effects on the catchment water balance because of localized soil moisture deficits resulting from substantial reduction in precipitation and throughfall reaching the ground surface, the redistribution of rainfall by throughfall and stemflow and the frequent contrast in drop size between precipitation falling onto the vegetation canopy and throughfall to the ground surface beneath the canopy.

The spatial distribution of precipitation may be changed in a number of ways as a result of passing through a vegetation canopy. It has been long established that generally in forested areas throughfall and dripping meltwater are concentrated at the edges of the tree crowns, whilst concentrated drip close to the trunk and stemflow itself often result in high values of infiltration and soil moisture recharge (cf. Specht, 1957; Voigt, 1960) and even the initiation of minor rills and channels in the surface (cf. Leopold *et al.*, 1964; Douglas, 1967; Jackson, 1975).

More recently, improvements in rainfall simulators and other instrumentation have enabled detailed studies of other types of vegetation cover, including grasses and agricultural crops. Finney (1984) investigated the possible paths taken by raindrops falling on Brussels sprouts, sugar beet and potatoes, i.e. they may fall between the leaves, their properties remaining unaltered; be intercepted and redirected as stemflow; be intercepted and coalesce, to fall subsequently as drip; or be intercepted and shattered by impact with the vegetation and then be redirected as small drops between the leaves. He found that as the plants matured and their interception area increased, the resulting decrease in throughfall was accompanied by an increased stemflow and leaf drip and a reduction in soil detachment except at leaf drip points.

Rainfall simulator experiments with tussock grasses showed the way in which the plant structure, with its convergent leaf arrays, directed intercepted rainfall towards the base of tussocks and sods and led De Ploey (1982) to the view that stemflow could play '... a major role in the process of runoff generation on slopes with steppe-like vegetation'.

3.4 Measuring interception

The normal method of measuring interception loss is to measure the precipitation above the vegetation layer (R), ground precipitation below the vegetation canopy (R_g) and in many cases stemflow (Q_s). The total interception loss may be calculated simply from

$$\sum I = R - (R_g + Q_s) \tag{3.1}$$

In the past this method has been used more for forest vegetation than for lower-order covers. Even then major problems arise because of the effects of the aerodynamic roughness of the vegetation cover on the catch of the canopy-level gauges (see Section 2.4.1) and the sampling difficulties imposed by the great spatial variability of throughfall and stemflow. To some extent these problems have been overcome by the use of aerodynamically shaped raingauges for the canopy-level measurement, a combination of randomly located trough and sheet gauges for throughfall measurement and spiral gutters sealed to the trunk, leading stemflow into a conveniently located collecting container. Recent work in the Amazonian rain forest, however, emphasized the magnitude of the sampling problems involved and indicated that previous calculations of throughfall and stemflow for that type of forest may be in error as a result (Lloyd and Marques, 1987).

For grasses and other lower vegetation other techniques may be possible. Small weighing lysimeters, for example, have been used to measure the wet-surface evaporation loss from heather (Hall, 1985; 1987), and in De Ploey's (1982) experiments *Molinia* stems protruded from the 100 cc measuring cylinders which were used to catch the stemflow. One of the most satisfactory methods of measuring grass interception losses was described by Corbett and Crouse (1968). They drove 25 cm diameter metal collars into the soil so that about 2.5 cm of the collar remained above the soil surface. When the grass height reached 5 cm, the soil surface inside the collar was sealed by first sifting on a fine layer of sand, to provide a uniform sloping surface for drainage, and then applying a latex emulsion which did not affect grass growth. A drain in the side of each collar carried combined throughfall and stemflow to a collecting container or measuring device.

Water-balance approaches have been used to measure indirectly the magnitude of interception loss. Some studies, for example, have utilized regular soil moisture measurements beneath the vegetation cover, together with precipitation measurements. In other cases, small instrumented catchments have been used to provide large-scale estimates of interception loss. Swank and Miner (1968), for example, reported that the effects of converting mature hardwoods on two experimental catchments in the southern Appalachians to eastern white pine was to reduce streamflow after 10 years by almost 10 cm. Since most of the water yield reduction occurred during the dormant season, it was attributed mainly to greater interception loss from white pine than from hardwoods. Increases in water yield, also attributable largely to interception effects, were reported by Pillsbury *et al.* (1962) and by Hibbert (1971), after conversion of chaparral scrub to grass.

Finally, there have been valuable attempts to quantify interception storage capacity and its individual components. Gamma-ray attenuation, for example, has been used to measure directly the amount of water held on a forest canopy (Calder and Wright, 1986). Herwitz (1985) determined the interception storage capacity of tropical rainforest leaf surfaces using a rainfall simulator and the interception storage capacity of the trunks and woody surfaces by immersing bark fragments in aqueous solutions. He combined these data with data on the leaf area index (LAI), i.e. leaf surface area/projected crown area, calculated with the help of large-scale aerial photographs, and the woody area index (WAI), i.e. woody surface area/projected crown area, to determine total interception storage capacity.

3.5 Modelling interception

Various approaches to modelling interception loss have been developed, although inevitably in view of the complexity of the interception process and the difficulty of establishing precise values for the major components and influencing factors (e.g. canopy and stem storage capacity, drip rates, stemflow, aerodynamic resistance, evaporation rates, etc.), most of the models suffer either from overgeneralization and simplification or from exacting and extensive data and processing demands.

The simplest models are those which incorporate empirical, regression-based expressions relating interception loss to precipitation characteristics. Horton (1919) was probably the first to propose

$$\sum I = \int_0^t E \, dt + S \tag{3.2}$$

for storms which saturate the vegetation canopy, where $\sum I$ is the amount of interception loss (in mm), E is the rate of evaporation of intercepted water and t is the duration of rainfall. S is the interception 'storage capacity' of the canopy, a term over which there is some confusion in the literature (Gash, 1988). As used here, it is the amount of water left on the canopy, in conditions of zero evaporation, after rainfall and drip have ceased, i.e. the minimum necessary to cover all the vegetation. This differs from its use, for example by Herwitz (1985), to signify the 'maximum storage capacity' of the vegetation canopy. Equation (3.2) can be elaborated, by considering separately evaporation before and after canopy saturation, to give

$$\sum I = \int_0^{t'} E \, dt + \int_{t'}^t E \, dt + S \tag{3.3}$$

where t' is the time taken for saturation of the canopy to occur.

Although Horton (1919) recognized that Eq. (3.2) was more logical, he concluded that it would often be more convenient to incorporate precipitation amounts rather than precipitation duration. Accordingly, there are many empirical models of interception loss which take the general form

$$\sum I = aP_g + b \tag{3.4}$$

where $\sum I$ is the amount of interception loss, P_g is the gross rainfall on the vegetation canopy and a and b are regression coefficients. Equation (3.4) can be used either to describe individual storm data or, if it is assumed that there is only one rainfall event per day, to describe daily interception loss as a function of daily gross rainfall (Gash, 1979).

Merriam (1960) incorporated an exponential expression to allow for the observed increase in storage with increased precipitation, giving

$$\sum I = S \left[1 - \exp\left(\frac{-P_g}{S}\right) \right] + E \, dt \tag{3.5}$$

and Jackson (1975) found that a semi-logarithmic curve fitted his data for tropical forest slightly better than other models, so that

$$\sum I = a + b \ln \bar{R} + c \ln t \tag{3.6}$$

where a, b and c are empirical coefficients and $\bar{R}$ is the average rate of rainfall during the storm.

Useful reviews of simple models such as these were provided by Zinke (1967), Jackson (1975), Gash (1979) and Massman (1983). They stressed that, while the models are easy to use, they do not always give satisfactory quantitative results when the coefficients are derived by regression against a specific set of data and that empirical results may not be valid for similar vegetation covers at other sites.

Models that are based on more fundamental physical reasoning tend to minimize many of the weaknesses of empirical models but usually require frequent (e.g. hourly) data inputs for rainfall and throughfall rates and for meteorologically based estimates of evaporation. Probably the most rigorous of such models is that developed by A. J. Rutter, which uses a digital computer program to solve the vegetation water-balance equation numerically.

The model, that was originally described by Rutter *et al.* (1971) and subsequently elaborated and generalized by Rutter *et al.* (1975) as a result of work in hardwood and coniferous forest stands, is predicated on the role of water storage on the vegetation canopy and stems. This store is added to by intercepted rainfall and depleted by evaporation, drip and drainage. The rates of evaporation and drip are assumed to vary with the amount of water on the canopy and, accordingly, the model is designed to calculate a running balance of rainfall, throughfall, evaporation and changes in canopy and stem storage. Evaporation from the wetted vegetation surfaces constitutes the interception loss.

The rate of input of water to the vegetation canopy is

$$(1 - p)R \tag{3.7}$$

where R is the rate of rainfall and p is the proportion of rain that falls through gaps in the canopy. The model assumes that when the depth of water (C) stored on the vegetation canopy equals or exceeds its storage capacity (S) evaporation will take place at the rate given by the Penman–Monteith equation (see Section 4.6.3 under 'The Penman–Monteith model'), i.e.

$$E_p = \frac{\Delta R_n + pc \, \delta e / r_a}{\lambda(\Delta + \gamma)} \tag{3.8}$$

where Δ is the rate of increase with temperature of the saturation vapour pressure of water at air temperature, R_n is net radiation, p and c are respectively the density and specific heat of air, δe is the vapour pressure deficit, r_a is the aerodynamic resistance (see Section 3.3), λ is the latent heat of vaporization of water and γ is the psychrometric constant.

When $C < S$ the evaporation is reduced proportionately so that

$$E = E_p \frac{C}{S} \tag{3.9}$$

The rate of drip drainage from the canopy is assumed to be a function of the storage capacity of the canopy and the depth of water stored so that

$$D = D' \exp(bC) \tag{3.10}$$

where D' and b were derived from observations as described by Rutter and Morton (1977). Subsequently the problems of deriving a satisfactory drip expression have been discussed and other expressions proposed to replace Eq. (3.10) (cf. Calder, 1977; Massman, 1980, 1983).

The storage capacity of the branch/stem system (S_t) is considered to be replenished by a constant proportion of rainfall (p_t) which is diverted to that part of the branch system that drains to the trunks. When the storage capacity has been completely filled depletion from this store will take place at the rate given by the Penman–Monteith equation, E_{pt}, which is assumed to be linearly related to E_p so that

$$E_{pt} = eE_p \tag{3.11}$$

where e is a constant for the forest stand; when the depth of water (C_t) on the branches and trunks is less than S_t the evaporative loss from them is

$$E_t = E_{pt} \frac{C_t}{S_t} \tag{3.12}$$

In contrast to the expression for drip from the canopy, the drainage of water in excess of S_t is assumed to be immediate.

The model was tested against observed interception losses for six forest types including both deciduous broad-leaved and evergreen coniferous species (Rutter *et al.*, 1975) and a graphical summary of the comparisons, indicating a very satisfactory model performance, is shown in Fig. 3.3.

An alternative approach is to integrate the mass balance equation analytically. In this context one of the most satisfactory attempts was the analytical model developed by Gash (1979) which, despite a number of simplifying assumptions, retains much of the physical reasoning of the more complex Rutter model. The Gash model calculates interception loss on a storm-by-storm basis and separately identifies the meteorological and biological controls of interception loss to give a framework within which results may be extrapolated more readily to other areas. The main simplifying assumptions are (Gash, 1979):

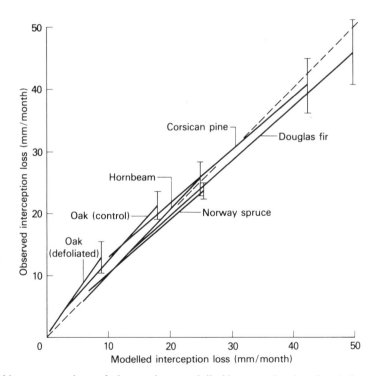

Figure 3.3 *Linear regressions of observed on modelled interception loss for six forest species, with standard deviations about the lines shown at the right-hand end of each. The lines are not extrapolated beyond the range of observed data. The dashed line has unit slope passing through the origin. (From an original diagram in Rutter et al., 1975. Blackwell Scientific Publications, Ltd.)*

(a) that it is possible to represent the real rainfall pattern by a series of discrete storms, separated by sufficiently long intervals for the canopy and trunks to dry,

(b) that the meteorological conditions prevailing during any wetting up of the canopy are similar to those prevailing during the storms and

(c) that there is no drip from the canopy during wetting up and that at the end of rainfall canopy storage reduces within about 30 minutes to the minimum value necessary for saturation.

Simplifying, the total interception loss during evaporation from a saturated canopy is

$$\sum I = \sum \left[\int_0^{t'} E \, dt + \frac{\bar{E}}{\bar{R}}(P_g - P'_g) \right] + nS \qquad \text{for } n \text{ storms} \qquad (3.13)$$

where $\bar{E}$ and $\bar{R}$ are the mean rates of evaporation and rainfall and P'_g is the amount of rain necessary to saturate the canopy. The interception loss during the wetting up of the canopy to saturation is

$$\sum I = n(1 - p - p_t)P'_g - nS \qquad \text{for } n \text{ storms} \qquad (3.14)$$

where p is the proportion of rain that falls through the canopy without striking a surface and p_t is the proportion that is diverted to the trunks as stemflow. The interception loss for small storms insufficient to saturate the canopy is

$$\sum I = (1 - p - p_t) \sum P_g \qquad \text{for } m \text{ storms} \qquad (3.15)$$

The interception loss from the trunks in storms that fill the storage is

$$\sum I = qS_t \qquad \text{for } q \text{ storms} \qquad (3.16a)$$

and for storms that do not fill the storage is

$$\sum I = p_t \sum P_g \qquad \text{for } n + m - q \text{ storms} \qquad (3.16b)$$

where S_t is the trunk storage capacity and q is the number of storms having sufficient rainfall to fill the trunk store.

The expression for total interception loss, derived by combining these four equations is

$$\sum I = (1 - p - p_t) \sum P'_g + \left(\frac{\bar{E}}{\bar{R}} \right) \sum (P_g - P'_g) + (1 - p - p_t) \sum P_g + qS_t + p_t \sum P_g$$
$$(3.17)$$

This model was applied to data from Thetford Forest in East Anglia and, as Fig. 3.4 indicates, produced satisfactory agreement between observed and modelled interception loss. In addition, both the Gash and Rutter models have been tested against data from Amazonian tropical forest (Lloyd *et al.*, 1988) with equally satisfactory results.

Inevitably research efforts continue to improve both the conceptual basis and the general applicability of models of interception loss. Many of these attempts have focused upon the type of drip/drainage expression used in the Rutter model; others have attempted to simplify both the structure and the data demands of the evaporation expression.

There is little doubt that drip expressions similar to Eq. (3.10) are unsatisfactory in the sense that they assume that the drip rate is determined solely by the amount of water

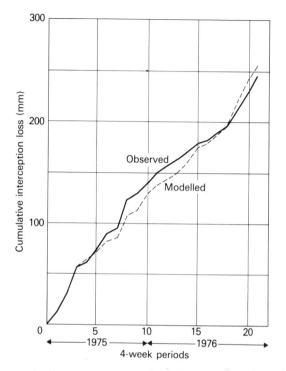

Figure 3.4 *Cumulative totals of observed and modelled interception loss for 21 four-week periods during 1975 and 1976 (from an original diagram by Gash, 1979).*

stored on the vegetation canopy. This means that they do not account explicitly for either the influence of rainfall rate upon the drip rate or for the physical dislodgement of previously intercepted rain by falling raindrops or for the multilayered nature of many vegetation canopies. Accordingly, Massman (1983) proposed a model for drip rate which assumes that, after some critical amount of water has been intercepted by leaf surfaces near the top of the canopy, further interception at that level would dislodge some of the stationary droplets which would then fall out of the tree or be intercepted at lower levels. In other words, during rainfall the drip rate may be controlled by water droplets cascading through the tree, alternatively falling from and impacting on successively lower surfaces (cf. Calder, 1977). As Massman (1983) observed, such a cascade process will probably be affected significantly by the momentum of the raindrops just above the canopy, one index of which will be the rainfall rate, so that drip rate may be expressed as

$$D = (D_o + d_oR)\left(\frac{C}{S}\right) \tag{3.18}$$

where D_o is an empirically determined constant and d_o varies with individual storm characteristics and requires both a computer and some extensive instrumentation for its calculation.

3.6 Interception losses from different types of vegetation

It would be expected, on the basis of the preceding discussions in this chapter, that interception loss will vary from one site to another in response to differences in vegetation and precipitation characteristics. The main vegetational effects will relate to differences in interception storage capacity from one vegetation type to another and in aerodynamic roughness and its implications for the aerodynamic resistance and the rate of evaporation from the wetted vegetation surface. The most important precipitation characteristics will be duration, frequency and intensity, apart from precipitation type, which will be discussed in the next section.

Because of the complexity of the interception process and the interrelationships between the vegetational and meteorological factors which determine the magnitude of interception losses, it is often difficult to make well-founded comparisons between published data on interception loss. It is clear, however, that in most cases, interception losses are greater from trees than from grasses or agricultural crops, although the reasons for this may vary with meteorological conditions. In the uplands of Britain, for example, where long-duration, low-intensity rain is common and vegetation surfaces are wet for considerable periods of time, the primary reason for the greater loss from trees is the much increased evaporation rate in wetted conditions. This is due to the greater aerodynamic roughness of the trees rather than because of their slightly higher interception storage capacity (Calder, 1979). In other conditions, however, the role of interception storage capacity in determining differences in interception loss between, say, trees and grass may be much more important. This would be the case where re-wetting is frequent, drying rapid and the duration of canopy wetting much shorter.

The values of interception loss for different vegetation covers which are quoted in this section must be interpreted, as far as possible, in the light of both the completeness of the measured data, where this is known, and also of weather conditions. For example, in some cases measurements were made of stemflow, in others an arbitrary allowance was made for this component and in still other cases it appears to have been ignored completely. Again the data presented in the literature are not always accompanied by an adequate analysis of meteorological conditions, particularly concerning the amount, duration, frequency, intensity and type of precipitation, all of which need to be known to permit a meaningful interpretation of the data.

3.6.1 *Woodlands*

Despite the fact that, in most cases, the leaf density is greater in deciduous than in coniferous forest, the bulk of the experimental evidence shows that interception losses are greater from the latter. Reviewing a broad range of Russian, European and American data, Rakhmanov (1966) suggested that coniferous forests, together with sparse woods and inhibited stands on peat bogs and other marshy terrain, intercept an average of 25–35 per cent of the annual precipitation compared with 15–25 per cent by broad-leaved forests. This contrast is illustrated for spruce and beech in Fig. 3.5. Both types of woodland show the reduction in relative importance of interception loss as rainfall amounts increase but, over the complete range of rainfall totals, interception loss is markedly greater from the spruce forests. One of the reasons for this contrast may be that, while water droplets remain clinging to separate spruce needles, they tend to run together on the beech leaves and so

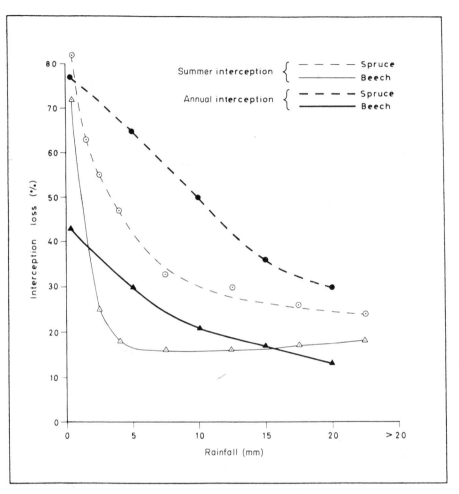

Figure 3.5 *Interception losses from spruce and beech forests (based on data from F. E. Eidmann, quoted by Penman, 1963, and from E. Hoppe quoted by Geiger, 1957).*

drop or flow on to twigs and branches (Geiger, 1957). It is also likely that the open texture of coniferous leaves allows freer circulation of air and consequently more rapid evaporation of the retained moisture. Observed annual interception losses from different coniferous species in British conditions have varied from 35 to 48 per cent and for hornbeam and oak were 36 and 18 per cent respectively (Law, 1956; Rutter *et al.*, 1975). For mixed evergreen forest in South Island, New Zealand, interception losses averaged 29 per cent (Pearce *et al.*, 1982) and variations from 12 to 54 per cent were quoted for tropical forest in Indonesia, China, Brazil and Puerto Rico by Franzle (1979) and Clegg (1963).

With regard to seasonal contrasts, winter and summer losses appear to be about the same from coniferous trees (Johnson, 1942; Wisler and Brater, 1959). Indeed, Fig. 3.6

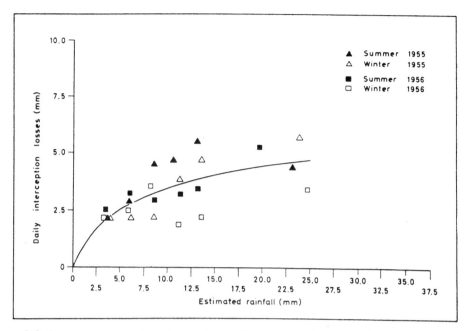

Figure 3.6 *Seasonal interception losses from sitka spruce (from an original diagram by Law, 1958).*

shows that, in northern England, winter losses from spruce may be slightly higher than summer losses. Expectedly, however, interception losses from deciduous trees are greatest during the period of full leaf. Lull (1964) quoted figures showing that, in northern hardwood and aspen–birch forests, interception losses with the trees in leaf were 15 and 10 per cent of the precipitation respectively, whereas with leafless trees the losses were 7 and 4 per cent. Similarly, Wisler and Brater (1959) suggested that summer losses for deciduous trees are two or three times greater than winter losses, although, in certain conditions, a considerable quantity of freezing rain may be stored on the twigs and branches and, being relatively immobile, is more susceptible to evaporation than is liquid rainwater at this time of year (Colman, 1953). In some climates, e.g. Mediterranean, rainfall rates will be higher in summer because of convective storms. This will give lower interception than in winter, when rainfall rates associated with long-duration, synoptic storms are lower (Gash, 1988).

A further important aspect of interception loss in wooded areas is that this often occurs at two or more levels within the vegetation cover. Precipitation is first intercepted by the upper canopy; some of the throughfall is then intercepted again by undergrowth, or by a layer of ground litter. Comparatively little is known about the importance of this secondary interception, although it will tend to increase with the amount of rainfall, because during light rains little or no throughfall will occur from the crown canopy, whereas during long, heavy storms throughfall will probably fill the interception storage capacity of the undergrowth or ground litter.

3.6.2 *Grasses and shrubs*

The total leaf area of a continuous cover of mature grass or shrub may closely resemble that of a closed canopy forest, so that the interception storage capacity may also be similar in magnitude to that of trees during the season of maximum development. Because of their higher aerodynamic resistance and shorter growing season, however, total annual interception loss from grasses is considerably less than from, say, deciduous woodland. Furthermore, in areas where grass is cut for hay or silage, or is heavily grazed in the field, interception losses will tend to be small, approaching zero in extreme cases (Lee, 1942).

Kittredge (1948) found that in California undisturbed grass of *Avena, stipa, lolium* and *bromus* species intercepted 26 per cent of an 826 mm seasonal rainfall, while in Missouri, bluegrass intercepted 17 per cent of the rainfall in the month before harvesting (Musgrave, 1938). In neither case was a correction made for stemflow, which was shown by Beard (1962) to account for between 35 and 45 per cent of total rainfall during a period when grass interception loss was about 13 per cent of the total rainfall. Work by Corbett and Crouse (1968), in which throughfall and stemflow were measured, showed that annual interception losses from *bromus* in southern California averaged about 8 per cent. Interception experiments with cut vegetation or with artificial sprinkling have given widely divergent results.

There is a paucity of data on interception by herbaceous and shrubby covers typical of the heaths and moorlands of Europe (Leyton *et al.*, 1967). This is an important omission since afforestation in such areas occurs at the expense of heather rather than grass. Measurements on heather in Scotland showed that between 35 and 66 per cent of precipitation penetrated the canopy, the amount increasing as precipitation increased, although no measurement was made of stemflow (Aranda and Coutts, 1963). Research at the Institute of Hydrology (Hall, 1985, 1987) showed that the aerodynamic resistance for heather is lower than for grass. During wet periods, therefore, interception losses from heather are likely to be much higher than those from grass. However, it has been shown that in dry periods transpiration losses from heather are significantly lower than those from grass (Wallace *et al.*, 1982). In regions of moderate annual rainfall (c. 1500 mm) the increased interception losses are likely to counterbalance the reduction in transpiration, in high rainfall areas interception losses will dominate and in drier areas the converse will be true (IH, 1985). Interception losses from mature chaparral cover in southern California averaged about 13 per cent of annual precipitation (Corbett and Crouse, 1968).

3.6.3 *Agricultural crops*

Again, data on the interception loss from agricultural crops are sparse in relation to those from forested areas. Figure 3.7 shows that interception by corn, soybeans and oats increases initially with increasing crop density. After a certain coverage has been attained, however, the subsequent increase of interception is slight, indicating that the approximate average interception by fully developed oats, soybeans and corn is 23, 35 and 40–50 per cent respectively. Since no measurements appear to have been made of stemflow, these figures would have to be reduced by an appropriate amount in order to represent the true interception loss. Other observations during the growing season for the same three crops showed that interception losses were about 7, 15 and 16 per cent of the total rainfall for

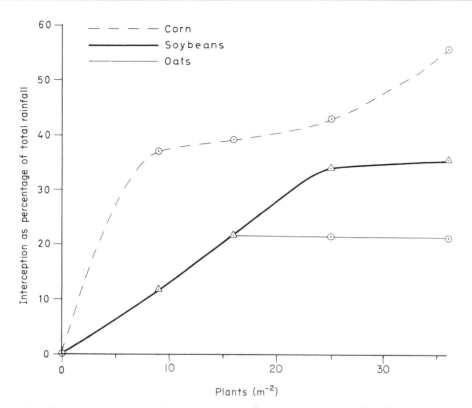

Figure 3.7 *Interception by three agricultural crops (based on data from Woollny quoted by Baver, 1956).*

oats, soybeans and corn respectively (Lull, 1964). Russian experiments showed that interception by spring wheat during the growing season was about the same as, or a little less than, that by forest in leaf for the same period, amounting to between 11 and 19 per cent of the total precipitation (Kontorshchikov and Eremina, 1963).

3.7 Interception of snow

Experimental evidence concerning the interception of snow has been frequently unsatisfactory and confusing and is largely restricted to woodland vegetation. Data reported by Lee (1980) indicated that conifers generally have a higher average interception storage capacity for snow than for rain. However, observations by J. Schubert and H. Hesselman, reported by Geiger (1957), showed that most of the snow falling on a forest reaches the forest floor. Schubert found that the ratio of snow outside to that inside the forest was 100:90, while Hesselman found the same depth of snow in a pine forest as in cuttings. Later work by West and Knoerr (1959) in the Sierra Nevada suggested that interception of snow in dense coniferous stands might amount to 8 per cent of the fall during the winter period.

There are both mechanical and thermodynamic factors that support the indication that snow interception losses may be relatively unimportant. For example, snow accumulating on vegetation surfaces is prone to large-scale mass release by rainwash and sliding under its own weight, frequently aided by wind-induced movement of the vegetation, and also the smaller-scale release of snow particles and meltwater drip. Such release mechanisms mean that in many areas snow remains on the vegetation cover for only a few days before falling to the ground, so that opportunities for evaporation are small. A typical sequence of events in Idaho is illustrated in the interceptograph trace in Fig. 3.8, which shows the continuous weight record of snow load on a tree (Satterlund and Haupt, 1970). On days 20 and 21 snow began and accumulated on the tree but as the storm turned to rain part of the intercepted snow was washed to the ground. On day 22 higher temperatures resulted in minor mass release, indicated by the irregular decline, and large mass release, indicated by the vertical trace. There was a slight decline on day 23 resulting from the evaporation of intercepted snow and then, on day 24, rain washed the remaining snow from the tree, leaving it bare and wet. On day 25 the liquid water evaporated, leaving the tree bare and dry. Similar patterns were observed during more recent tree weighing experiments in the Institute of Hydrology's field site at Aviemore, Scotland (IH, 1985). Even when snow remains on the vegetation cover for long periods of time, the energy available for evaporation and sublimation is minimal (cf. Hoover and Leaf, 1967; Miller, 1965, 1967), and in some areas, such as north-west Europe, when there *is* a transfer of water it tends to be to the snow cover, in the form of condensation, rather than away from it, in the form of evaporation or sublimation (Penman, 1963).

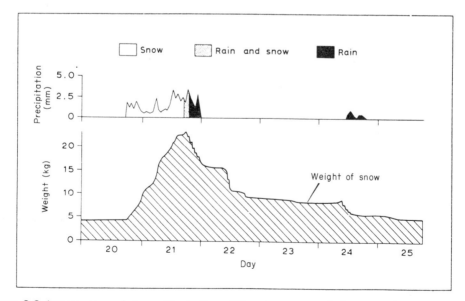

Figure 3.8 *Interceptograph trace illustrating different aspects of the snow-interception water balance. See text for explanation. (From an original diagram by D. R. Satterlund and H. F. Haupt, Water Resources Research, **6**, p. 650. © 1970 American Geophysical Union.)*

Even so, some investigators have suggested that the interceptions of snow and rain are quantitatively similar (Johnson, 1942; Rowe and Hendrix, 1951), while Satterlund and Eschner (1965), having defined the energy sources and vapour pressure deficits at the canopy level, concluded that these could account for significantly greater losses from intercepted snow than from a snow cover on open ground.

Lull (1964) suggested that there is a marked contrast between the interception of falling snow in coniferous and deciduous woodland, a view seemingly supported by a number of measurements of the depth of snow accumulation under different vegetation covers (Eschner and Satterlund, 1963; Lull and Rushmore, 1960; Schomaker, 1968; Lee, 1980).

More recently gamma-ray attenuation and tree-weighing experiments in Scotland (IH, 1987) have shown that spruce canopies can hold in excess of 28 mm water equivalent of snow and that evaporation rates from snow-covered canopies can be as high as evaporation rates from rain-wetted canopies, especially when the snow is melting during incursions of tropical maritime air. Furthermore, although sublimation rates are lower than evaporation rates, sublimation is important in snow interception from Scottish upland coniferous forest because the canopy capacity for snow may be an order of magnitude greater than for water and because of the long periods of subzero temperatures during which sublimation can take place. The combination of high evaporation rates, large storage capacity and sublimation effects indicates a potential for significant snow interception losses, and indeed losses of 86–90 per cent of total snowfall were observed for individual storms (IH, 1987).

Many of the discrepancies evident in the above discussion undoubtedly result from the difficulties of measuring snowfall, particularly in view of its tendency to drift at the edge of pronounced barriers such as forests, and in forest clearings, i.e. the very locations in which measurements are normally made. Encouragingly, more recent investigations (cf. IH, 1985, 1987) have placed a much needed emphasis on the aerodynamic processes involving transport of snow through and from the forest canopy to the site of final deposition and on the energy and water fluxes above the snow cover. As already indicated, preliminary results are exciting and appear to reverse a number of the long-held assumptions outlined above.

3.8 Interception in urban areas

In most urban areas interception by vegetation is comparatively unimportant. A significant percentage of precipitation may, however, be held by and evaporated from the surfaces of buildings, although reliable experimental data are virtually non-existent. In extreme cases, where water is led from a roof into a storage container and thence evaporated, the interception loss may be approximately 100 per cent, although normally such water is led into a drainage system or into the subsoil via sewers and storm drains.

3.9 Horizontal interception

In conditions of high atmospheric humidity, it is known that some types of vegetation, particularly trees, are able to extract moisture from the air, which would not otherwise have reached the ground as precipitation. Water droplets are formed on leaves, twigs and branches by direct condensation, often in such quantities that they fall to the ground like rainfall. This phenomenon is commonly experienced under trees during fog (Kerfoot, 1968), particularly when windspeeds are relatively high, and may be utilized for agricultural purposes. In parts of Japan, for example, forested belts are maintained to filter out a large percentage of the water from fogs moving in off the sea (Ekern, 1964). This process is also observed, however, in agricultural crops and has been held responsible for the lodging of grain crops which often occurs in calm, misty conditions (Penman, 1967).

In the sense that this *horizontal interception* or *fog drip* represents measurable precipitation beneath the vegetation canopy where none is recorded in the open, Kittredge (1948) suggested that it could be regarded as negative interception. Reference to the three main components of interception as defined in Section 3.1, however, indicates that although interception *loss* in these conditions is negative, throughfall and stemflow are positive, and it is these two components of interception that assume the greatest hydrological significance.

Experimental evidence has frequently emphasized that horizontal interception is essentially an edge effect and that its importance decreases markedly away from the borders of, say, a forested plot or area of relatively taller vegetation (Geiger, 1957; Oberlander, 1956). The border nature of the phenomenon is also apparent from results of various experiments with fog gauges. Nagel (1956), for example, used a fog gauge in which a wire gauze cylinder was placed above the funnel of a normal raingauge, on the basis that the gauze would intercept horizontally driven cloud and fog droplets in a manner similar to that of natural vegetation surfaces. It was found that during a period of 12 months fog drip from the 'tablecloth' of orographic cloud on Table Mountain was 3294 mm, compared with a measured rainfall of 1940 mm. Because of edge effects it is impossible to extrapolate these results to yield a value of fog drip for the vegetation cover on the mountain, but even an order of magnitude reduction of the 1354 mm excess measured fog drip would leave an impressive contribution to the water balance of Table Mountain (Penman, 1963). Similar difficulties arise when attempting to relate measurements of fog drip to natural horizontal interception on Mount Wellington, Tasmania (Twomey, 1957). In this case the fog gauge caught 1117 mm in a 10-day period during February, in comparison with the 104 mm catch of an adjacent raingauge.

Although it is difficult to quantify natural horizontal interception or fog drip, systematic experiments with artificial leaves showed that fog drip depends on the characteristics of both the fog and the intercepting surface, and led Merriam (1973) to look forward to identifying the characteristics of each in a way that would '... permit a reliable quantitative evaluation of fog drip potential'. Meanwhile we may deduce that at the windward edge of areas of upstanding vegetation, and on high crests and ridges, in foggy localities, rainfall may be exceeded by a factor of two or three (Kittredge, 1948). Certainly, it has been suggested that in those parts of Tasmania where soil and vegetation types are typical of a considerably higher rainfall than that actually measured, the anomalies can probably be attributed to the interception of cloud water by vegetation (Twomey, 1957). In North America, too, such additions to the normal precipitation are important in determining plant distributions along the west central coast (Lull, 1964) and, as Reynolds (1967) observed, a real gain in precipitation has also been demonstrated in Israel (Waisel, 1960), Hawaii (Ekern, 1964) and Chile (Kummerow, 1962).

3.10 Surface storage

Surface storage, or *depression storage* as it is commonly known, is an aspect of water losses that is allied closely to interception. It is, in fact, another example of precipitated moisture, most of which is returned to the atmosphere by evaporation before taking an active part in the land-bound portion of the hydrological cycle. Surface storage comprises the water retained in hollows and depressions in the ground surface during and after rainfall. This water is then either evaporated directly, or is used by vegetation, or else infiltrates into the soil.

Although surface storage is commonly thought of as a small-scale phenomenon related to minor depressions and puddles, it may also assume considerable importance on a larger scale where topographical conditions are particularly favourable. On the Canadian prairies, for example, glaciation created a great deal of surface storage in most of the catchment areas (Stitchling and Blackwell, 1957).

References

Aranda, J. M. and J. R. H. Coutts (1963) Micrometeorological observations in an afforested area in Aberdeenshire: rainfall characteristics, *J. Soil Sci.*, **14**: 124–33.

Baver, L. D. (1956) *Soil physics*, 3rd edn, John Wiley and Sons, New York.

Beard, J. S. (1962) Rainfall interception by grass, *Journal of South African Foresters*, **42**: 12–15.

Burgy, R. H. and C. R. Pomeroy (1958) Interception losses in grassy vegetation. *Trans. AGU*, **39**: 1095–100.

Calder, I. R. (1977) A model of transpiration and interception loss from a spruce forest in Plynlimon, central Wales, *J. Hydrol.*, **33**: 247–65.

Calder, I. R. (1979) Do trees use more water than grass?, *Water Services*, **83**: 11–14.

Calder, I. R. and I. R. Wright (1986) Gamma ray attenuation studies of interception from Sitka spruce: some evidence for an additional transport mechanism, *WRR*, **22**: 409–17.

Clegg, A. G. (1963) Rainfall interception in a tropical forest, *Caribbean Foresters*, **24**: 75–9.

Colman, E. A. (1953) *Vegetation and watershed management*, Ronald, New York.

Corbett, E. S. and R. P. Crouse (1968) Rainfall interception by annual grass and chaparral . . . losses compared. *Research Paper PSW-48*, USFS, Berkeley.

De Ploey, J. (1982) A stemflow equation for grasses and similar vegetation, *Catena*, **9**: 139–52.

Douglas, I. (1967) Natural and man-made erosion in the humid tropics of Australia, Malaysia and Singapore, *IASH Publ.*, **75**: 17–30.

Ekern, P. C. (1964) Direct interception of cloud water on Lanaihale, Hawaii, *Proc. SSSA*, **28**: 419–21.

Eschner, A. R. and D. R. Satterlund (1963) Snow deposition and melt under different vegetative covers in central New York, *Research Note NE-13*, USFS, 6 pp.

Finney, H. J. (1984) The effect of crop covers on rainfall characteristics and splash detachment, *Journal of Agricultural Engineering Research*, **29**: 337–43.

Franzle, O. (1979) The water balance of the tropical rain forest of Amazonia and the effects of human impact, *Applied Sciences and Development*, **13**: 88–117 (Institute for Scientific Cooperation, Tubingen, FRG).

Gash, J. H. C. (1979) An analytical model of rainfall interception by forests, *QJRMS*, **105**: 43–55.

Gash, J. H. C. (1988) Personal communication.

Geiger, R. (1957) *The climate near the ground* (Transl.), Harvard University Press, Massachusetts.

Goodell, B. C. (1963) A reappraisal of precipitation interception by plants and attendant water loss, *J. Soil and Water Cons.*, **18**: 231–4.

Hall, R. L. (1985) Further interception studies of heather using a wet surface weighing lysimeter system, *J. Hydrol.*, **81**: 193–210.

Hall, R. L. (1987) Processes of evaporation from vegetation of the uplands of Scotland, *Trans. Roy. Soc. Edinburgh: Earth Sciences*, **78**: 327–34.

Helvey, J. D. (1967) Interception by eastern white pine, *WRR*, **3**: 723–9.

Herwitz, S. R. (1985) Interception storage capacities of tropical rainforest canopy trees, *J. Hydrol.*, **77**: 237–52.

Hewlett, J. D. and A. R. Hibbert (1961) Increase in water yield after several types of forest cutting, *Bull. IASH*, **6**: 5–17.

Hibbert, A. R. (1967) Forest treatment effects on water yield, in *Forest hydrology*, W. E. Sopper and H. W. Lull (eds), Pergamon, Oxford, pp. 527–43.

Hibbert, A. R. (1971) Increases in streamflow after converting chaparral to grass, *WRR*, **7**, 71–80.

Hirata, T. (1929) *Contributions to the problem of the relation between forest and water in Japan*, Imperial Forest Experimental Station, Meguro, Tokyo, 41 pp.

Hoover, M. D. and C. F. Leaf (1967) Process and significance of interception in Colorado subalpine forest. in *Forest hydrology*, W. E. Sopper and H. W. Lull (eds), Pergamon, Oxford, pp. 213–24.

Horton, R. E. (1919) Rainfall interception, *Mon. Wea. Rev.*, **47**: 603–23.

IH (1982) *Research Report 1978–81*, Institute of Hydrology, Wallingford.

IH (1985) *Research Report 1981–84*, Institute of Hydrology, Wallingford.

IH (1987) Report for the period 1 April 1985–31 March 1986, in *The Natural Environment Research Council Report for 1985/86*, Swindon, NERC, pp. 61–78.

Jackson, I. J. (1975) The relationships between rainfall parameters and interception by tropical forest, *J. Hydrol.*, **24**: 215–38.

Johnson, W. M. (1942) The interception of rain and snow by a forest of young Ponderosa pine, *Trans. AGU*, **23**: 566–70.

Kerfoot, O. (1968) Mist precipitation on vegetation, *Forest Abstracts*, **29**: 8–20.

Kittredge, J. (1948) *Forest influences*, McGraw-Hill, New York.

Kontorshchikov, A. S. and K. A. Eremina (1963) Interception of precipitation by spring wheat during the growing season, *Soviet Hydrol.*, **2**: 400–409.

Kummerow, J. (1962) Quantitative measurements of fog in the Fray Jorge National Park, *Forest Abstracts*, **24**: 4576.

Law, F. (1956) The effect of afforestation upon the yield of water catchment areas, *Journal of British Waterworks Association*, **38**: 489–94.

Law, F. (1958) Measurement of rainfall, interception and evaporation losses in a plantation of Sitka spruce trees, *Proc. IASH Gen. Assoc. of Toronto*, **2**: 397–411.

Lee, C. H. (1942) Transpiration and total evaporation, Ch. 8 in *Hydrology*, O. E. Meinzer (ed), McGraw-Hill, New York.

Lee, R. (1980) *Forest hydrology*, Columbia University Press, New York.

Leopold, L. B., M. G. Wolman and J. P. Miller (1964) *Fluvial processes in geomorphology*, Freeman, San Francisco.

Leyton, L., E. R. C. Reynolds and F. B. Thompson (1967) Rainfall interception in forest and moorland, in *Forest hydrology*, W. E. Sopper and H. W. Lull (eds), Pergamon, Oxford, pp. 163–78.

Lloyd, C. R. and A. de O. Marques (1987) Spatial variability of throughfall and stemflow measurements in Amazonian rain forest, *Agricultural Forestry Meteorology*, **42**: 63–73.

Lloyd, C. R., J. H. C. Gash, W. J. Shuttleworth and A. de O. Marques (1988) The measurement and modelling of rainfall interception by Amazonian rain forest, *Agricultural Forestry Meteorology*, **43**, 277–94.

Lull, H. W. (1964) Ecological and silvicultural aspects, Sec. 6 in V. T. Chow (ed), *Handbook of applied hydrology*, McGraw-Hill, New York.

Lull, H. W. and F. M. Rushmore (1960) Snow accumulation and melt under certain forest conditions in the Adirondacks, *NE Forest Experimental Station Paper 138*, USDA.

McMillan, W. D. and R. H. Burgy (1960) Interception loss from grass, *JGR*, **65**: 2389–94.

Massman, W. J. (1980) Water storage on forest foliage: a general model, *WRR*, **16**: 210–16.

Massman, W. J. (1983) The derivation and validation of a new model for the interception of rainfall by forests, *Agric. Met.*, **28**: 261–86.

Merriam, R. A. (1960) A note on the interception loss equation, *JGR*, **65**: 3850–1.

Merriam, R. A. (1973) Fog drip from artificial leaves in a fog wind tunnel, *WRR*, **9**: 1591–8.

Miller, D. H. (1965) The heat and water budget of the earth's surface, *Advances in Geophysics*, **11**: 175–302.

Miller, D. H. (1967) Sources of energy for thermodynamically-caused transport of intercepted snow from forest crowns, in *Forest hydrology*, W. E. Sopper and H. W. Lull (eds), Pergamon, Oxford, pp. 201–11.

Moore, C. J. (1976) Eddy flux measurements above a pine forest, *QJRMS*, **102**: 913–18.

Murphy, C. E. and K. R. Knoerr (1975) The evaporation of intercepted rainfall from a forest stand: An analysis by simulation, *WRR*, **11**: 273–80.

Musgrave, G. W. (1938) Field research offers significant new findings, *Soil Conservation*, **3**: 210–14.

Nagel, J. F. (1956) Fog precipitation on Table Mountain, *QJRMS*, **82**: 452–60.

Oberlander, G. T. (1956) Summer fog precipitation on the San Francisco peninsula, *Ecology*, **37**: 851–2.

Olszewski, J. L. (1976) Relation between the amount of rainfall reaching the forest floor and the amount of rainfall over a mixed deciduous forest, *Phytocoenosis*, **5**: 127–56 (Warszawa).

Patric, J. H. (1966) Rainfall interception by mature coniferous forests of southeast Alaska, *J. Soil and Water Cons.*, **21**: 229–31.

Pearce, A. J., L. K. Rowe and J. B. Stewart (1980) Nighttime, wet canopy evaporation rates and the water balance of an evergreen mixed forest, *WRR*, **16**: 955–9.

Pearce, A. J., L. K. Rowe and C. L. O'Loughlin (1982) Hydrologic regime of undisturbed mixed evergreen forests, South Nelson, New Zealand, *J. Hydrol. (N.Z.)*, **21**: 98–116.

Penman, H. L. (1963) *Vegetation and hydrology*, CAB, Farnham Royal.

Penman, H. L. (1967) In discussion of J. Delfs, Interception and stemflow in stands of Norway spruce and beech in West Germany, in *Forest hydrology*, W. E. Sopper and H. W. Lull (eds), Pergamon, Oxford, pp. 179–85.

Pillsbury, A. F., R. E. Pelishek, J. F. Osborn and T. E. Szuszkiewicz (1962) Effects of vegetation manipulation on the disposition of precipitation on chaparral-covered watersheds, *JGR*, **67**: 695–702.

Rakhmanov, V. V. (1958) Are the precipitations intercepted by the tree crowns a loss to the forest?, *Botanicheskii Zhurnal*, **43**: 1630–3. (Trans. PST Cat. No. 293, Office Tech. Serv. US Dept. Commerce, Washington, D.C.).

Rakhmanov, V. V. (1966) *Role of forests in water conservation*, Goslesbumizdat, Moscow, 1962; Translated and edited by A. Gourevitch and L. M. Hughes, Israel Program for Scientific Translations Ltd, Jerusalem.

Reynolds, E. R. C. (1967) The hydrological cycle as affected by vegetation differences, *JIWES*, **21**: 322–30.

Rowe, P. B. and T. M. Hendrix (1951) Interception of rain and snow by second growth of Ponderosa pine, *Trans. AGU*, **32**: 903–8.

Rutter, A. J. (1963) Studies in the water relations of *Pinus sylvestris* in plantation conditions, *Journal of Ecology*, **51**: 191–203.

Rutter, A. J. (1967) An analysis of evaporation from a stand of Scots pine, in *Forest hydrology*, W. E. Sopper and H. W. Lull (eds), Pergamon, Oxford, pp. 403–17.

Rutter, A. J. and A. J. Morton (1977) A predictive model of rainfall interception in forests: III. Sensitivity of the model to stand parameters and meteorological variables, *Journal of Applied Ecology*, **14**: 567–88.

Rutter, A. J., K. A. Kershaw, P. C. Robins and A. J. Morton (1971) A predictive model of rainfall interception in forests, I. Derivation of the model from observations in a plantation of Corsican pine, *Agric. Met.*, **9**: 367–84.

Rutter, A. J., A. J. Morton and P. C. Robins (1975) A predictive model of rainfall interception in forests: II. Generalization of the model and comparison with observations in some coniferous and hardwood stands, *Journal of Applied Ecology*, **12**: 367–80.

Satterlund, D. R. and A. R. Eschner (1965) The surface geometry of a closed conifer forest in relation to losses of intercepted snow, *Research Paper NE-34*, USFS.

Satterlund, D. R. and H. F. Haupt (1970) The disposition of snow caught by conifer crowns, *WRR*, **6**: 649–52.

Schomaker, C. E. (1968) Comparison of snow interception by a hardwood and a conifer forest, *Research in the Life Sciences*, **16**: 35–43.

Shaw, E. M. (1988) *Hydrology in practice*, 2nd edn, Van Nostrand Reinhold, Wokingham, 539 pp.

Singh, B. and G. Szeicz (1979) The effect of intercepted rainfall on the water balance of a hardwood forest, *WRR*, **15**: 131–8.

Specht, R. L. (1957) Dark Island Heath (Ninety-Mile Plain, S. Australia). IV. Soil moisture patterns produced by rainfall, interception and stem-flow, *Australian Journal of Botany*, **5**: 137–50.

Stewart, J. B. (1977) Evaporation from the wet canopy of a pine forest, *WRR*, **13**: 915–21.

Stewart, J. B. and A. S. Thom (1973) Energy budgets in pine forest, *QJRMS*, **99**: 154–70.

Stitchling, W. and S. R. Blackwell (1957) Drainage area as a hydrologic factor on the glaciated Canadian prairies, *Proc. IASH Gen. Assoc. of Toronto*. **3**: 365–76.

Swank, W. T. and N. H. Miner (1968) Conversion of hardwood-covered watersheds to white pine reduces water yield, *WRR*, **4**: 947–54.

Thom, A. S. and H. R. Oliver (1977) On Penman's equation for estimating regional evaporation, *QJRMS*, **103**: 345–57.

Thorud, D. B. (1967) The effect of applied interception on transpiration rates of potted Ponderosa pine, *WRR*, **3**: 443–50.

Twomey, S. (1957) Precipitation by direct interception of cloud-water, *Weather*, **12**: 120–2.

Voigt, G. K. (1960) Distribution of rainfall under forest stands, *Forest Science*, **6**: 2–10.

Waisel, Y. (1960) Fog precipitation by trees, *La-Yaaran*, **9**: 29.

Wallace, J. S., J. M. Roberts and A. M. Roberts (1982) Evaporation from heather moorland in North Yorkshire, England, in *Hydrological research basins and their use in water resources planning*, Proceedings of International Symposium, Bern, September, pp. 397–405.

West, A. J. and K. R. Knoerr (1959) Water losses in the Sierra Nevada, *Journal of American Waterworks Association*, **51**: 481–8.

Whitmore, J. S. (1961) Agrohydrology, *Technical Report 22*, South African Department of Water Affairs.

Wilson, E. M. (1983) *Engineering hydrology*, 3rd edn, Macmillan, London.

Wisler, C. O. and E. F. Brater (1959) *Hydrology*, 2nd edn, John Wiley and Sons, New York.

Zinke, P. J. (1967) Forest interception studies in the United States, in *Forest hydrology*, W. E. Sopper and H. W. Lull (eds), Pergamon, Oxford, pp. 137–61.

4. Evaporation

4.1 Introduction and definitions

Evaporation is the process by which a liquid is changed into a gas. In the context of the hydrological cycle, this involves the conversion to water vapour and the return to the atmosphere of the solid or liquid precipitation that reaches the earth's surface. Indirectly the most important form of evaporation is probably that which takes place from the seas and oceans, since this forms the main source of all the water on the land areas and is the principal factor in the large-scale transfer of water and water vapour between the oceans and the continents. Other important forms include the evaporation of intercepted moisture from vegetation surfaces which was discussed in Chapter 3, the evaporation from bare soil surfaces and from open water surfaces such as streams, rivers and lakes, and transpiration whereby water vapour escapes from within plants, principally via the leaves. Direct evaporation from falling precipitation can be ignored by the hydrologist, who is concerned only with the amount of precipitation that reaches the earth's surface.

The combined processes of evaporation and transpiration constitute *total evaporation*, which is often referred to by the somewhat clumsy term *evapotranspiration*, or sometimes as *consumptive use*, thereby emphasizing that the necessary consumption of water in the production of plant material represents a 'use' rather than a 'loss'. Total evaporation is obviously a major component of the hydrological cycle, accounting for the disposal of nearly 100 per cent of precipitation in arid regions, about 75 per cent in humid regions and 100 per cent globally.

4.2 The process of evaporation

The process of evaporation can occur naturally only if there is an input of *energy* either directly from the sun or indirectly from the atmosphere itself, and is controlled by the rate at which the water vapour produced can diffuse away from the earth's surface by means of molecular and turbulent *diffusion processes* (Shuttleworth, 1979).

In much simplified terms the molecules comprising a given mass of water, no matter whether this is a large lake or a very thin layer covering an individual soil grain only a fraction of a millimetre in diameter, are in constant motion. Adding heat to the water causes the molecules to become increasingly energized and to move more rapidly, the result being an increase in the distance between liquid molecules and an associated weakening of the forces between them. At high temperatures, therefore, more of the molecules near the water surface will tend to fly off into the lower layers of the overlying air. At the same time, of course, water vapour molecules in these lower air layers are also in continual motion, and some of these will penetrate into the underlying mass of water. The rate of evaporation at any given time will, therefore, depend on the number of molecules leaving the water surface less the number of returning molecules. If this is a negative quantity, i.e. if more molecules are returning to the water surface than are leaving it, condensation is said to be taking place.

Generally speaking, the evaporation of water from a given surface is greatest in warm, dry conditions and least in cold, humid conditions, because when the air is warm, the saturation vapour pressure (E) of water is high and when the air is dry, the actual vapour pressure (e) of the water in the air is low. In other words, in warm dry conditions the saturation deficit ($E - e$) is large and conversely in cold moist conditions it is small. There is thus an underlying relationship between the size of the saturation deficit and the rate of evaporation.

Since the process of evaporation involves the net movement of water vapour molecules from an evaporating surface into the overlying air, it will eventually lead to the saturation of the lowest layers of the overlying air and the consequent cessation of evaporation when ($E - e$) is zero. In absolutely calm conditions this situation would soon occur. Normally, however, diffusion processes resulting from turbulence or convection mix the lowest layers with the overlying air, thereby effectively reducing the water vapour content and permitting further evaporation to take place. Clearly, the stronger the wind, the more vigorous and the more effective will be the turbulent action in the air; and the greater the difference between the surface and overlying air temperatures, the greater will be the effect of convection.

The plant transpiration system provides a particular example of the evaporation process. As illustrated in Fig. 4.1, it consists first of the *stomata*, the leaf orifices through which water vapour escapes to the atmosphere. Lenticels are occasional holes in bark tissue which perform a similar function on a much smaller scale; in addition, some transpiration takes place through the cuticle of leaves (Hewlett, 1982; Hewlett and Nutter, 1969). Second, there are the *spongy leaf cells*, from the surface of which vaporization takes place, the thin film of water covering the cell surfaces being supplied from inside the cells. These supplies of water are in turn replaced by the translocation of water through the *xylem*, which is a low-resistance hydraulic conductor, from the roots. Finally, the *root system* obtains water by absorption at those surfaces that are in contact with soil moisture.

Like the other components of total evaporation, transpiration depends upon a supply of energy. A simplified view, outlined by Hewlett (1982) and Hewlett and Nutter (1969) in Fig. 4.1, is that radiant and sensible heat energy supplied to the leaves and stems of plants causes evaporation within the leaf and that, when the resulting water vapour has diffused through the stoma and has been carried away by turbulent mixing, the resulting water loss produces a water deficit within the leaf cells. This water deficit represents a suction force or potential which is transmitted from cell to cell right through the system from the leaf to the root and is capable of drawing up moisture through even the highest trees against the force of gravity.

The water potential, negative in a transpiring plant cell, may be considered as

$$\psi = O + P + Z \tag{4.1}$$

where ψ is total potential, O is osmotic potential, P is cell wall pressure potential and Z is the gravity potential. In an actively transpiring plant, the osmotic potential of the cell sap in root and leaf will be strongly negative, cell wall pressure will approximate zero and the gravity potential will be positive downwards. Some workers consider that the osmotic potential, developed at the root surfaces because of the concentration difference between

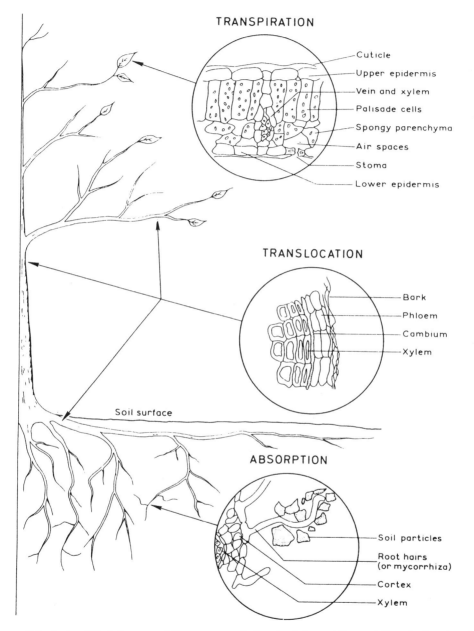

TRANSPIRATION

- Cuticle
- Upper epidermis
- Vein and xylem
- Palisade cells
- Spongy parenchyma
- Air spaces
- Stoma
- Lower epidermis

TRANSLOCATION

- Bark
- Phloem
- Cambium
- Xylem

Soil surface

ABSORPTION

- Soil particles
- Root hairs (or mycorrhiza)
- Cortex
- Xylem

Figure 4.1 *A simplified diagram of the movement of water into and through the plant system (from an original diagram by Hewlett, 1982, and Hewlett and Nutter, 1969).*

the soil solution and the sap solution, is the primary energy source for the transpiration process (Eagleson, 1970), capable of developing a pressure difference of the order of 10 atmospheres (Hendricks and Hansen, 1962). Certainly, the rate of supply of water to the leaves may exceed the rate at which moisture is removed from the leaves by transpiration, in which case the stomatal cavities fill with moisture and liquid may fall from the leaves. This process, *guttation*, is common with some types of plant such as willow (Eagleson, 1970).

Transpiration is thus a flow process through a conducting medium of varying resistance, the soil–plant–atmosphere continuum (SPAC) (Philip, 1966), involving both chemical and phase changes. A simplified representation of the process as a series of driving forces and resistances was attributed by Hewlett and Nutter (1969) to Van den Honert who, in 1948, suggested that

$$V = \frac{\psi_{soil} - \psi_{root}}{R_{cortex}} = \frac{\psi_{root} - \psi_{leaf}}{R_{xylem}} = \frac{\psi_{leaf} - \psi_{air}}{R_{stoma}} \qquad (4.2)$$

where V is the flow of transpiration water towards the leaf, ψ is the total potential at various levels and R is the resistance offered at each level. This equation applies only if transpiration is steady through time, which, of course, it seldom is.

Most physicists studying the transpiration process ascribed 'primary' control to the stomatal apparatus because the resistance is largest in the vapour pathway from the parenchyma cell through the stoma (Hewlett and Nutter, 1969). Other evidence, however, suggests that in some conditions it is the root resistance that is the most important link in the transpiration flow (cf. CSIRO, 1969; see also Section 4.5.3 under 'Root control').

4.3 Evaporation from free water surfaces

The basic components of the evaporation process from a free water surface were first expressed in quantitative terms by Dalton (1802), who suggested that if other factors remain constant, evaporation is proportional to the windspeed and the vapour pressure deficit, i.e. the difference between saturation vapour pressure at the temperature of the water surface and the actual vapour pressure of the overlying air. Dalton's law, although never expressed by the author in mathematical terms, has formed the starting point of much of the subsequent work on evaporation. In practice a number of factors, both meteorological and physical, affect the rate of free water evaporation, although it is often difficult to assess the relative importance of each of them.

4.3.1 *Meteorological factors*

Solar energy
The dual role of energy and diffusion processes in governing the evaporation process has already been stressed. The change in state of water from a liquid to a gas involves the expenditure of about 2.47×10^6 J/kg at 10°C (or, in more familiar units, approximately 590 calories per gram of water at ordinary field temperatures). Solar energy will therefore set the broad limits and govern the main variations in the rate of evaporation. Indeed Linsley *et al.* (1958) suggested that evaporation losses from a free water surface represent

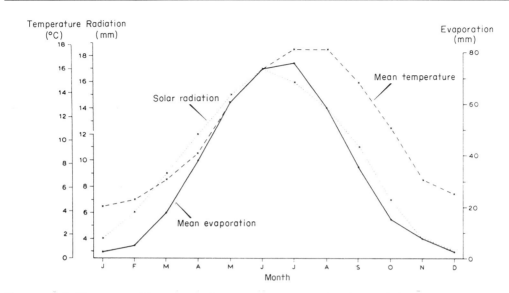

Figure 4.2 *Mean monthly evaporation, temperature and solar radiation (mm evaporation equivalent) for Camden Square, London (evaporation and temperature data from Miller, 1953).*

essentially 'solar evaporation', and this is emphasized in Fig. 4.2 which shows that the relationship of evaporation with solar radiation is closer than that with temperature.

Humidity
It has been noted that the rate of evaporation is proportional to the difference between the actual humidity and the saturated humidity at given temperatures. Actual vapour pressure varies only slightly from time to time throughout the day (see Fig. 4.3), except with a change of air mass, i.e. because of its different water vapour content each air mass affecting the British Isles will tend to have a different evaporating capacity, other conditions, e.g. temperature, being equal. In contrast, relative humidity is more variable (see Fig. 4.3) and as the relative humidity of the air over the evaporating surface increases, the net transfer of water molecules from the evaporating surface is gradually reduced, although even at 100 per cent relative humidity some evaporation normally takes place.

Since relative humidity increases as the air temperature falls, even though the water vapour content of the air remains constant, a decrease in temperature may result in a decrease in the rate of evaporation. Thus in cold weather evaporation may be smaller than in warm weather simply because the overlying air is able to hold only a small amount of water vapour below saturation level.

Diffusion processes
Diffusion of water molecules from a free water surface into a completely still layer of air will reduce as the air layer approaches saturation point and as more of the water molecules in the air reenter the water surface. Air movement is, therefore, necessary to remove the lowest moist air layers in contact with the water surface and to mix them with the upper drier layers. In fact, absolutely calm air, even above a water surface, is very rarely

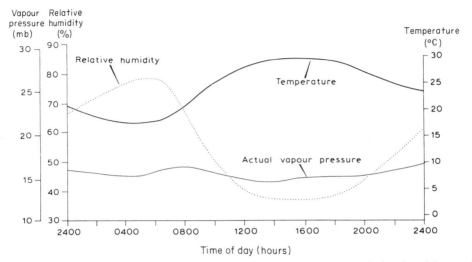

Figure 4.3 *An example of the diurnal variation of vapour pressure, relative humidity and air temperature (from an original diagram in ASCE, 1949).*

experienced in natural conditions, so that the rate of evaporation is almost always influenced to some extent by turbulent air movement. The degree of turbulence is closely related to wind velocity and surface roughness, although this frictional turbulence may be enhanced by convective turbulence where there is a suitable gradient in mean air temperature away from the evaporating surface.

The relationship between windspeed and evaporation holds good only to a certain point: beyond a critical value further increase in windspeed leads to no further increase in evaporation. This critical windspeed will vary as other conditions influencing evaporation change, but the associated turbulence will diffuse water molecules sufficiently rapidly to enable evaporation to proceed at the maximum rate governed by the existing energy and humidity conditions. Even if diffusion proceeded more rapidly, the rate of evaporation would not be increased. In these circumstances wind does not *cause* evaporation but permits a given rate of evaporation to be maintained.

4.3.2 *Physical factors*

The rate of evaporation from water surfaces exposed to identical meteorological conditions may vary as a result of physical differences between the water bodies concerned.

Salinity

For example, evaporation decreases by about 1 per cent for each 1 per cent increase in salinity because of the reduced vapour pressure of the saline water. Accordingly evaporation from sea water with an average salinity of about 35 parts per thousand is some 2 to 3 per cent less than evaporation from fresh water, although this effect is normally small enough to be discounted when comparing evaporation rates from different 'fresh' water bodies.

Water depth

The effect of water depth upon the seasonal distribution of evaporation (although not upon annual totals) may be quite considerable. The seasonal temperature regime of a shallow lake will normally approximate closely to the seasonal air temperature regime, so that maximum rates of evaporation will occur during the summer and minimum rates during the winter, as illustrated by curve A in Fig. 4.4. Large, deep lakes, however, not only have a much higher capacity for heat storage than small water bodies but in the middle and higher latitudes they normally experience a marked thermal stratification which also affects evaporation from their surfaces. During the spring and summer, heat entering the water surface is mixed downward for only a limited distance by wind action. This results in the formation of an upper layer of water, the epilimnion, in which temperatures are relatively uniform and higher than those in the rest of the water body. With cooling of the surface water during autumn and its associated increase in density, the cooled water sinks and is replaced by warmer water from below. In ice-free conditions this turnover may continue throughout the winter as a result of wind action (Zumberge and Ayers, 1964). In any case the turnover results in approximately uniform temperatures and densities throughout the depth of water so that turbulence resulting from wind action can easily activate deep mechanical mixing. There is, therefore, a slow release by the deep water body of stored heat during the autumn and winter months which means that a supply of heat energy in excess of that received directly from the sun is made available for evaporation at that time of year. The net result of this heat storage on relative air and water temperatures is that water temperatures are lower than air temperatures during the summer and higher than air temperatures during the winter. Since, according to Dalton's law, evaporation is proportional to the difference between saturation vapour pressure at the temperature of the water surface and the actual vapour pressure of the overlying air, it will be apparent that the highest rates of evaporation from deep water bodies should occur during the winter, as shown by curve B in Fig. 4.4. Furthermore, at that season diffusion processes will be rapid, as a result of convectional activity, encouraged by the temperature gradient, whereas during the summer the colder water will tend to cool and stabilize the air immediately above it and so inhibit diffusion.

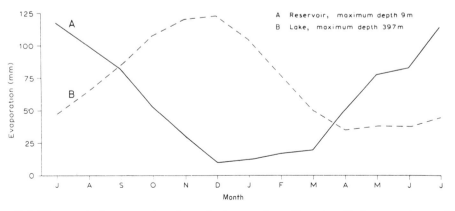

Figure 4.4 *The trend of evaporation through the year from Kempton Park reservoir near London (curve A) and Lake Superior (curve B) (based on data from Lapworth, 1956, and Hickman, 1940).*

Size of surface

With constant windspeed, the evaporation rate is related to the size of the evaporating surface and to the relative humidity (cf. Fig. 4.5). As air moves across a large lake there will be a reciprocal decrease in the rate of evaporation as the fully adjusted boundary layer or 'vapour blanket' increases in thickness. The larger the lake, the greater will be the total reduction in the depth of water evaporated, although of course the total volume of water evaporated may well increase with the size of the water surface.

In the case of a continuous water surface, e.g. the oceans, the humidity of the air will be largely independent of the distance it has travelled, except in coastal areas. In these conditions, therefore, evaporation will be uniform over much larger areas and will be closely related to the amount of heat energy available. At the other extreme, small evaporating surfaces such as evaporimeters and pans exert little influence on the temperature or humidity of the overlying air. The small amount of water vapour which leaves the surface, even with high rates of evaporation, is quickly diffused so that a continuous high rate of evaporation is maintained.

The magnitude of differences in evaporation rates from different-sized evaporating surfaces will be considerably affected by the humidity of the 'incoming' air (see Fig. 4.5). If this is initially high it can be modified only slightly, even after a lengthy passage across a large lake. There will thus be little difference in the evaporation rates from large and small water surfaces. If, on the other hand, the initial humidity of the air is very low, it may be considerably increased while crossing a large lake, so that the proportional decrease in the rate of evaporation as compared to that from a small water surface may be much more significant.

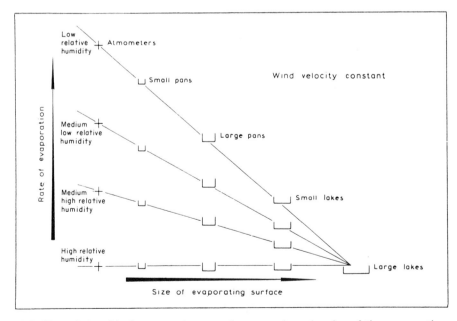

Figure 4.5 *The relationship between the rate of evaporation, the size of the evaporating surface and the relative humidity (from an original diagram by Thornthwaite and Mather, 1955).*

4.3.3 *Evaporation from snow*

Although reliable data are sparse, it is widely believed that evaporation from water in the solid state, i.e. ice and snow, is relatively insignificant. The main meteorological considerations were summarized adequately by Linsley *et al.* (1958) and by Gray *et al.* (1970). Of particular importance is the fact that since snow and ice melt at 0°C, and since evaporation can occur only when the vapour pressure of the overlying air is less than that of the snow surface, evaporation from snow itself will cease when the dew point rises to 0°C (although some evaporation of meltwater may take place) and that as temperatures rise above freezing, the rate of snowmelt must exceed the rate of evaporation. On this basis Linsley *et al.* suggested that, contrary to a widespread belief, fohn-type winds do not result in high evaporation rates and that there is probably an upper limit of about 5 mm/day (water equivalent) for evaporation from a snow surface, although most experimental data indicate evaporation rates considerably below this. Other considerations are that snow surfaces have a high and variable albedo, which depends on a number of factors such as age, water content and depth, and that since, in general, approximately $2.82–2.85 \times 10^6$ J/kg (675–680 cal/g) are required to sublimate snow whereas the latent heat of fusion is only about 3.35×10^5 J/kg (80 cal/g) the mass cf water melted is substantially greater than the amount evaporated (Gray *et al.*, 1970). The limited energy normally available for snow evaporation has already been referred to in the discussion of snow interception (see Section 3.7).

Unfortunately, for several practical reasons it has been difficult in the past to substantiate theory by measurement. Interesting and valuable results have been obtained from time to time, particularly by Croft (1944), Clyde (1931), Baker (1917), Kittredge (1953), Garstka *et al.* (1958) and the US Army Corps of Engineers (1956). Kittredge's work in the western Sierra Nevada was very comprehensive and led him to the conclusion that the outstanding feature of snow evaporation in this area was the 'small magnitude' of the measured losses. Indeed, the data reported by various investigators were fairly consistent on this point and indicate that, during the spring months, evaporation averages about 25 mm/month or less (Linsley *et al.*, 1949).

Interestingly, however, recent measurements using gamma-ray attenuation and tree weighing have shown that evaporation rates from snow-covered forest canopies can be as high as evaporation rates from rain-wetted canopies when the snow is melting (IH, 1987). The same work also indicated that although sublimation rates are lower than evaporation rates, sublimation may be quantitatively significant during prolonged periods of subzero temperatures (see Section 3.7).

4.3.4 *Amounts and variations of free water evaporation*

Since evaporation from free water surfaces is largely determined by the amount of solar energy available, both seasonal and areal variations of evaporation are closely related to similar variations in the duration and intensity of solar radiation. A few selected examples will suffice to illustrate this point.

Smith (1965) analysed evaporation pan data obtained between 1935 and 1961 at Harrogate (54°N) and the mean monthly totals for this 26 year period are plotted in Fig. 4.6. Curve A shows that the highest rates are experienced in June and July and the lowest

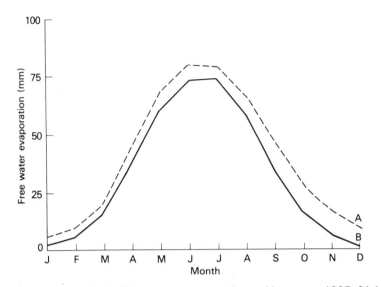

Figure 4.6 *Mean monthly totals of free water evaporation at Harrogate, 1935–61 (curve A), and Camden Square, London, 1885–1938 (curve B) (based on data from Smith, 1965, and Meteorological Office, 1938).*

in December and January and that there is a fairly uniform graduation between these two extremes. It should be noted that 299 mm (61.4 per cent) of evaporation occur in the four 'summer' months from May to August and only 44 mm (9.1 per cent) occur in the four 'winter' months from November to February. Long-period (1885–1938) monthly evaporation totals from the evaporation pan at Camden Square, London, are also plotted for comparison and it will be seen that curve B is almost identical in shape to curve A although, in this case, May to August evaporation represents 68.6 per cent and November to February evaporation only 4.4 per cent of the mean annual total.

In most hydrological investigations, variations of free water evaporation are of less interest than areal variations of total evaporation from a vegetation-covered surface. Reliable data sets which may be used as a basis for mapping water surface evaporation are therefore sparse. For a small land area such as the British Isles the pattern is likely to reflect a simple latitudinal response to solar radiation by maritime influences, particularly the advection of moist oceanic air from the south-west which would depress evaporation totals over much of Wales and south-west England. For a larger land mass, such as North America, one would again expect the latitudinal distribution of energy availability to dominate, with perhaps a more subtle overlay of the influence of continentality. This is largely borne out by Fig. 4.7, which shows a map of mean annual lake evaporation in the conterminous USA.

4.4 Evaporation from soils

The rate of evaporation from a soil surface will be governed by the same meteorological factors that govern the evaporation loss from a free water surface since soil evaporation is

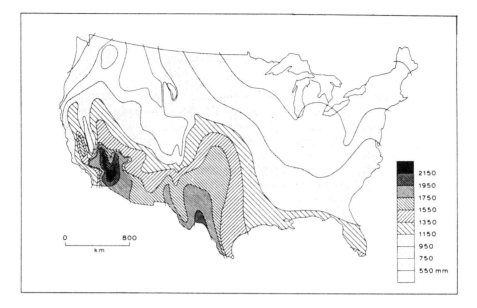

Figure 4.7 *Mean annual lake evaporation over the conterminous USA (based on an original diagram by Geraghty et al., 1973).*

merely the evaporation of films of water surrounding the soil grains and filling the spaces between them. However, in the case of free water evaporation, the supply of moisture is always, by definition, so plentiful that it exerts no limiting influence on the rate of water loss; in other words, what Horton called the 'evaporation opportunity' is always 100 per cent. On the other hand, evaporation from soils is often less than evaporation from a free water surface, not because the climatic conditions are different but because there is not a sufficient supply of water in the soil to be evaporated; in other words, the 'evaporation opportunity' is less than 100 per cent. Thus the most important additional factors affecting evaporation from soils are those that influence the evaporation opportunity.

4.4.1 *Soil moisture content*

The moisture content of the surface layers of the soil exerts the most direct influence on evaporation opportunity. With saturated soils the evaporation opportunity may be greater than 100 per cent because the countless minute irregularities of the soil surface comprise a larger evaporating surface than that of an 'identical' area of water. The finer the soil texture and the smoother the surface, the closer will the evaporation opportunity approximate to 100 per cent at saturation. Early investigators found that the relationship between soil moisture content and evaporation was quite close (cf. Fortier, 1907; Harris and Robinson, 1916; Whitney and Cameron, 1904; Widtsoe, 1914; Zumberge and Ayers, 1964);

evaporation decreases rapidly as the surface moisture content falls until, with a dry soil surface, it is zero. This pattern has since been confirmed by many experiments: Veihmeyer and Brooks (1954), for example, reported the data illustrated in Fig. 4.8, which shows the relationship between open water and soil evaporation during a dry period after irrigation. The soil was at field capacity on 9 July and, although for the first few days the rate of evaporation was high, the soil curve soon flattens out and by the end of the period illustrated total evaporation from the water surface had exceeded total evaporation from the soil surface by 372 mm.

Normally it is the moisture content of the few centimetres of surface soil that plays the decisive role; the subsoil may be saturated but because of the slow movement of soil moisture, it may have a negligible effect on the rate of evaporation from the surface. There will, of course, be some movement of water vapour from the subsoil to the surface, but this will normally represent only a minute fraction of the total evaporation loss. The way in which rainfall replenishes the soil moisture from above therefore assumes considerable importance. If other conditions remain constant, evaporation will be greater from a soil

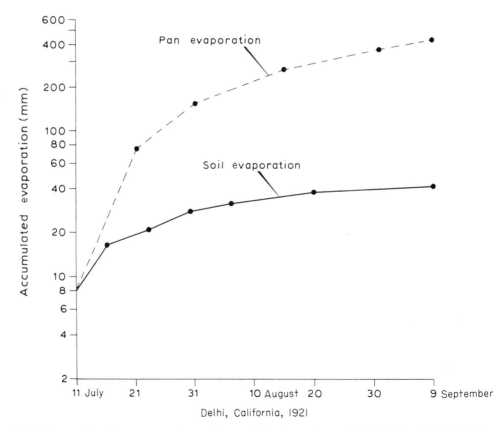

Figure 4.8 *Evaporation from open water and from soil during a dry period after irrigation (based on data in Veihmeyer and Brooks, 1954).*

surface frequently wetted by intermittent showers than from a soil surface thoroughly soaked by the same quantity of rain falling in one storm.

4.4.2 *Soil capillary characteristics*

In areas where rainfall is not a frequent source of soil moisture replenishment, and especially in desert areas, evaporation opportunity tends to vary in relation to those soil factors that affect the complex processes of heat and water vapour transfer from the evaporating front at some depth below the soil surface (Menenti, 1984) and the capillary rise of moisture stored in the soil layers immediately beneath the surface. The upward movement of soil moisture by capillarity is conditioned partly by the size and arrangement of the soil particles. In fine-textured soil, capillary movement is effective over larger vertical distances, e.g. a metre or so. In coarse-textured soil it may be only a few centimetres. However, the speed of movement tends to vary inversely with the height of capillary lift so that, in neither case where special conditions prevail, does the supply of water to the soil surface by capillary activity increase significantly the total amount of evaporation.

One such special condition is the presence of shallow groundwater. Numerous experiments (cf. Keen, 1927; Parshall, 1930; Tanner, 1957; Veihmeyer and Brooks, 1954) have shown that soil evaporation is at a maximum when the water table is at the surface and decreases, quite rapidly at first, as the water table retreats downwards. However, after a depth of about one metre has been reached, further decline in water table height is accompanied by only a slight change in the rate of evaporation. This is illustrated in Fig. 4.9, which is based on figures of evaporation from waterlogged soils at Davis, California

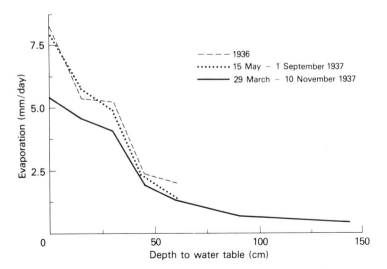

Figure 4.9 *The relationship between soil evaporation and water table depth (based on data in Veihmeyer, 1938).*

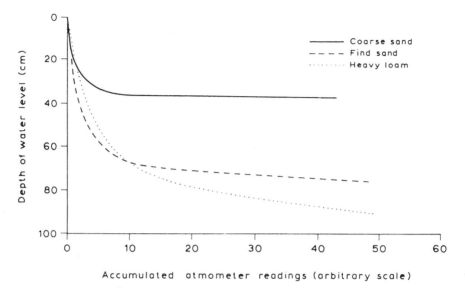

Figure 4.10 *Graphs of declining water table level plotted against cumulative evaporation. The flattening out of each curve indicates the level at which capillary movement of water to the soil surface becomes insignificant (from an original diagram by Keen, 1927).*

(Veihmeyer, 1938). Naturally, the critical depth beyond which evaporation is only slightly affected will depend largely on the capillary characteristics of the soil and will, therefore, tend to vary with soil texture and particle size. It will be greatest in the fine-textured silts and loams and least in coarse sands. Classic experiments on this phenomenon were reported by Keen (1927) who found that capillary lift contributed insignificantly to total evaporation losses after the water table had receded to about 35 cm below the surface in coarse sand; 70 cm below the surface in fine sand; and 85 cm below the surface in heavy loam soil (see Fig. 4.10).

4.4.3 *Other factors*

Soil colour will tend to affect evaporation because darker soils, having a low albedo, will absorb more heat than lighter soils. The resulting increase in surface temperature may modify the evaporation rate significantly. Milly (1984a, 1984b) discussed the complex relationship between evaporation from a soil surface and the distribution of temperature in the soil profile and showed how variations of soil temperature in time and depth cause corresponding variations in the coefficients of water transport. Finally, in many locations, cf. the British Isles, discussion of evaporation from bare soil may be largely hypothetical, since extensive areas completely devoid of vegetation, e.g. ploughed land, are usually found only during the colder months when, in any case, climatic conditions favour only a low rate of evaporation. During the spring and early summer months, however, and again in the autumn and early winter, climatic conditions may favour considerable evaporation losses from soil partially covered by immature, widely spaced crops whose presence will tend to reduce soil evaporation by shading and sheltering the surface and by increasing the relative humidity of the lower air layer.

4.4.4 *Amounts and variations of soil evaporation*

The operation of the additional factors affecting soil evaporation, which have been discussed in the preceding sections, mean that both seasonal and areal variations of this component of total evaporation are far less rhythmic and predictable than variations of free water evaporation.

An analysis, by Penman and Schofield (1941), of typical British conditions showed that although annual evaporation is relatively stable, it does tend to be higher in wet years and lower in dry years, the difference being largely accounted for by differences in the totals of summer evaporation. In dry summers soil evaporation is considerably less than the evaporation from a free water surface, but in summers when re-wetting is frequent soil evaporation will be much higher.

Similarly, the areal pattern of annual soil evaporation will vary considerably from year to year depending on the interaction between meteorological factors, on the one hand, and the wetness of the surface layers of the soil, on the other—basically the difference in this context between demand and supply. So numerous are the variables involved that a meaningful map of soil evaporation could have been drawn in the past only from accurate drain gauge or similar measurements. Recent advances in remote sensing techniques, however, now permit estimates of instantaneous evaporation rates which may be used to determine the spatial variability of evaporation from bare soil using multiband radiometer data (Rice and Jackson, 1985; Menenti, 1984). An alternative approach using C-band radar to sense surface soil moisture conditions in conjunction with a soil water transfer model was described by Prevot *et al.* (1984).

4.5 Evaporation from vegetation covers

The description of evaporation in Section 4.2 illustrated the complexity of that process in respect of evaporation from a vegetation-covered surface. In particular, transpiration is dependent upon a sequence of water-moving processes. For example, water at a point in the soil profile moves under the influence of a moisture gradient towards the root hairs, is there absorbed and is subsequently translocated through the plant root and stem system to the leaves at rates determined by the various resistances imposed by the plant system. It is finally vaporized in the stomatal cavities before passing through the stomatal apertures to the atmosphere. Soil, plant and atmosphere thus form parts of a continuous flow path of varying resistance in which water moves at varying rates and undergoes both chemical and phase changes.

Not surprisingly, the full understanding and representation of evaporation from vegetated surfaces have proved elusive, and hydrologists and climatologists have resorted to stratagems of simplification in order to derive numerical values of evaporation for use in hydrological analysis and modelling. The most important simplification has undoubtedly been the development of the concept of *potential* evaporation.

4.5.1 *Potential evaporation and actual evaporation*

An essential distinction between potential evaporation and the evaporation that *actually* takes place from a vegetated surface is that the concept of potential evaporation involves a supply of moisture, either from the soil or from the atmosphere as precipitation, which is at all times sufficient to meet the demands of the transpiring vegetation cover. It was defined by Thornthwaite (1944) as 'the water loss which will occur if at no time there is a deficiency of water in the soil for use of vegetation'. Subsequent modifications to this concept have limited its application to the evaporation that takes place from a continuous and unbroken green vegetation cover; cf. Penman *et al.* (1956) who defined potential evaporation as the '. . . evaporation from an extended surface of short green crop, actively growing, completely shading the ground, of uniform height and not short of water'.

If it is assumed that the term 'extended' means of sufficient extent to minimize advectional influences, that the effects of vegetation height are excluded by the term 'short', the effects of shape and roughness by 'uniform' and the role of soil moisture movement by the term 'never short of water', potential evaporation may be regarded as a climatological response which will not be affected by the movement of water through the soil and which will be affected by plant behaviour and plant type only insofar as these affect colour (and therefore albedo) and stomatal closure (Ward, 1971).

Equally clearly, a concept so restrictingly defined is likely to be of only limited practical significance. For example, although the Penman concept suggests that if all the restrictive conditions are fulfilled potential evaporation represents the maximum possible evaporative loss from a vegetation-covered surface, these conditions are unlikely to be fulfilled except, perhaps, for a very large surface of close-mown grass in a humid environment. In other conditions theoretical argument and experimental evidence indicate that the shape and height of the vegetation cover and the supply of large-scale advective energy affect the transpiration rate in such a way that, even in the humid conditions of Britain or the Netherlands, the actual rate of evaporation under these conditions of optimum water supply can exceed potential evaporation. Furthermore, it was shown in Section 3.3 that the evaporation of water intercepted by the vegetation surface may greatly exceed the rate of transpiration from the same vegetation cover.

Realistically, therefore, the concept of potential evaporation should embrace the evaporation of intercepted moisture as proposed, for example, by van Wijk and de Vries (1954) who defined *wet-surface evaporation* as the maximum evaporation rate from a wet surface of similar shape, colour and dimensions as the crop under consideration. This extended notion of potential evaporation was also espoused by van Bavel (1966) who suggested that 'when the surface is wet and imposes no restriction upon the flow of water vapor, the potential value is reached'. As Lee (1980) observed, however, this definition, by ignoring advective influences, permits potential evaporation to vary widely with local microclimate and thereby removes entirely the generality of the concept.

Indeed, such are the ambiguities surrounding the various notions of a climatologically determined maximum rate of evaporation from a vegetated surface that it may be argued that the simple concept of potential evaporation, as originally advanced by Thornthwaite (1944) and Penman *et al.* (1956), has outlived its usefulness. Certainly it represents only one of the three main conditions in which evaporative losses occur from a vegetation cover, i.e. (a) high evaporation from a wetted vegetation cover, irrespective of soil moisture status, (b) the Thornthwaite–Penman conditions when the ready availability of moisture to the plant root system imposes no restriction on the evaporation rate from an unwetted vegetation cover and (c) the condition where moisture stress within the soil–plant–atmosphere system, frequently resulting from inadequate soil moisture supply, reduces the *actual* evaporation loss from an unwetted vegetation cover below the climatologically dictated loss.

The term potential evaporation is used in Chapter 3 to refer to condition (a) and in this chapter to refer to conditions (a) *and* (b); i.e. it is recognized that there is a range of conditions of surface wetness and/or adequate soil moisture supply that will permit evaporation to take place at a rate determined largely by atmospheric variables. Actual evaporation may take place at the potential rate if the appropriate surface wetness and/or soil moisture conditions obtain; otherwise actual evaporation will take place at a lower rate and may be greatly influenced by plant and soil conditions.

From the foregoing discussion it will be clear that the factors affecting evaporation from vegetation covers are numerous and will vary in importance depending on surface wetness and soil moisture conditions. For convenience they may be categorized as meteorological, plant and soil factors although, in practice, this distinction is sometimes difficult to maintain.

4.5.2 *Meteorological factors*

The energy and humidity factors discussed in Sections 4.3.1 and 4.4 in the context of free water and soil evaporation apply also to the evaporation of water from vegetation covers. One of the significant differences, however, between vegetation covers, on the one hand, and free water and soil surfaces, on the other, is in their aerodynamic roughness and its impact on the role of turbulence and diffusion processes in evaporation. It has been emphasized that evaporation from vegetation comprises the evaporation of intercepted water from the wet surface of the vegetation and transpiration through the stomatal apertures in the leaves. In simple terms the magnitude of each of these component processes is a response to the combined operation both of meteorological variables, which encourage the vaporization of liquid water on or in the leaf and the flow of the resultant water vapour into the enveloping air, and also the resistances that discourage that vapour flow. The two principal resistances are the aerodynamic or boundary layer resistance (r_a), which is a measure of the resistance encountered by water vapour moving from the outer surface of the vegetation cover into the surrounding atmosphere, and the surface or stomatal resistance (r_s), which is a physiological resistance imposed by the vegetation itself and which will be considered further under the heading 'vegetational factors' in Section 4.5.3.

The aerodynamic resistance, which commonly varies between 10 and 100 s/m, depends solely on the physical properties of the vegetation cover insofar as these affect aerodynamic roughness and zero plane displacement. Vegetation height will clearly be important since the coefficient of turbulent exchange increases by a factor of 2 with a change in vegetation height from 10 to 90 cm and by a factor of more than 5 with a change from a short cut surface at about 2 or 3 cm (as implied in the Penman definition of potential evaporation?) to a vegetation height of 90 cm (Rijtema, 1968). Expectedly, therefore, the aerodynamic resistance for trees is normally found to be an order of magnitude less than for grass because trees are not only taller but also present a relatively

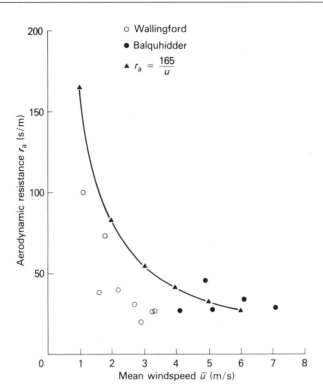

Figure 4.11 *Theoretical and observed relationships between aerodynamic resistance and mean windspeed (based on an original diagram in Hall, 1987).*

rougher surface to the wind and are more efficient in generating the forced eddy convection which, in most meteorological conditions, is the dominant mechanism of vertical water vapour transport (Calder, 1979).

It was seen in Section 4.3.1 under 'Diffusion processes' that eddy diffusion is dependent on windspeed as well as on surface roughness. Clearly, therefore, r_a should be inversely related to windspeed in a predictable way, cf. the line of theoretical fit in Fig. 4.11. The point data for wetted grass at Wallingford and Balquhidder also shown in this diagram confirm that there is indeed an inverse relationship between r_a and u but indicate that this does not take the form predicted by classical eddy-diffusion theory. Hall (1987) quoted several studies including Finnigan (1979), Denmead (1984) and Grant *et al.* (1986), which showed that large-scale turbulence in the form of intermittent energetic gusts are a major mechanism in water vapour transport and weaken the classically predicted relationship between r_a and mean windspeed.

4.5.3 *Vegetational factors*

Increasingly during the past decade or so hydrological interest has focused on plant factors that may influence the rate of evaporation from vegetation covers and Federer (1975) drew attention to the shift in American research towards the study of stomatal influences in particular. Earlier discussion of these questions had frequently been inconclusive and confused partly because of the paucity of adequate measured data and partly because of the absence of a sufficiently robust conceptual framework within which ideas could be developed. Discussion here will concentrate on three main vegetational influences on evaporation, i.e. albedo, stomatal control and root water uptake.

Albedo
Penman's definition of potential evaporation referred to a 'green' crop, but of course not all plants remain green and not all green plants are the same colour as measured by reflectivity or albedo. Since albedo materially influences the energy balance at the evaporating surface its effect on evaporation may be significant. Typical values for selected grass, agricultural crops and trees are shown in Table 4.1, but in each case measurements range about the value shown and in addition albedo varies with sun angle and wetness of the vegetation surface, as illustrated in Fig. 4.12.

Table 4.1 Typical values of albedo for selected surface conditions (based on data compiled by Oke, 1978).

Surface	Condition	Albedo
Soil	Dark, wet Light, dry	0.05 . . . . . . 0.40
Grass	Long (1 m) Short (2–3 cm)	0.16 . . . . . . 0.26
Crops		0.18–0.25
Forest Deciduous Coniferous	Bare Leafed	0.15 . . . . . . 0.20 0.05–0.15
Water	Small zenith angle Large zenith angle	0.03–0.10 0.10–1.00
Snow	Old Fresh	0.40 . . . . . . 0.95

Stomatal control
Water vapour diffuses through the leaf stomata into the atmosphere in response to a given atmospheric vapour pressure deficit, at a rate that depends partly on the size of the vapour pressure deficit, partly on vegetation type, insofar as this may affect the number and configuration of the stomata, and partly also on the complex interaction of a number of past and present environmental variables including soil moisture deficit, light intensity and, to a lesser extent, the aerodynamic or boundary layer resistance. Stomatal or surface resistance (r_s) is of major importance in the evaporation process because it is usually an order of magnitude greater than aerodynamic resistance, commonly varying between 100 and 1000 s/m (Lee, 1980); it therefore offers significant possibilities for artificial control of plant evaporation by either biochemical or genetical modification (Federer, 1975).

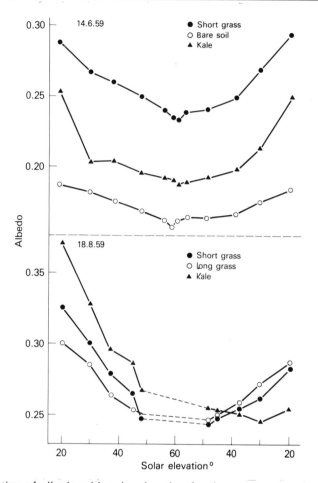

Figure 4.12 *Variation of albedo with solar elevation for three crops showing enhanced albedo values after overnight dew on 18 August (based on an original diagram by Monteith and Szeicz, 1961).*

In general, the stomatal control of vapour diffusion is associated with the maintenance of leaf turgidity, However, the factors affecting stomatal resistance are varied and an understanding of the trigger mechanisms has emerged only slowly. These were discussed by Federer (1975) and Sharkey and Ogawa (1987), and include the pumping of potassium ions into the guard cells (cf. Fischer, 1971), the effect of carbon dioxide concentration and the role of abscisic acid (cf. Zabadal, 1974). The effect of light appears to differ little among species, with stomata achieving full opening in one-tenth of full sunlight (cf. Turner and Begg, 1973). On sunny days, therefore, r_s on exposed leaves decreases rapidly at sunrise, remains at a minimum value (r_{sm}) all day if the water supply to the leaf is adequate and increases at sunset (Federer, 1975).

Physiological factors may be very important causes of interspecies variation in r_s. The low rate of transpiration from heather, for example, is caused by its large r_s, 50–170 s/m according to Wallace *et al.* (1984) and Miranda *et al.* (1984), which results from the stomata occurring only in a groove, the edge of which is lined with fine hairs, on the abaxial side of the leaves so that free movement of water vapour is impaired (Hall, 1987).

As transpiration occurs, the leaf water potential declines, thus causing increased inflow of water from regions of higher potential in the stems, roots and soil (Federer, 1975). With continued desiccation a critical leaf water potential (Turner, 1974) will be reached at which stomata begin to close, thereby limiting further water loss. This critical value may depend on plant type and history and on leaf location, i.e. leaves in the upper canopy may have lower values. Certainly it has been found that leaves in the lower canopy experience earlier stomatal closure (Stevenson and Shaw, 1971), implying a preferential water supply to the more exposed leaves in the upper canopy (Federer, 1975).

When the soil is sufficiently moist maximum values of r_s are maintained throughout the day, leaf water potential remains above the critical value and evaporation is controlled by meteorological factors. However, when evaporation from the leaves exceeds the rate of moisture supply to them through the soil–stem system, the critical leaf water potential is reached, stomata begin to close, r_s increases above the minimum value and water shortage reduces actual evaporation below the potential rate via the mechanism of stomatal control. Exactly *when* this occurs is still not clear but, as Federer (1975) observed, 'it is the classic question of availability of soil water stated in terms of stomatal control'. It illustrates clearly the difficulty of distinguishing between meteorological, plant and soil influences on evaporation, to which attention was drawn earlier at the end of Section 4.5.1. The divergence of evidence appears to be wide, ranging from work that showed stomatal control of evaporation in wet soil conditions (cf. Shepherd, 1972; McNaughton and Black, 1973) to other studies in which evaporation appeared not to be reduced below the potential rate until *very* low values of soil moisture content had been attained (cf. Veihmeyer, 1972; Ritchie, 1973).

This apparently conflicting evidence is not surprising. It has been shown that the soil–plant–atmosphere system is characterized by great variability of moisture content in space and time, and one would therefore expect that different vegetation types, growing in different soil and atmospheric conditions, will limit evaporation at different values of soil moisture status. Indeed, it might be argued that much more remarkable is the similarity of transpiration from different forest species noted by Roberts (1979) and explained in terms of a possible similarity both of r_{sm} and also of the relationship between r_s and atmospheric humidity. Roberts further suggested that similar r_{sm} values in forests might result from tree species being genetically adapted to similar soil moisture conditions.

There is also a sense in which the integration of the complex heterogeneity of, say, water potentials and surface resistances of individual leaves imposes a certain homogeneity on the system as a whole. Thus evaporation from a vegetation canopy covering a given area of ground (A_g) is the sum of evaporation from the greater area of all the individual leaf surfaces (A_l), i.e.

$$E_c = \bar{E}_l \left(\frac{A_l}{A_g} \right) \tag{4.3}$$

where E_c is the combined evaporation from the canopy, $\bar{E}_l$ is the mean evaporation from the leaves and the ratio A_l/A_g is the leaf area index which commonly varies between 5 and

10 (Hewlett and Nutter, 1969). Similarly, if $\bar{r}_s$ is the mean surface resistance of the leaves then the surface resistance of the canopy r_c may be regarded as

$$r_c = \bar{r}_s\left(\frac{A_g}{A_l}\right) \qquad (4.4)$$

which was described by Lee (1980) as '... the overall physiological control of transpiration'. Simply dividing mean r_s for the canopy by the leaf area index represents a gross simplification which makes no allowance for microclimatological interaction between the individual leaves but it is a commonly employed technique (cf. Brun *et al.*, 1973; Szeicz *et al.*, 1973). Smith *et al.* (1985) found that r_c for well-irrigated wheat varied significantly with the leaf area index and vapour pressure deficit, as shown in Fig. 4.13.

Work at the Institute of Hydrology (IH, 1979) led to the derivation of a semi-empirical equation for spruce forest:

$$r_c = 74.5 \frac{1 - 0.3\cos\{2P_i[(D - 222)/365]\}}{1 - 0.45\delta e} \qquad (4.5)$$

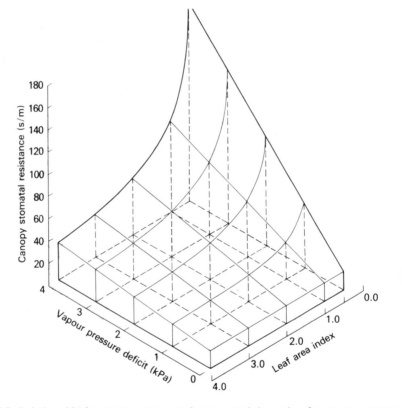

Figure 4.13 *Relationship between canopy resistance and the ratio of vapour pressure deficit and leaf area index derived using infrared thermometry and associated meteorological data (based on an original diagram in Smith* et al., *1985).*

where D is the day number of the year and δe is vapour pressure deficit (in kPa). This work was subsequently developed and refined (e.g. Stewart, 1988).

Continued experimentation and development of artificially applied transpiration suppressants confirms the perception that vegetation factors play an important role in determining evaporation losses from vegetation covers since, clearly, successful application of this technique will depend upon a correct understanding of the evaporation process and particularly of the plant physiological factors that affect it. An early review was provided by Davenport *et al.* (1969). In terms of foliar sprays the three main possibilities are, first, the use of reflecting materials that decrease the heat load on the leaf; secondly, the application of film-forming materials that slow down the escape of water vapour from the leaves by decreasing their permeability; and, third, the use of substances that increase stomatal resistance either by closing or by narrowing the stomatal apertures. The most promising approach appears to be through the modification of stomatal aperture, using either phenylmercuric acetate (PMA) or decenylsuccinic acid (cf. Kreith *et al.*, 1975). Both have been demonstrated as effective for a range of plants under controlled conditions. For large natural areas, aerial spraying represents the only economically viable application technique but has the disadvantage that transpiration suppressant is delivered mainly to the upper leaf surface whereas the stomata are normally located in the underside of the leaf. Where stomata are more amenably distributed, as in red pine, impressive reductions in soil water depletion have been recorded after application of antitranspirants (Waggoner and Bravdo, 1967). The three major questions, however, that still remain to be answered (Kreith *et al.*, 1975) concern the period of effectiveness of the application, its cost-effectiveness, including its effect on growth since a barrier to water vapour movement in one direction will be a barrier to carbon dioxide movement in the other, and its environmental integrity.

Finally Fig. 4.14 may be used to summarize much of the preceding discussion of meteorological and vegetational factors which has focused attention upon the relationship between evaporation from a vegetation cover and the aerodynamic and surface resistances associated with that cover. The diagram shows calculated evaporation rates

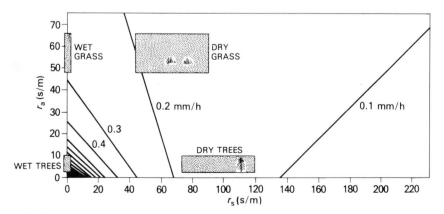

Figure 4.14 *Calculated evaporation rates as a function of r_a and r_s for typical cool summer daytime conditions in Britain. See text for explanation (based on an original diagram by Calder, 1979).*

for meteorological conditions typical of a British cool summer day (i.e. net radiation = 200 W/m²; vapour pressure deficit = 5 mbar; air temperature = 10°C). Note particularly that the evaporation rates from grass and coniferous trees are similar in dry conditions but that when the vegetation surfaces are wet evaporation rates differ substantially, increasing by a factor of 1.4–1.8 for grass but by a factor of 5–15 for the trees (Calder, 1979).

Root control

The review of stomatal control in the foregoing section drew attention to the necessity of adequate soil–plant water supply to the leaves to maintain leaf water potential at supercritical levels. Most contributors to the discussion, however, appear to treat the roots as inert, uniform sinks for soil moisture and it was encouraging, therefore, to find this passivity questioned vigorously by Monteith (1985) at a major ASAE conference on 'Advances in Evapotranspiration' when he urged that:

> . . . it is time that soil physicists and root physiologists got together with microclimatologists and leaf physiologists to do some comprehensive work on the soil–plant–atmosphere continuum. . . . To understand how canopy resistance changes with time when water is limiting, we need to know much more about the rate at which roots move towards water and water moves towards roots.

A commonly accepted model of the role of the root system in the evaporation process was advanced by Penman (1969) who argued that physical theory supports the findings of Veihmeyer and others (referred to in preceding subsection) that evaporation may be maintained at the potential rate through a wide range of reduction of soil moisture content. Penman suggested that this was a likely outcome, especially where roots ramify throughout the soil. The fact, however, that the Veihmeyer hypothesis breaks down so frequently in field conditions led Penman to the formulation of the 'root constant' concept (Pearl et al., 1954; Penman, 1949) which suggests that because moisture, on the whole, moves relatively slowly through the soil, the readily available moisture is effectively restricted to that water in the immediate proximity of the root system (Pearl et al., 1954). It will, therefore, be limited in amount partly by the depth of the soil available and partly by the soil type. The root constant was the term given to the available moisture so defined by Penman (1949) and may be described as the maximum soil moisture deficit that can be built up without checking transpiration. As the term itself implies, the root constant was considered to be primarily a plant characteristic, although it may be modified by other factors such as soil type, and its value for grass was estimated by Penman at between 7 and 12 cm, although at lower values by others (cf. *New Scientist*, 1961). This means, in effect, that some 7–12 cm of soil moisture can be removed by grass before transpiration falls below the potential rate; after this reduction below field capacity it is assumed that transpiration falls to about one-tenth of the potential rate. Since the main factor affecting the root constant is the depth of rooting, its value should vary with the type of plant and may be as much as 25–30 cm or more for some trees.

Visser (1964, 1966) described the flow of soil moisture to plant roots during drying and evolved a formula for this flow. It was concluded that the extraction of soil moisture by the roots tends to cause a dry cylinder around each root and that the conductivity of this relatively dry cylinder will considerably increase the resistance for liquid flow. As did the Penman root constant concept, the Visser model implied that the extent and efficiency of the root system should help to determine the total amount of moisture available to the plant cover and the rapidity with which, in drying conditions, the rate of actual evaporation falls below the potential rate. Thus, the root–surface ratio, the ratio of root surface to soil volume, should be an important factor. This ratio, which normally decreases sharply with depth, is influenced by the ability of the root system to extend into moist soil and to dry out additional cylinders around the new roots. In addition, fungal growths on the roots of many trees serve to increase the surface area of absorbing roots and may increase the availability of soil water (Hewlett and Nutter, 1969).

Numerous experiments have investigated the relationship between root range and evaporation, and some of these were reviewed by Penman (1963) and Douglass (1967). Generally it was suggested that a forest will transpire more than a pasture because trees, on the whole, are more deeply rooted than grass. Increasingly, however, evidence is accumulating that the relationship between root density and water uptake is not as simple as these earlier views suggested. Certainly field experimentation reported by McGowan *et al.* (1984) and McGowan and Tzimas (1985) showed that the capture of soil water by crop roots is not solely dependent upon root development and soil moisture potential, as assumed by Penman and Visser and subsequently by workers such as Hillel (1980) and Taylor and Kleeper (1978), but upon the *differences* of potential that can develop between soil and root. Measurements during and after the great British drought of 1975–6 on three consecutive crops of winter wheat grown in the same field showed that the crop with the largest root system, indeed an unusually large root system, grown in the second drought year, 1976, was the least efficient in extracting soil moisture and so failed to dry the soil as thoroughly as the crops in 1975 and 1977. Plant water potential data showed that this restricted use of the available soil moisture was associated with '. . . failure to make any significant osmotic adjustment, leading to premature loss of leaf turgor and stomatal closure' (McGowan *et al.*, 1984).

The importance of osmotic adjustment in determining the efficiency of plant water use emerged largely as a result of pioneering work by Hsaio *et al.* (1976) and Turner (Begg and Turner, 1976; Jones and Turner, 1978). Much remains to be understood, but it is already clear that the Penman-type root constant, even for the same species growing on the same soil, is *not* constant but can vary significantly from year to year. Accordingly, the root constant should be reinterpreted (McGowan and Tzimas, 1985) as '. . . the condition where the maximum possible capture rate by the complete root system becomes less than the evaporative demand on the leaves', and it should be recognized that this condition is more clearly associated with the minimum xylem potential that can develop in the root system, in response to osmotic adjustment, than with root density.

It will be seen in a later section (Section 4.6.3 under 'The Penman–Monteith model') that this has important implications for attempts to estimate actual evaporation from vegetation covers on the basis of values of soil moisture status where, as with the MORECS model, such attempts rely explicitly on the root constant concept or on a presumed relationship between canopy resistance and soil moisture deficit.

4.5.4 *Soil factors*

Numerous attempts have been made to relate evaporation from vegetation covers to soil conditions and particularly to soil moisture status. It will be clear, however, that apart from the factors discussed in Section 4.4, the influence of soil moisture conditions on evaporation from vegetated surfaces is so intricately enmeshed with vegetational 'control' exerted via the stomatal and root systems discussed in Sections 4.5.2 and 4.5.3 that it cannot easily or even helpfully be disentangled.

The correlations between soil moisture potential and evaporation which have been explored through the medium of drying curves and the root constant can be seen to represent a 'black-box' substitute for the detailed understanding of the complex soil–plant–water relationships which has only recently begun to emerge from sophisticated experimental investigation of aerodynamic and physiological resistances to water movement through the soil–plant–atmosphere continuum.

Like many black-box approximations, however, drying curves such as those illustrated in Fig. 4.15 have enabled significant advances to be made in the understanding, measurement and modelling of evaporation from vegetated covers, and they continue to form the basis of much of the ongoing research reported at the 1985 ASAE conference on 'Advances in Evapotranspiration', to which earlier reference has been made.

The relationships summarized by the curves in Fig. 4.15 differ widely, from the Veihmeyer model at one extreme to the Thornthwaite model at the other. Veihmeyer (1927, 1972) and Veihmeyer and Hendrickson (1927, 1955) reported 'pot' experiments with young prune trees which demonstrated that water losses were the same when the soil moisture content had been reduced almost to the wilting point, as when the soil was at field capacity; i.e. throughout the period of depletion of the soil moisture storage available to the plant between wilting point and field capacity, evaporation takes place at the potential rate (cf. line A in Fig. 4.15). This view was reaffirmed by Halkias *et al.*

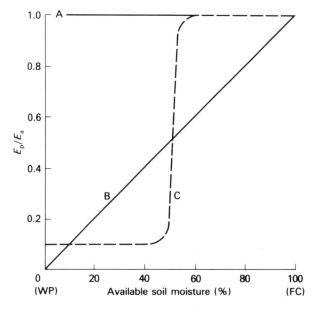

Figure 4.15 *The relationship between actual and potential evaporation through the range of drying from field capacity (FC) and wilting point (WP) according to Veihmeyer and Hendrickson (line A), Thornthwaite (line B) and Penman (curve C). See text for explanation.*

(1955), and Linsley *et al.* (1958) suggested that the total amount of the stored soil moisture so available varies with soil type from about 4 cm for each 100 cm depth in a sand to 17 cm or more for each 100 cm depth in a clay loam. Similarly, Gardner and Ehlig (1963) reported experiments with birdsfoot trefoil, cotton and pepper, which showed that there was little variation in transpiration rates as the soil moisture content decreased until plants began to wilt; thereafter, there was a virtually linear relationship between water content and transpiration rate.

At the other extreme is the view that evaporation decreases as soil moisture content decreases throughout the range of drying. This was argued, for example, by Kramer (1952) and Lassen *et al.* (1952). Later, Thornthwaite (1954) discussed the rate of soil moisture depletion in a sandy loam and noted that, although at first evaporation occurred at approximately the potential rate, '... within a week when three centimetres of water have been lost from the soil, the depressing effect on evaporation has become significant. Within three weeks the soil moisture has been reduced to a point where the evapotranspiration is only about 25 per cent of the potential rate'. Thornthwaite and Mather (1955) eventually assumed that there would be a linear reduction of actual evaporation rate below the potential rate as a function of residual soil moisture storage expressed as a fraction of soil moisture storage at field capacity; i.e. if 50 per cent of the field capacity moisture storage remains, actual evaporation will be 50 per cent of potential evaporation (cf. line B in Fig. 4.15). This assumption has formed a basic premise of several successful water-balance studies, e.g. Carter (1958) and van Hylckama (1956).

Between these two extremes there are several intermediate models, one of the best known of which is that of the root constant proposed by Penman (1949) and discussed earlier (cf. curve C in Fig. 4.15).

Each of these simplifying assumptions has played a valuable role in the past. Some will undoubtedly continue to do so in the future. All must now be viewed, however, in the light of our better understanding of the physiology of plant–water relations and assessed accordingly.

4.5.5 *The components of evaporation from vegetation covers*

The components of total evaporation from a vegetation-covered surface, i.e. bare soil or free water evaporation, transpiration and the evaporation of water intercepted by the vegetation surfaces, have been fully described and discussed in this and the preceding chapter. Throughout, the relationships of these principal evaporation components have been implied without *directly* addressing the question 'Which component is the most important?' The answer to this question is clearly conditional upon the availability of moisture for the evaporation process and specifically will vary depending on whether evaporation is taking place from a wetted or a non-wetted vegetation cover.

In the case of wetted vegetation it is no longer a matter of debate that the evaporation of intercepted water will normally take place at a rate much higher than the transpiration of water by means of stomatal diffusion, i.e. that the interception component will dominate the evaporation total. Furthermore, it seems likely that the evaporation of intercepted water will be relatively more important where surface wetting by precipitation occurs predominantly at night rather than during the day (Pearce *et al.*, 1980). Even in high-rainfall areas, however, vegetation surfaces are normally dry more often than they are wet, and the relative importance of interception and transpiration components over a calendar period such as a year will reflect the relative duration of wet and dry vegetation conditions as well as the interaction of other associated soil, vegetational and atmospheric conditions. For example, Shuttleworth (1988) reported that, in an area of Amazonian rainforest, average evaporation over two years was within 5 per cent of potential evaporation. However, in wet months average evaporation exceeded potential estimates by about 10 per cent and fell below such estimates by at least this proportion in dry months.

Expectedly, therefore, empirical evidence varies widely even within geographically restricted areas. For coniferous forested areas in Britain, for example, Calder (1976) reported interception loss (I) of 1570 mm and transpiration loss (T) of 930 mm over a three-year period in the forested headwaters of the River Severn in mid Wales during which precipitation (P) was 5464 mm. Law (1956, 1958) found average annual figures of $I = 371$ mm, $T = 340$ mm and $P = 984$ mm for the Hodder catchment in the Yorkshire Pennines and Gash and Stewart (1977) observed $I = 213$ mm, $T = 353$ mm and $P = 595$ mm near Thetford in East Anglia. Finally, Hall (1987) presented modelled data for the Monachyle catchment in Scotland which showed annual values for heather of $I = 199$ mm, $T = 91$ mm and $P = 2632$ mm and for grass of $I = 101$ mm, $T = 72$ mm and $P = 2632$ mm. In contrast, the 1983–85 Amazon rainforest data reported by Shuttleworth (1988) were $I = 683$ mm, $T = 2065$ mm and $P = 5492$ mm.

In terms of the comparison between soil evaporation and the other components of evaporation, experimental evidence suggests that, in general, soil evaporation is the least important. For example, where the vegetation cover is continuous, or almost so, it will severely limit opportunities for direct evaporation from the ground surface. This was partially substantiated by Russian experiments reported by Penman (1963) on the use of water by stands of aspen which showed transpiration exceeded soil evaporation by a factor of between 2.1 and 4.3. Again in areas of shallow groundwater, transpiration probably exceeds soil evaporation. Veihmeyer (1938), for example, showed that when the water table was some 120 cm below the ground surface, losses of water by transpiration were approximately 100 times greater than the surface evaporation from the soil.

Where the vegetation cover is not continuous, however, or where it has a relatively small plant mass, energy-balance considerations suggest that the relationship between the components of evaporation may differ from that outlined above. This was indicated by Ritchie (1972) and in work by Baumgartner (1967), in which comparisons between forest, meadow grass and cultivated crops showed soil evaporation accounting for 10, 25 and 45 per cent respectively of the total evaporation loss. Similarly, Thornthwaite and Hare (1965) suggested that, from the standpoint of energetics, soil evaporation in an open-row crop will contribute considerably to the total water loss and King (1961) reported that, in corn crops in the mid-west of the USA, soil evaporation was about 30 per cent of total evaporation in dry years and as much as 50 per cent in wet years. The UK Institute of Hydrology (IH, 1987) reported the initiation of a study of the detailed processes of evaporation from sparse dryland crops typical of those grown in many low rainfall areas of Africa and the Middle East. In these vegetation conditions, direct soil evaporation may on occasions, e.g. after rainfall, account for twice the water loss resulting from transpiration (cf. Fig. 4.16a).

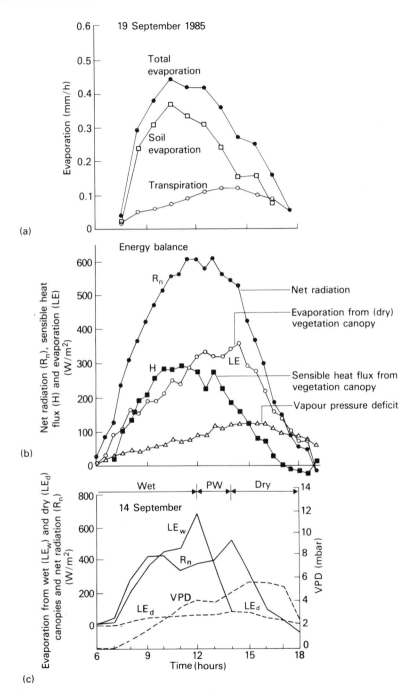

Figure 4.16 (a) *Diurnal trends in evaporation from millet (from an original diagram in IH, 1987),* (b) *energy budget components for a young Douglas fir forest in British Columbia (from an original diagram by McNaughton and Black, 1973) and (c) a hardwood forest in Quebec (from an original diagram by Singh and Szeicz, 1979).*

4.5.6 *Amounts and variations of evaporation from vegetation covers*

Diurnal and seasonal variations of potential evaporation (E_p) are by definition closely related to energy availability and are therefore largely predictable. In normal circumstances, however, *potential* values of evaporation are not achieved because of the resistances imposed on water vapour movement through the soil–plant–atmosphere continuum. Typical patterns of diurnal and seasonal water loss are shown in Figs 4.16 and 4.17.

For example, the energy budget components in Fig. 4.16b were measured for a young Douglas fir forest at Blaney in British Columbia (McNaughton and Black, 1973) and indicate that although evaporation accounted for a substantial proportion of the energy available, the sensible heat flux from the comparatively dry canopy was numerically very similar. Figure 4.16c shows comparable data for a hardwood forest in Canada whose canopy was initially wet and then dried progressively through the day. The graphs confirm

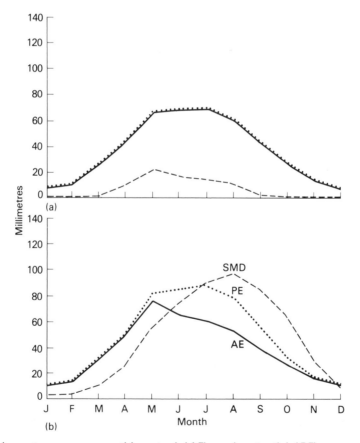

Figure 4.17 *Long-term mean monthly actual (AE) and potential (PE) evaporation and soil moisture deficit (SMD) for (a) MORECS grid square 83 (Lake District) and (b) MORECS grid square 141 (Suffolk). (Based on diagrams in Shawyer and Westcott, 1987. By permission of the Controller of Her Majesty's Stationery Office.)*

the high rates of evaporation from the wetted canopy which exceed the net radiation supply at times, indicating that the wetted canopy acts as an important sink for advected energy from either the overflowing air or the canopy space itself (Singh and Szeicz, 1979). Finally, the long-term mean monthly estimated evaporation data for two 40 km grid squares in England (Fig. 4.17) indicate the close similarity between actual and potential evaporation in areas of low soil moisture deficit (cf. Lake District) and the marked divergence of the two in areas of high soil moisture deficit (cf. Suffolk).

Annual estimates of E_p for the stations with long periods of data, cf. the Radcliffe Observatory, Oxford, for 1881–1966 and the Edgbaston Observatory, Birmingham, for 1900–68, were analysed at the Institute of Hydrology (Rodda *et al.*, 1976) and by Takhar and Rudge (1970), and indicate considerable year-to-year variations, between 430 and 560 mm for Oxford, and a period of significantly low values centred on 1930 at both stations.

The value of mapping the spatial variation of potential and actual evaporation from vegetated surfaces for a variety of climatological, water-balance and resource evaluation procedures has long been recognized. Gurnell (1981) listed a sample of more than 20 published mapping studies ranging in scale from a single English county (cf. Foyster, 1973) to the world (cf. Baumgartner and Reichel, 1975).

Other examples are shown in Figs 4.18 and 4.19. Figure 4.18 emphasizes the overriding control of available energy in determining E_p for Europe. The comparatively regular pattern of isolines, increasing southwards, is broken only by the mountains of central and southern Europe, where relatively lower evaporation rates prevail. Figure 4.19 shows the distribution of E_p and E_a for Britain. The derivation of the E_p map, which is based on a variety of sources, was detailed by Ward (1976). The pattern of isolines shows a broad latitudinal decrease from south to north and also a marked decrease away from the coast. Both characteristics reflect the availability of net radiation, although the higher coastal values are probably also a reflection of the additional drying power associated with higher windspeeds in coastal areas. Conversely, reduced values in the main upland areas are largely associated with low radiation availability. Again, the derivation of the E_a map was discussed by Ward (1976). Based on measured data from a large number of river basins, the map shows that over most of England and Wales actual evaporation values lie between 400 and 499 mm, despite the much wider variation in the potential values. A recent example of evaporation mapping was provided by Schulze (1985) who discussed the problems of mapping regional E_p in areas of sparse climatological data.

Various authors, including Monteith (1985), have drawn attention to the potentially important role of remotely sensed data in the preparation of evaporation maps for a wide range of scales from individual fields to major river basins. Some of the detailed possibilities in this respect were reviewed by Hatfield (1983, 1985), and examples of the use of GOES imagery were discussed by Leith and Solomon (1985) and of Meteosat and NOAA thermal IR imagery by Seguin *et al.* (1985). Figure 4.20 shows an example of daily evaporation mapping at a field/farm scale in the Netherlands which was based on infrared line scanning (IRLS) reported by Soer (1980) and Nieuwenhuis (1981).

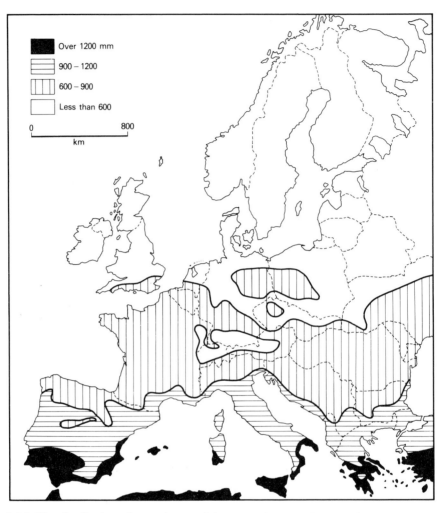

Figure 4.18 *The distribution of annual potential evaporation over Europe (from an original map by Mohrmann and Kessler, 1959).*

4.6 Modelling evaporation

4.6.1 *Modelling and measurement*

Modelling and estimation of evaporation, especially of total evaporation from a vegetation-covered surface, is the normal method of quantifying evaporation amounts. In this respect evaporation contrasts strongly with some other hydrological variables, such as runoff or rainfall, for which quantities are normally obtained by direct measurement. This is not because measurement of evaporation is impossible or even in many cases difficult. Indeed, there are numerous measurement techniques that have been extensively reviewed in the literature. These include the traditional approaches using atmometers,

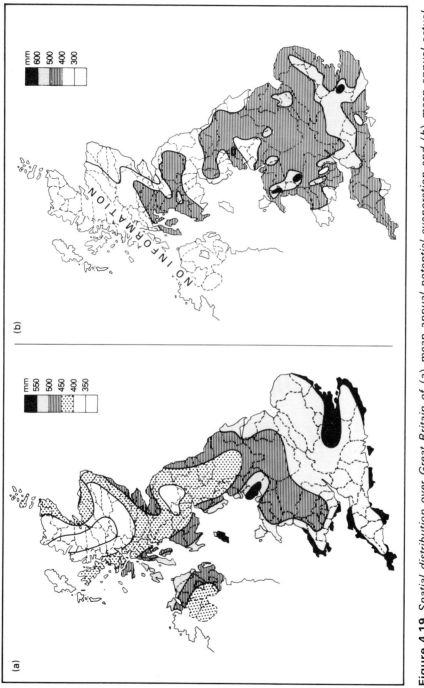

Figure 4.19 Spatial distribution over Great Britain of (a) mean annual potential evaporation and (b) mean annual actual evaporation (based on original diagrams in Ward, 1981).

Figure 4.20 *Daily evaporation rates for an area near Losser, the Netherlands, mapped from remotely sensed crop surface temperatures. (Reprinted by permission of the publisher from Estimation of regional evapotranspiration and soil moisture conditions using remotely sensed crop surface temperatures, by G. J. R. Soer, Remote Sensing of Environment, **9**, pp. 27–45. © 1980 Elsevier Science Publishing Co. Inc.)*

evaporimeters and evaporation pans (cf. Shaw, 1983), percolation gauges and lysimeters (cf. Rodda *et al.*, 1976; Shaw, 1983) and soil moisture status (cf. Shuttleworth, 1979). In addition there are less widely used techniques involving measurement of the ascent of sap within the plant stem (cf. Federer, 1975; Jordan and Kline, 1977; NWSCA, 1987), the use of whole-plant evaporation chambers (cf. Stark, 1967; CSIRO, 1985) and macroscale measurements of the catchment water balance (cf. Ward, 1981; Woo and Waylen, 1985). Finally, there have been important advances in recent years in the development of instruments, described at an early stage by McIlroy (1971), to make routine measurements of evaporation. The Hydra (Shuttleworth *et al.*, 1988), for example, is a compact portable instrument comprising an ultrasonic anemometer which, together with humidity and temperature sensors, is linked to a microprocessor to enable analysis of measurements as they are made in the field.

Much of the requisite instrumentation for such measurements lacks standardization, is demanding in terms of maintenance and routine observing techniques and is expensive. In addition, all these approaches to the measurement of evaporation from water, soil or vegetated surfaces, together with the theoretical approaches discussed in the remainder of this chapter, suffer from the overriding problem of point sample representativeness, and most of them suffer also from the distorting effects of advection which, as Fig. 4.5 indicates, is inversely proportional to the size of sample evaporating surface. Often, therefore, it is difficult to know what these measurements mean in terms of a valid representation of areal evaporation at a field, catchment or regional scale. However, Gash (1986) derived a simple method of estimating the areal sample of micrometeorological measurements as a function of measurement height and aerodynamic roughness.

4.6.2 *Theoretical equations*

Earlier discussions in this chapter have indicated that the physics of the evaporation process relate primarily to the provision at the evaporating surface of sufficient energy for the latent heat of vaporization and the operation above the evaporating surface of diffusion processes to provide a means for removing the water vapour produced by evaporation. There have been many attempts to calculate evaporation from equations based upon physical theory. Inevitably, the two main approaches have related to turbulent transfer (diffusion) processes and energy-balance processes.

Turbulent transfer approach
Air, like any other fluid, can have a *laminar* or a *turbulent* motion. In the first case, it moves in straight lines or along smooth, regular curves in one direction. In the second, air particles follow irregular, tortuous, fluctuating paths, evidenced in the cross-currents and gusts of wind with intervening brief lulls. True laminar flow is observed in the lowest few millimetres of the atmosphere; above this, friction between the air and the ground surface induces eddies and whirls and other irregular air movements. This is the planetary boundary layer where the depth and strength of turbulence is largely dependent upon the roughness of the ground surface and the strength of the wind, and increases as these two factors increase. It is also influenced greatly by the stability of the air, which varies diurnally. Normally, during the daytime, the lapse rate increases, reaching a maximum in the early afternoon. The associated convectional activity and the increased buoyancy of the air together reinforce the degree of turbulence and deepen the planetary boundary layer to as much as 1–2 km. At night, on the other hand, the ground surface cools and lapse conditions are normally replaced by an inversion of temperature. This decreases the buoyancy of the air, damps down convectional activity and effectively suppresses the turbulent motion of the air until, on calm nights, an almost laminar flow is experienced. As would be expected, this diurnal variation of turbulent activity is greatest with clear skies and scarcely apparent in completely overcast conditions (Sutton, 1962).

The greater the intensity of turbulence, the more effectively are water vapour molecules dispersed or diffused upwards into the atmosphere. When, however, the level of turbulence is low, as in the evening, the consequent reduction in the mixing and diffusion of water vapour through the lower layers of the atmosphere leads to greatly increased relative humidity in a thin layer of air immediately above the ground surface. Subsequent cooling may then result in the formation of a shallow layer of ground fog.

The motion of the air in the boundary layer will tend to establish a uniform moisture content at all heights, provided that no water vapour is added or removed. If water vapour *is* added to the bottom of the boundary layer by evaporation it will be dispersed and diffused upwards, but as long as the evaporation continues the moisture content of the air will be highest at the ground surface and lowest at the top of the turbulent layer. This moisture gradient will persist until evaporation ceases. A reverse gradient will develop if water vapour is removed at the bottom of the turbulent layer by condensation in the form of dew or frost. It follows from this that, with a constant intensity of mixing, an increase in the rate of evaporation will be reflected in an increased moisture gradient in the boundary layer. Alternatively, with a constant rate of evaporation, variation of moisture gradient will reflect changes in the intensity of mixing. Accordingly, it should be possible to determine the rate of evaporation from any surface by reference to the moisture gradient and the intensity of turbulent mixing. Measurements are needed of the moisture content and the windspeed at a minimum of two known heights within the turbulent layer.

Within the turbulent transfer approach three main lines of development have been based upon the work of Prandtl (1925), Taylor (1922, 1935) and Richardson (Richardson, 1926; Richardson and Stommel, 1948). Associated with these different approaches, numerous evaporation formulae have been developed. Since, in each case, the rate of evaporation is determined on the basis of small humidity and windspeed differences over a narrow height range within the boundary layer, the frequency and the accuracy of the instrumental observations must be very high. Certainly, for most hydrological problems where replication is desirable and where the use of existing standard climatological data would enable the derivation of long-term evaporation estimates, the turbulent transfer approach makes unacceptable demands upon instrumental and data accuracy.

Energy-balance approach
Instead of concentrating attention upon the upward diffusion of water vapour from the evaporating surface, the energy-balance approach emphasizes that, in order for water to change in state from a liquid to a gas, a certain amount of energy (approximately 2.47×10^6 J/kg or 590 cal/g at normal air temperatures) is needed. If the total amount of radiation received by a given area could be determined and if all this radiation were used in evaporation, it would then be a simple matter to calculate the depth of water evaporated.

However, only a part of the incident radiation is used for evaporation (H_e). Some is reflected back into the atmosphere (R_e). Of the remainder that is absorbed, some is returned as long-wave radiation from the soil or water surface (R_a), some is used to heat the overlying air (H_a) and some to heat the soil or water body or the biomass (H_b). Finally, a negligible part is used in plant growth to produce carbohydrate (H_c) (Pearl *et al.*, 1954). An energy balance equation can, therefore, be written in the following terms, where R represents the total incoming radiation:

$$R - R_e - R_a = H_e + H_a + H_b + H_c \qquad (4.6)$$

Net radiation income = radiation expenditure

The energy-balance approach thus involves two main steps: first, to determine the total radiation income and, second, to apportion its outgo under the various headings outlined above in order that the amount used in evaporation may be determined. The application of the energy-balance approach to the estimation of evaporation from water surfaces was first suggested by Angstrom (1920), and later formulae were developed by Bowen (1926), McEwen (1930), Richardson (1931) and Cummings (1936).

The measurement of the convective fluxes, which would be necessary to solve Eq. (4.6) is exceedingly difficult. Some simplification is possible if the principle of similarity is invoked. This asumes that an eddy is non-discriminatory with regard to the property being transported (Oke, 1978); i.e. it will carry heat, water vapour, momentum, carbon dioxide, etc., with equal facility. Such simplification provides the basis for the application of the turbulent transfer approach (discussed in the preceding subsection), which assumes incorrectly that $K_v = K_m$, and the Bowen ratio method, whose main advantage is that it assumes correctly that $K_v = K_h$.

The Bowen ratio is the ratio of the sensible and latent heat fluxes:

$$\beta = \frac{H_a + H_b}{H_e}, \qquad \text{often simplified to } \beta = \frac{H_a}{H_e} \qquad (4.7)$$

Typical average values of β are (Oke, 1978) 0.1 for tropical oceans, 0.1–0.3 for tropical wet jungles, 0.4–0.8 for temperate forests and grassland, 2.0–6.0 for semi-arid areas and 10.0 or more for deserts.

The energy used in actual evaporation is then given by

$$H_e = \frac{H_a + H_b}{1 + \beta}$$

or

$$\frac{H_e}{H_a + H_b} = \frac{1}{1 + \beta} \qquad (4.8)$$

The solution of this equation demands accurate measurements of radiation, soil heat flux and vertical profiles of temperature and humidity, and although these would be unremarkable measurements in a research situation they impose substantial limitations for the application of the energy-balance approach to the routine calculation of evaporation. Attention has therefore focused on ways in which evaporation may be estimated from empirical or semi-empirical models of the evaporating system, in which the model input comprises readily available, routine meteorological measurements.

4.6.3 *Empirical and semi-empirical models*

Scores of empirical and semi-empirical evaporation models have been developed. A small minority of these are now used on a routine basis, but the majority are simply 'hydrological museum pieces' (Rodda *et al.*, 1976). Probably the most widely used, especially where data are limited, is the model developed by C. W. Thornthwaite, but almost as well known is that of H. L. Penman, especially in the form as subsequently modified and developed by J. L. Monteith and coworkers and now known as the 'Penman–Monteith' model.

The Thornthwaite model
Thornthwaite's formula for the estimation of potential evaporation was presented and modified in a number of papers (Thornthwaite, 1944, 1948, 1954). He began by

estimating a monthly heat index i by means of the equation:

$$i = \left(\frac{t}{5}\right)^{1.514}$$
(4.9)

where t is the mean temperature of the month under consideration expressed in °C. Twelve monthly indexes are then added together to give an annual heat index I. Monthly potential evaporation is then calculated as

$$E_p = 1.6b\left(\frac{10t}{I}\right)^a$$
(4.10)

where E_p is monthly potential evaporation in cm, t is the mean monthly temperature in °C, a is a cubic function of I and b is a factor to correct for unequal day length between months.

The only factors taken into account are mean air temperature and hours of daylight. The method therefore appears to be extremely empirical and somewhat complex, and has been criticized on both grounds.

However, Thornthwaite's choice of an empirical model was quite deliberate, as also was his choice of mean air temperature as the main variable influencing potential evaporation (Thornthwaite and Hare, 1965). He justified the selection of mean air temperature on the grounds that there is a fixed relationship between that part of the net radiation which is used for heating and that part which is used for evaporation when conditions are suited to evaporation at the potential rate, ie. when the soil is continuously moist (Thornthwaite, 1954). This means, in effect, that although the Thornthwaite model is empirical, it nevertheless estimates potential evaporation by an indirect reference to the radiation balance at the evaporating surface.

The Thornthwaite model has been adopted more widely than any other. Paradoxically, it is probably because it expresses E_p simply as a function of mean air temperature and day length, two quantities that are independent of the rate of evaporation, that it is applicable over a wide range of climatological conditions (Sibbons, 1956). Certainly, experimental and documentary evidence has justified its application in almost every climatological area (cf. Carter, 1954, 1955, 1956; Cochrane, 1959; Howe, 1956; van Hylckama, 1956, 1958; Mather, 1954; Rao, 1958; Sanderson, 1950; Thornthwaite, 1953). This is not to say that, in all of these cases, the Thornthwaite model has permitted a continuously precise and accurate estimation of monthly or short-term E_p; it would be unrealistic to expect this to be so. It is likely, however, that in each case the Thornthwaite approach has provided a convenient and easily applicable method whose proven weaknesses have given it some superiority over other less well-tried techniques. Over very short periods of time, when mean temperature is not a suitable measure of incoming radiation, and in areas like the British Isles, where advection effects resulting from frequent air mass changes lead to frequent rapid changes in mean air temperature and humidity, it is likely that the Thornthwaite approach will be relatively unsuccessful. Even so, during the summer months, when E_p is of greatest significance, more stable air mass conditions tend to prevail, thereby strengthening the relationship between temperature, the radiation balance and evaporation.

It is unfortunate that the enthusiasm with which the Thornthwaite model has been adopted in so many different conditions has often been allowed to obscure

Thornthwaite's own caution that the chief obstacle to the development of a rational evaporation model '. . . is the lack of understanding of why $[E_p]$ corresponding to a given temperature is not everywhere the same' (Thornthwaite, 1948) and the significance of that caveat for the severe underestimation of evaporation rates in dry climates (Monteith, 1985).

The Penman–Monteith model

Although imposing greater data demands than the Thornthwaite approach, the Penman–Monteith evaporation model has become firmly established, not only as a routine method of quantifying evaporation but also as the basis for most of the significant conceptual advances in evaporation research that have taken place in recent decades.

Penman (1948) devised a model for potential evaporation which combined the turbulent transfer and the energy-balance approaches. This was later restated by Penman in slightly modified forms (1952, 1954, 1956, 1963). Basically, there are three equations. The first equation is a measure of the drying power of the air. This increases with a large saturation deficit, indicating that the air is dry, and with high windspeeds. The first equation is, therefore, derived from the basic pattern of the turbulent transfer approach and takes the form:

$$E_a = 0.35(e_a - e_d)\left(1 + \frac{u}{100}\right) \qquad \text{mm/d} \qquad (4.11)$$

where e_a is the saturation vapour pressure of water at the mean air temperature, e_d is the saturation vapour pressure of water at the dew point temperature, or the actual vapour pressure at the mean air temperature, and u is the windspeed at a height of two metres above the ground surface.

The second equation provides an estimate of the net radiation available for evaporation and heating at the evaporating surface and takes the form:

$$H = A - B \qquad \text{mm/d} \qquad (4.12)$$

where A is the short-wave incoming radiation and B is the long-wave outgoing radiation, as estimated in the following expressions:

$$A = (1 - r) R_a\left(0.18 + 0.55\frac{n}{N}\right) \qquad \text{mm/d} \qquad (4.13)$$

$$B = \sigma T_a^4(0.56 - 0.09\sqrt{e_d})\left(0.10 + 0.90\frac{n}{N}\right) \qquad \text{mm/d} \qquad (4.14)$$

where R_a is the theoretical radiation intensity at the evaporating surface in the absence of an atmosphere, expressed in evaporation units; r is the reflection coefficient of the evaporating surface, n/N is the ratio of actual/possible hours of bright sunshine; σT_a^4 is the theoretical backradiation which would leave the area in the absence of an atmosphere, T_a being the mean air temperature in kelvin and σ being Stefan's constant; and e_d is as in Eq. (4.11).

Penman assumed that the heat flux into and out of the soil, which usually represents about two per cent of the total incoming energy (Pearl *et al.*, 1954), is small enough to be ignored, so he simply divided the net radiation between heating the air and evaporation. The proportion of it used in evaporation is then estimated by combining Eqs (4.11) and

(4.12) to give

$$E = \left(\frac{\Delta}{\gamma} H + E_a \right) \Big/ \left(\frac{\Delta}{\gamma} + 1 \right) \quad \text{mm/d} \tag{4.15}$$

where Δ is the slope of the saturation vapour pressure curve for water at the mean air temperature. This relationship assumes equality of the coefficients of water vapour and convective heat transfer, a requirement that is well met in windy and unstable conditions (Cole and Green, 1963) but which may be less certain in conditions of strong insolation and light winds (Milthorpe, 1960). The ratio Δ/γ is dimensionless and makes allowance for the relative significance of net radiation and evaporativity in total evaporation. At 10, 20 and 30°C the values of Δ/γ are 1.3, 2.3 and 3.9 respectively (Penman, 1963); i.e. during those periods, such as the summer months, when totals of evaporation are significantly high, the net radiation term is given more weight than the evaporativity term. Furthermore, in humid areas, H is usually greater than E_a. H therefore tends to be the dominant term in the equation. The quantity E estimated in Eq. (4.15) will vary according to the value assigned to the reflection coefficient r in Eq. (4.13). If $r = 0.25$, E will represent E_p from an extended short green crop whereas, if $r = 0.05$, E will represent E_o, the evaporation from an extended sheet of open water.

Modifications to this model have been suggested from time to time, many of them concerned with the radiation terms in Eqs (4.13) and (4.14). Glover and McCulloch (1958), for example, proposed the alteration of Eq. (4.13) thus:

$$A = (1 - r)R_a \left[0.29 \cos (\text{lat}) + 0.52 \frac{n}{N} \right] \quad \text{mm/d} \tag{4.16}$$

based upon regression analysis of values of radiation and hours of sunshine to give a better relationship over the range of latitude from 0 to 60°. Later modifications, based on work by L. P. Smith at the Meteorological Office, were described in MAFF (1967).

If all or part of the available energy represented in Eq. (4.12) is measured directly then even short-period estimates of E_p may be tolerably accurate. Several techniques may be employed. First, measured incoming radiation may be substituted for Eq. (4.13) (Makkink, 1957). A second option is to replace the whole of the expression for H, i.e. Eq. (4.12), by measured net radiation (Cole and Green, 1963). Third, in addition to net radiation, measurements may be made of the heat flow through the soil. In this way the complete energy balance is measured and errors caused in the estimation of E_p by the neglect of heat storage may be obviated. This, in turn, might correct a large part of the seasonal discrepancy which has been noted between measured and estimated results using the Penman model (Ward, 1963) and which has been attributed to the fact that a larger proportion of the incoming radiation will be required to heat the soil in the spring than in the autumn, so that the proportion available for evaporation will be correspondingly lower.

However, the insertion of measured radiation values in place of the Penman radiation terms will result in an improved estimate of E_p only if the network of radiation instruments is sufficiently dense to enable a representative sampling of vegetation type having different albedo and surface temperatures. Otherwise the gain in accuracy of the point measurements tends to be lost in the extrapolation to an areal estimate.

Much of the discussion earlier in this chapter of evaporation from vegetated surfaces focused on the important roles of aerodynamic resistance (r_a) and surface resistance (r_s) in accounting for variations of evaporation between vegetation covers that differ in terms of wetness, structure and physiology. Neither resistance appears to be incorporated in the Penman model of the evaporation process, although interestingly Thom and Oliver (1977) showed that the wind function in Eq. (4.11) implicitly incorporates some reasonable assumption about both. Their explicit incorporation was proposed by Penman and Schofield (1951) and was subsequently presented in its most familiar form by Monteith (1965), who treated the canopy as a single extensive isothermal leaf so that

$$\lambda E = \frac{\Delta H + \rho c (e_a - e_d)/r_a}{\Delta + \gamma [1 + (r_s/r_a)]} \tag{4.17}$$

where ρ is the density of the air, c is the specific heat of the air and all other terms are as previously defined.

In theory this modification should permit incorporation of the important aerodynamic and physiological influences of the vegetation cover on evaporation as well as the largely meteorological influences on which the original Penman model focused. In practice, however, although the Penman–Monteith model has played a most valuable role in the development of conceptual understanding of the complex evaporation process from wetted and unwetted vegetation surfaces, it has not been used widely to determine values of evaporation. This was admitted by Monteith (1985) when he noted that the model had '. . . been used mainly as a diagnostic tool for estimating r_s when λE is known, rather than as a prognostic tool for estimating λE when r_s is assumed'.

The main problem, of course, lies in the difficulty of obtaining adequate measurement of the vegetational factors, and especially of r_s which is a complex function of many climatological and biological factors including radiation, saturation deficit, soil moisture status and biomass characteristics. Much valuable research has been conducted by the Institute of Hydrology on the dependence of r_a and r_s on environmental variables using the Penman–Monteith equation. Work in Thetford Forest, for example, showed how surface resistance values were highly dependent on the atmospheric humidity deficit and less so on the input of solar radiation (IH, 1985; Stewart, 1988). Other similar work has been done on bracken (IH, 1979) and heathland vegetation (IH, 1985). Table 4.2 shows one set of values for r_a and r_s as determined by Szeicz et al. (1969) for sites in southern England.

One important prognostic use of the Penman–Monteith model has been its incorporation in the UK Meteorological Office rainfall and evaporation calculation system

Table 4.2 Values of r_a and r_s (s/m) for selected sites in southern England (based on data in Szeicz et al., 1969).

Month	Open water		Pine forest		Potatoes		Lucerne	
	r_a	r_s	r_a	r_s	r_a	r_s	r_a	r_s
June	125	0	3	98	61	113	60	64
September	125	0	3	120	38	110	52	60

(MORECS). This system uses daily meteorological data to produce weekly estimates of evaporation, soil moisture deficit and effective rainfall for 40×40 km grid squares in Great Britain. Since its introduction in 1978, MORECS has undergone a number of changes, including extensive revisions in 1981 and 1983. A useful description and evaluation of the calculation of evaporation and soil moisture extraction was given by Thompson (1982) and Gardner and Field (1983). A modified version of the Penman–Monteith equation is used to calculate daily potential and actual evaporation and in the latter case the canopy resistance is adjusted in relation to the magnitude of the soil moisture deficit. Net radiation is calculated from sunshine duration and a correction factor has been introduced into the model to allow for the difference between surface temperature and air temperature as suggested by Monteith (1981).

MORECS models soil moisture extraction using a two-layer moisture reservoir. When the soil is at field capacity, layer X contains 40 per cent (X_{max}) and layer Y 60 per cent of the available soil water. Water in X is freely available to the vegetation cover; water in Y is not and becomes increasingly difficult to extract as the moisture status is reduced. This is achieved by increasing the canopy resistance in relation to the decrease in moisture content in Y. Moisture is extracted from Y only if X is empty and rainfall replenishes X first and then Y only when X is full. As Gardner and Field (1983) observed, the value of X_{max} is important (a) because it defines a soil moisture deficit threshold rather similar to Penman's 'root constant' (see Section 4.5.3 under 'Root control') and (b) because being 40 per cent of the available soil water it defines the total available water in the soil, i.e. $2.5X_{max}$. Different X_{max} values are allocated to different surface covers ranging from bare soil to forest, and for each surface cover three values of X_{max} are used to represent soils having low, medium and high available water capacity, e.g. for grass the X_{max} values are 37.5, 50.0 and 62.5 mm, representing total available water (or maximum permitted deficits) of 94.0, 125.0 and 156.0 mm respectively for soils of low, medium and high water availability.

Results from the revised MORECS models have been encouraging, although further improvements to the models for evaporation and soil moisture depletion will undoubtedly be made. Specimen MORECS results are shown in Fig. 4.17 (Section 4.5.6).

The role of advection

Much of the discussion of evaporation in the literature (including that presented in this chapter) tends to be imprecise about the role of advection. Reference was made in the definition of potential evaporation, for example, to an 'extensive' evaporating surface and to the assumption that a sufficiently extensive surface would obviate advectional effects. Implicit in such terminology is the recognition that evaporation is influenced strongly by vapour pressure deficit and that, in turn, vapour pressure deficit is linked to the available energy and the canopy resistance. Conceptually it is possible to move in opposite directions from this position.

First, it may be recognized that with heterogeneous surfaces and contrasting itinerant air masses the linkage between vapour pressure deficit, available energy and canopy resistance may be broken when air with greatly different characteristics is imported from another area. In other words, when an air mass moves across the boundary between two surfaces of different wetness a horizontal gradient of saturation deficit is produced, leading to 'advectional enhancement' (McNaughton et al., 1979) if the air moves from a drier to a wetter surface and the saturation deficit increases and to advectional depression when the movement is in the opposite sense and the saturation deficit increases. As McNaughton *et al.* (1979) observed, although 'some significant effect of advection on local evaporation rates is the rule rather than the exception', the same methods, including the Penman–Monteith model, are normally used to estimate evaporation when advective enhancement is large as at other times. In fact the use of the Penman–Monteith model would not be unreasonable if values of canopy resistance were known to sufficient accuracy. At present they are not, although research in this area continues.

Second, it may be recognized that with an 'infinite' rather than an extensive surface, surface exchange rates control the behaviour of the atmosphere and give rise to an 'equilibrium' evaporation rate (Shuttleworth, 1979) which may be stated as

$$\lambda E_{eq} = H\left(\frac{\Delta}{\Delta + \gamma}\right) \tag{4.18}$$

Equilibrium evaporation thus represents the lower limit of evaporation from a wet surface. This was recognized by Slatyer and McIlroy (1961) and formalized much later by McNaughton (1976a, 1976b). Doubt persists, however, about the extent to which $\Delta/(\Delta + \gamma)$ does, in fact, represent an equilibrium rate. Priestley and Taylor (1972), in discussion of their own proposed simplification of the Penman model which eliminates the term involving windspeed and atmospheric humidity deficit, noted that the average rate of evaporation from vegetation freely supplied with water was $1.26[\Delta/(\Delta + \gamma)]$. This higher value has not been satisfactorily explained although the additional energy implied has been ascribed to the entrainment of relatively warm, dry air downwards through the upper surface of the planetary boundary layer (Monteith, 1985).

In a number of contributions, Morton (1978, 1983, 1985) has used the argument that the Penman–Monteith model, which serves as a 'cause' of evaporation from a small moist surface, is, because of a scale-based feedback mechanism, an 'effect' of evaporation from a large vegetation covered suface to develop a complementary relationship between actual and potential evaporation on an areal basis. His Complementary Relationship Areal Evapotranspiration (CRAE) model is described most clearly in Morton (1985) and assumes that given a similar input of radiation, the potential evaporation from a drier area, where actual evaporation is low, is higher than from a wetter area, where the actual evaporation may be quite high. It then follows that the sum of actual and potential evaporation is constant, i.e. there is a complementary relationship between them. Following Bouchet (1963), Brutsaert and Stricker (1979) and Morton (1978, 1983, 1985) quantified this relationship and used it to estimate the regional actual evaporation from estimates of potential evaporation. However, tests of the model for lowland tropical sites by the Institute of Hydrology (IH, 1985) were not encouraging.

References

Angstrom, A. (1920) Applications of heat radiation measurement to the problems of evaporation from lakes and the heat convection at their surfaces, *Geografisca Ann.*, **2**: 237–52.

ASAE (1985) *Advances in evapotranspiration*, Proceedings of National Conference, Chicago, December 16–17, 1985, American Society of Agricultural Engineers, St Joseph, Mich., 453 pp.

ASCE (1949) *Hydrology handbook*, American Society of Civil Engineers, New York.

Baker, F. S. (1917) Some field experiments on the evaporation from snow surfaces, *Mon. Wea. Rev.*, **45**: 363–6.

Baumgartner, A. (1967) Energetic bases for differential vaporization from forest and agricultural lands, in *Forest hydrology*, W. E. Sopper and H. W. Lull (eds), Oxford, Pergamon, pp. 381–9.

Baumgartner, A. and E. Reichel (1975) *The world water balance: mean annual global, continental and maritime precipitation, evaporation and runoff*, Elsevier Scientific, New York.

Begg, J. E. and N. C. Turner (1976) Crop water deficits, *Advances in Agronomy*, **28**: 161–217.

Bouchet, R. T. (1963) Evapotranspiration réelle et potentielle, signification climatique, in *General Assembly IAHS, Berkeley: Committee for Evaporation, IAHS Publ.*, **62**: 134–42.

Bowen, I. S. (1926) The ratio of heat losses by conduction and by evaporation from any water surface, *Physical Review*, **27**: 779–87.

Brun, L. J., E. T. Kanemasu and W. C. Powers (1973) Estimating transpiration resistance, *Agron. J.*, **65**: 326–8.

Brutsaert, W. H. and H. Stricker (1979) An advection-aridity approach to estimate actual regional evapotranspiration, *WRR*, **15**: 443–50.

Calder, I. R. (1976) The measurement of water loss from a forested area using a 'natural' lysimer, *J. Hydrol.*, **30**: 311–25.

Calder, I. R. (1979) Do trees use more water than grass?, *Water Services*, **83**: 11–14.

Carter, D. B. (1954) Climates of Africa and India according to Thornthwaite's 1948 classification, *Publ. in Climatol.*, **7**: 455–79.

Carter, D. B. (1955) The water balance of the Lake Maracaibo basin, *Publ. in Climatol.*, **8**: 209–27.

Carter, D. B. (1956) The water balance of the Mediterranean and Black Seas, *Publ. in Climatol.*, **9**: 123–74.

Carter, D. B. (1958) The average water balance of the Delaware basin, *Publ. in Climatol.*, **11**: 249–70.

Clyde, G. D. (1931) Snow-melting characteristics, *Utah Agricultural Experimental Station Bull.*, **231**.

Cochrane, N. J. (1959) Observable evapotranspiration in the basin of the River Thames, *QJRMS*, **85**: 57–9.

Cole, J. A. and M. J. Green (1963) Measurements of net radiation over vegetation and of other climatic factors affecting transpiration losses in water catchments, *IASH Committee for Evaporation, Publ.*, **62**: 190–202.

Croft, A. R. (1944) Evaporation from snow, *Bull. Amer. Met. Soc.*, **25**: 334–7.

CSIRO (1969) *Annual Report 1968–69*, Commonwealth Scientific and Industrial Research Organisation, Griffith, NSW, 75 pp.

CSIRO (1985) *Research Report 1983–1984*, Division of Irrigation Research, Commonwealth Scientific and Industrial Research Organisation, Griffith, NSW, 154 pp.

Cummings, N. W. (1936) Evaporation from water surfaces, *Trans. AGU*, **17**: 507–9.

Dalton, J. (1802) Experimental essays on the constitution of mixed gases, *Manchester Literary and Philosophical Society Memo.*, **5**: 535–602.

Davenport, D. C., R. M. Hagan and P. E. Martin (1969) Antitranspirants research and its possible application in hydrology, *WRR*, **5**: 735–43.

Denmead, O. T. (1984) Plant physiological methods for studying evapotranspiration: problems of telling the forest from the trees, *Agric. Water Manag.*, **8**: 167–89.

Douglass, J. E. (1967) Effects of species and arrangement of forests on evapotranspiration, in *Forest hydrology*, W. D. Sopper and H. W. Lull (eds), Pergamon, Oxford, pp. 451–61.

Eagleson, P. S. (1970) *Dynamic hydrology*, McGraw-Hill, New York, 462 pp.

Federer, C. A. (1975) Evapotranspiration, *Reviews of Geophysics and Space Physics*, **13**: 442–5.

Finnigan, J. J. (1979) Turbulence in waving wheat II. Structure of momentum transfer, *Boundary-Layer Met.*, **16**: 213–36.

Fischer, R. A. (1971) Role of potassium in stomatal opening in the leaf of *Vicia faba*, *Plant Physiol.*, **47**: 555–8.

Fortier, S. (1907) Evaporation losses in irrigation, *Engng. News*, **58**: 304–7.

Foyster, A. M. (1973) Application of the grid square technique to mapping of evapotranspiration, *J. Hydrol.*, **19**: 205–26.

Gardner, C. M. K. and M. Field (1983) An evaluation of the success of MORECS, a meteorological model, in estimating soil moisture deficits, *Agric. Met.*, **29**: 269–84.

Gardner, W. R. and C. F. Ehlig (1963) The influence of soil water on transpiration by plants, *JGR*, **68**, 5719–24.

Garstka, W. U., L. D. Love, B. C. Goodell and S. A. Bertle (1958) *Factors affecting snowmelt and streamflow*, US Department of the Interior and USDA Forest Service.

Gash, J. H. C. (1986) A note on estimating the effect of a limited fetch on micrometeorological evaporation measurements, *Boundary-Layer Met.*, **35**: 409–13.

Gash, J. H. C. and J. B. Stewart (1977) The evaporation from Thetford Forest during 1975, *J. Hydrol.*, **35**: 385–96.

Geraghty, J. J., D. W. Miller, F. van der Leeden and F. L. Troise (1973) *Water atlas of the United States*, Water Information Center Inc., New York.

Glover, J. and J. S. G. McCulloch (1958) The empirical relation between solar radiation and hours of sunshine, *QJRMS*, **84**: 172–5.

Grant, R. H., G. E. Bertolini and L. P. Herrington (1986) The intermittent vertical heat flux over a spruce forest canopy, *Boundary-Layer Met.*, **35**: 317–30.

Gray, D. M., G. A. McKay and J. M. Wigham (1970) Energy, evaporation and evapotranspiration, in *Handbook on the principles of hydrology*, D. M. Gray (ed.), sec. III, National Research Council of Canada, Ottawa.

Gurnell, A. M. (1981) Mapping potential evapotranspiration: the smooth interpolation of isolines with a low density station network, *Applied Geography*, **1**: 167–83.

Halkias, N. A., F. J. Veihmeyer and A. H. Hendrickson (1955) Determining water needs for crops from climatic data, *Hilgardia*, **24**: 207–33.

Hall, R. L. (1987) Processes of evaporation from vegetation of the uplands of Scotland, *Trans. Roy. Soc. Edinburgh: Earth Sciences*, **78**: 327–34.

Harris, F. S. and J. S. Robinson (1916) Factors affecting the evaporation of moisture from the soil, *J. Agric. Res.*, **7**: 439–61.

Hatfield, J. L. (1983) Evapotranspiration obtained from remote sensing methods, in D. E. Hillel (ed.), *Advances in irrigation*, vol. 2, Academic Press, New York, pp. 395–416.

Hatfield, J. L. (1985) Estimation of regional evapotranspiration, in *Advances in Evapotranspiration*, ASAE, pp. 357–65.

Hendricks, D. W. and V. E. Hansen (1962) Mechanics of evapotranspiration, *Proc. ASCE, J. Irrig. Drainage Div.*, **88**(IR2): 67–82.

Hewlett, J. D. (1982) *Principles of forest hydrology*, University of Georgia Press, Athens, 183 pp.

Hewlett, J. D. and W. L. Nutter (1969) *An outline of forest hydrology*, University of Georgia Press, Athens, 137 pp.

Hickman, H. C. (1940) Evaporation experiments, in Hydrology of the Great Lakes: A Symposium, *Trans. ASCE*, **105**: 807–49.

Hillel, D. (1980) *Fundamentals of soil physics*, Academic Press, New York, 413 pp.

Howe, G. M. (1956) The moisture balance in England and Wales based on the concept of potential evapotranspiration, *Weather*, **11**: 74–82.

Hsaio, T. C., E. Acevedo, E. Fereres and D. W. Henderson (1976) Water stress, growth and osmotic adjustment, *Phil. Trans. Roy. Soc. London, B*, **273**: 479–500.

IH (1979) *Research Report 1976–78*, Institute of Hydrology, NERC, 124 pp.

IH (1985) *Research Report 1981–84*, Institute of Hydrology, NERC, 86 pp.

IH (1987) Report for the period 1 April 1985–31 March 1986, in *The Natural Environment Research Council Report for 1985/86*, NERC, Swindon, pp. 61–78.

Jones, M. M. and N. C. Turner (1978) Osmotic adjustment in leaves of sorghum in response to water deficits, *Plant Physiol.*, **61**: 122–6.

Jordan, C. F. and J. R. Kline (1977) Transpiration of trees in a tropical rainforest, *Journal of Applied Ecology*, **14**: 853–60.

Keen, B. A. (1927) The limited role of capillarity in supplying water to the plant roots, *Proceedings and Papers of the 1st International Congress on Soil Science*, **1**: 504–11.

King, K. M. (1961) Evaporation from land surfaces, in *Proceedings of Hydrology Symposium No. 2, Evaporation*, Queen's Printer, Ottawa, pp. 55–80.

Kittredge, J. (1953) Influence of forests on snow in the Ponderosa-subar pine-fir zone of the central Sierra Nevada, *Hilgardia*, **22**: 1–96.

Kramer, P. J. (1952) Plant and soil water relations on the watershed, *Journal Forestry*, 50, 92–5.

Kreith, F., Taori and J. E. Anderson (1975) Persistence of selected antitranspirants, *WRR*, **11**: 281–6.

Lapworth, C. F. (1956) Evaporation from the water surface of a reservoir, Cyclostyled copy of a paper given before the International Union of Geodesy and Geophysics.

Lassen, L., H. W. Lull and B. Frank (1952) Some plant–soil–water relations in watershed management, *Circular 910*, USDA.

Law, F. (1956) The effect of afforestation upon the yield of water catchment areas, *Journal of British Waterworks Association*, **38**: 489–94.

Law, F. (1958) Measurement of rainfall, interception and evaporation losses in a plantation of Sitka spruce trees, *Proc. IASH Gen. Assoc. of Toronto*, **2**: 397–411.

Lee, R. (1980) *Forest hydrology*, Columbia University Press, New York, 349 pp.

Leith, R. M. and S. I. Solomon (1985) Estimation of precipitation, evapotranspiration and runoff using GOES, in *Advances in evapotranspiration*, ASAE, pp. 366–76.

Linsley, R. K., M. A. Kohler and J. L. H. Paulhus (1949) *Applied hydrology*, McGraw-Hill, New York.

Linsley, R. K., M. A. Kohler and J. L. H. Paulhus (1958) *Hydrology for engineers*, McGraw-Hill, New York.

McEwen, G. F. (1930) Results of evaporation studies, *Scripps Institute Oceanographic Technical Series*, **2**: 401–15.

McGowan, M. and E. Tzimas (1985) Water relations of winter wheat: the root system, petiolar, resistance and development of a root abstraction equation, *Experimental Agriculture*, **21**: 377–88.

McGowan, M., P. Blanch, P. J. Gregory and D. Haycock (1984) Water relations of winter wheat 5. The root system and osmotic adjustment in relation to crop evaporation, *J. Agric. Sci.* **102**: 415–25.

McIlroy, I. C. (1971) An instrument for continuous recording of natural evaporation, *Agric. Met.*, **9**: 93–100.

McNaughton, K. G. (1976a) Evaporation and advection I: Evaporation from extensive homogeneous surfaces, *QJRMS*, **102**: 181–91.

McNaughton, K. G. (1976b) Evaporation and advection II: Evaporation downwind of a boundary separating region having different surface resistances and available energies, *QJRMS*, **102**: 193–202.

McNaughton, K. G. and T. A. Black (1973) A study of evapotranspiration from a Douglas fir forest using the energy balance approach, *WRR*, **9**: 1579–90.

McNaughton, K. G., B. E. Clothier and J. P. Kerr (1979) Evaporation from land surfaces, in *Physical hydrology: New Zealand experience*, D. L. Murray and P. Ackroyd (eds), Hydrological Society, Wellington, N.Z., pp. 97–119.

MAFF (1967) Potential Transpiration, *Technical Bulletin 16*, Ministry of Agriculture, Fisheries and Food, HMSO, 17 pp.

Makkink, G. F. (1957) Ekzameno de la formulo de Penman, *Neth. J. Agric. Sci.*, **5**: 290–305.

Mather, J. R. (ed.) (1954) The measurement of potential evapotranspiration, *Publ. in Climatol.*, **7**: 225 pp.

Menenti, M. (1984) Physical aspects and determination of evaporation in deserts applying remote sensing techniques, *Report 10* (special issue), Institute of Land and Water Management Research (ICW), Wageningen, 202 pp.

Meteorological Office (1938) *British Rainfall*, 145 pp.

Miller, A. A. (1953) *The skin of the earth*, Methuen, London.

Milly, P. C. D. (1984a) A linear analysis of thermal effects on evaporation from soil, *WRR*, **20**: 1075–85.

Milly, P. C. D. (1984b) A simulation analysis of thermal effects on evaporation from soil, *WRR*, **20**: 1087–98.

Milthorpe, F. L. (1960) The income and loss of water in arid and semi-arid zones, in *Plant–water relationships in arid and semi-arid conditions*, UNESCO, Paris, pp. 9–36.

Miranda, A. C., P. G. Jarvis and J. Grace (1984) Transpiration and evaporation from heather moorland, *Boundary-Layer Met.*, **28**: 227–43.

Mohrmann, J. C. J. and J. Kessler (1959) Water deficiencies in European agriculture, *Publication 5*, International Institute of Land Reclamation and Improvement.

Monteith, J. L. (1965) Evaporation and environment, *Proceedings of Symposium on Experimental Biology*, **19**: 205–34.

Monteith, J. L. (1981) Evaporation and surface temperature, *QJRMS*, **107**: 1–27.

Monteith, J. L. (1985) Evaporation from land surfaces: progress in analysis and prediction since 1948, in *Advances in evaporation*, ASAE, pp. 4–12.

Monteith, J. L. and G. Szeicz (1961) The radiation balance of bare soil and vegetation. *QJRMS*, **87**: 159–70.

Morton, F. I. (1978) Estimating evapotranspiration from potential evaporation: practicality of an iconoclastic approach, *J. Hydrol.*, **38**: 1–32.

Morton, F. I. (1983) Operational estimates of areal evapotranspiration and their significance to the science and practice of hydrology, *J. Hydrol.*, **66**: 1–76.

Morton, F. I. (1985) The complementary relationship areal evapotranspiration model: how it works, in *Advances in evapotranspiration*, ASAE, pp. 377–84.

New Scientist (1961) Making irrigation in Britain a more exact science, *New Scientist*, **11**: 11.

Nieuwenhuis, G. J. A. (1981) Application of HCMM satellite and airplane reflection and heat maps in agrohydrology, *Advances in Space Research*, **1**: 71–86.

NWSCA (1987) Measuring tree transpiration, in *Streamland 58*, Ministry of Works and Development, National Water and Soil Conservation Authority, Wellington, N.Z., 4 pp.

Oke, T. R. (1978) *Boundary layer climates*, Methuen, London, 372 pp.

Parshall, R. L. (1930) Experiments to determine the rate of evaporation from saturated soils and river bed sands, *Trans. ASCE*, **94**: 961–99.

Pearce, A. J., L. K. Rowe and J. B. Stewart (1980) Nighttime, wet canopy evaporation rates and the water balance of an evergreen mixed forest, *WRR*, **16**, 955–9.

Pearl, R. T., *et al.* (1954) The calculation of irrigation need, *Technical Bulletin 4*, Ministry of Agriculture and Fisheries.

Penman, H. L. (1948) Natural evaporation from open water, bare soil and grass, *Proc. Roy. Soc., A*, **193**: 120–45.

Penman, H. L. (1949) The dependence of transpiration on weather and soil conditions, *J. Soil Sci.*, **1**: 74–89.

Penman, H. L. (1952) Experiments on the irrigation of sugar beet, *J. Agric. Sci.*, **42**: 286–92.

Penman, H. L. (1954) Evaporation over parts of Europe, *IASH General Association, Rome*, **3**: 168–76.

Penman, H. L. (1956) Evaporation: an introductory survey, *Neth. J. Agric. Sci.*, **4**: 9–29.

Penman, H. L. (1963) *Vegetation and Hydrology*, Farnham Royal, Commonwealth Agricultural Bureaux.

Penman, H. L. (1969) The role of vegetation in soil water problems, in *Water in the unsaturated zone*, P. E. Rijtema and H. Wassink (eds), IASH-UNESCO, pp. 49–61.

Penman, H. L. and R. K. Schofield (1941) Drainage and evaporation from fallow soil at Rothamsted, *J. Agric. Sci.*, **31**: 74–109.

Penman, H. L. and R. K. Schofield (1951) Some physical aspects of assimilation and transpiration, *Proceedings of Symposium Society on Experimental Biology*, **5**.

Penman, H. L., *et al.* (1956) Discussions of evaporation etc., *Neth. J. Agric. Sci.*, **4**: 87–97.

Philip, J. R. (1966) Plant water relations: some physical aspects, *Annual Review of Plant Physiology*, **17**: 245–68.

Prandtl, L. (1925) Bericht über Untersuchungen zur Turbulenz, *Zeitschrift für angewandte Mathematik und Mechanik*, **5**: 136–9.

Prevot, L., R. Bernard, O. Traconet, D. Vidal-Madjar and J. L. Thony (1984) Evaporation from a bare soil evaluated using a soil water transfer model and remotely sensed surface moisture data, *WRR*, **20**, 311–16.

Priestley, C. H. B. and R. J. Taylor (1972) On the assessment of surface heat flux and evaporation using large scale parameters, *Mon. Wea. Rev.*, **100**: 81–92.

Rao, B. (1958) The water balance of the Ohio River basin, *Bull. Amer. Met. Soc.*, **39**: 153–4.

Rice, R. C. and R. D. Jackson (1985) Spatial distribution of evaporation from bare soil, in *Advances in evapotranspiration*, ASAE, pp. 447–53.

Richardson, B. (1931) Evaporation as a function of insolation, *Trans. ASCE*, **95**: 996–1019.

Richardson, L. F. (1926) Atmospheric diffusion shown on a distance–neighbour graph, *Proc. Roy. Soc., A*, **110**: 709–37.

Richardson, L. F. and H. Stommel (1948) Note on the eddy diffusion in the sea, *J. Meterol.*, **5**: 238–40.

Rijtema, P. E. (1968) On the relation between transpiration, soil physical properties and crop production as a basis for water supply plans, *Technical Bulletin 58*, Institute of Land and Water Management Research.

Ritchie, J. T. (1972) Model for predicting evaporation from a row crop with incomplete cover, *WRR*, **8**: 1204–13.

Ritchie, J. T. (1973) Influence of soil water status and meteorological conditions on evaporation from a canopy, *Agron. J.*, **65**: 893–7.

Roberts, J. M. (1979) Evaporation from some British forest trees—environmental and plant controls, *Publication 527* WMO Symposium on Forest Meteorology, World Meteorological Organization, pp. 227–34.

Rodda, J. C., R. A. Downing and F. M. Law (1976) *Systematic hydrology*, Newnes-Butterworths, London, 399 pp.

Sanderson, M. (1950) Some Canadian developments in agricultural climatology, *Weather*, **5**: 381–8.

Schulze, R. E. (1985) Regional potential evaporation mapping in areas of sparse climatic data, in *Problems of regional hydrology*, H. E. Muller and K.-R. Nippes (eds), 5th Report of IGU Working Group on the IHP, Kirchzarten, Verl. Beitr. Zur Hydrologie, pp. 373–86.

Seguin, B., J. C. Mandeville, Y. Kerr and J. P. Guinot (1985) A proposed methodology for daily ET mapping using thermal IR satellite imagery, in *Advances in evapotranspiration*, pp. 385–92.

Sharkey, T. D. and T. Ogawa (1987) Stomatal responses to light, in *Stomatal function*, E. Zeiger, G. D. Farquhar and I. R. Cowan (eds), Stanford University Press, Stanford, Calif., 503 pp.

Shaw, E. M. (1983) *Hydrology in practice*, Van Nostrand Reinhold, London, 569 pp.

Shawyer, M. S. and P. Westcott (1987) The MORECS climatological dataset—a history of water balance variables over Great Britain since 1961, *Met. Mag.*, **116**: 205–11.

Shepherd, W. (1972) Some evidence of stomatal restriction of evaporation from well-watered plant canopies, *WRR*, **8**: 1092–5.

Shuttleworth, W. J. (1979) *Evaporation*, Report 56, Institute of Hydrology NERC, 61 pp.

Shuttleworth, W. J. (1988) Evaporation from Amazonian rain forest, *Proc. Roy. Soc. London, B*, **233**: 321–46.

Shuttleworth, W. J., J. H. C. Gash, C. R. Lloyd, C. J. Moore and J. S. Wallace (1988) An integrated micrometeorological system for evaporation measurement, *Agricultural Forest Meteorology*, **143**, 295–317.

Sibbons, J. L. H. (1956) The climatic approach to potential evapotranspiration, *Advances of Science*, **13**: 354–6.

Singh, B. and G. Szeicz (1979) The effect of intercepted rainfall on the water balance of a hardwood forest, *WRR*, **15**: 131–8.

Slatyer, R. O. and I. C. McIlroy (1961) *Practical microclimatology*, CSIRO and UNESCO.

Smith, K. (1965) A long-term assessment of the Penman and Thornthwaite potential evapotranspiration formulae, *J. Hydrol.*, **2**: 277–90.

Smith, R. C. G., H. D. Barrs, J. L. Steiner (1985) Relationship between wheat yield and foliage temperature: theory and its application to infrared measurements, *Agricultural Forest Meteorology*, **36**: 129–43.

Soer, G. J. R. (1980) Estimation of regional evapotranspiration and soil moisture conditions using remotely sensed crop surface temperatures, *Remote Sensing of Environ.*, **9**: 27–45.

Stark, N. (1967) The transpirometer for measuring the transpiration of desert plants, *J. Hydrol.*, **5**: 143–57.

Stevenson, K. R. and R. H. Shaw (1971) Effects of leaf orientation on leaf resistance to water vapour diffusion in soybean leaves, *Agron. J.*, **63**: 327–9.

Stewart, J. B. (1988) Modelling surface conductance of pine forest, *Agricultural Forest Meteorology*, **43**: 19–35.

Sutton, O. G. (1962) *The challenge of the atmosphere*, Hutchinson, London.

Szeicz, G., G. Endrodi and S. Tajchman (1969) Aerodynamic and surface factors in evaporation, *WRR*, **5**: 380–94.

Szeicz, G., C. H. M. van Bavel and S. Takami (1973) Stomatal factor in the water use and dry matter production by sorghum, *Agric. Met.*, **12**: 361–89.

Takhar, H. S. and A. J. Rudge (1970) Evaporation studies in standard catchments, *J. Hydrol.*, **11**: 329–62.

Tanner, C. B. (1957) Factors affecting evaporation from plants and soils. *J. Soil and Water Cons.*, **12**: 221–7.

Taylor, G. I. (1922) Diffusion by continuous movements, *Proceedings of London Mathematics Society*, **20**: 196–212.

Taylor, G. I. (1935) Statistical theory of turbulence, *Proc. Roy. Soc., A*, **151**: 494–512.

Taylor, H. M. and B. Klepper (1978) The role of rooting characteristics in the supply of water to plants, *Advances in Agronomy*, **30**: 99–128.

Thom, A. S. and H. R. Oliver (1977) On Penman's equation for estimating regional evaporation, *QJRMS*, **103**: 345–58.

Thompson, N. (1982) A comparison of formulae for the calculation of water loss from vegetated surfaces, *Agric. Met.*, **26**: 265–72.

Thornthwaite, C. W. (1944) A contribution to the Report of the Committee on Transpiration and Evaporation, 1943–44, *Trans. AGU*, **25**, 686–93.

Thornthwaite, C. W. (1948) An approach towards a rational classification of climate, *Geographical Review*, **38**: 55–94.

Thornthwaite, C. W. (1953) Climate and scientific irrigation in New Jersey, *Publ. in Climatol.*, **6**: 3–8.

Thornthwaite, C. W. (1954) A reexamination of the concept and measurement of potential evapotranspiration, *Publ. in Climatol.*, **7**.

Thornthwaite, C. W. and F. K. Hare (1965) The loss of water to the air, *Meteorological Monographs*, **6**: 163–80.

Thornthwaite, C. W. and J. R. Mather (1955) The water balance, *Publ. in Climatol.*, **8**: 1–86.

Turner, N. C. (1974) Stomatal behaviour and water status of maize, sorghum, and tobacco under field conditions. II. At low soil water potential, *Plant Physiol.*, **53**: 360–5.

Turner, N. C. and J. E. Begg (1973) Stomatal behaviour and water status of maize, sorghum, and tobacco under field conditions. I. At high soil water potential, *Plant Physiol.*, **51**: 31–6.

US Army Corps of Engineers (1956) *Snow hydrology*, USACE, N. Pacific Division, Portland, Oreg.

van Bavel, C. H. M. (1966) Potential evaporation: the combination concept and its experimental verification, *WRR*, **2**: 455–67.

van Hylckama, T. E. A. (1956) The water balance of the earth, *Publ. in Climatol.*, **9**: 57–117.

van Hylckama, T. E. A. (1958) Modification of the water balance approach for basins within the Delaware Valley, *Publ. in Climatol.*, **11**: 271–91.

van Wijk, W. R. and D. A. de Vries (1954) Evapotranspiration, *Neth. J. Agric. Sci.*, **2**: 105–18.

Veihmeyer, F. J. (1927) Some factors affecting the irrigation requirements of deciduous orchards, *Hilgardia*, **2**: 125–284.

Veihmeyer, F. J. (1938) Evaporation from soils and transpiration, *Trans. AGU*, **19**: 612–15.

Veihmeyer, F. J. (1972) The availability of soil moisture to plants: results of empirical experiments with fruit trees, *Soil Sci.*, **114**: 268–94.

Veihmeyer, F. J. and F. A. Brooks (1954) Measurements of cumulative evaporation from bare soil, *Trans. AGU*, **35**: 601–7.

Veihmeyer, F. J. and A. H. Hendrickson (1927) Soil moisture conditions in relation to plant growth, *Plant Physiol.*, **2**: 71–82.

Veihmeyer, F. J. and A. H. Hendrickson (1955) Does transpiration decrease as the soil moisture decreases?, *Trans. AGU*, **36**: 425–8.

Visser, W. C. (1964) Moisture requirements of crops and rate of moisture depletion of the soil, *Technical Bulletin 32*, Institute for Land and Water Management Research.

Visser, W. C. (1966) Progress in the knowledge about the effect of soil moisture content on plant production, *Technical Bulletin 45*, Institute for Land and Water Management Research, 44 pp.

Waggoner, P. E. and B. A. Bravdo (1967) Stomata and the hydrologic cycle, *Proceedings of National Academy of Science,* **57**: 1096–102.

Wallace, J. S., J. Roberts and W. J. Shuttleworth (1984) A comparison of methods for estimating aerodynamic resistance of heather (*Calluna vulgaris* (L.) Hull) in the field, *Agricultural Forest Meteorology,* **32**: 289–305.

Ward, R. C. (1963) Observations of potential evapotranspiration (PE) on the Thames floodplain, 1959–60, *J. Hydrol.,* **1**: 183–94.

Ward, R. C. (1971) Measuring evapotranspiration: a review, *J. Hydrol.,* **13**: 1–21.

Ward, R. C. (1976) Evaporation, humidity and the water balance, in *The climate of the British Isles,* T. J. Chandler and S. Gregory (eds), Longman, London, pp. 183–98.

Ward, R. C. (1981) River systems and river regimes, in *British rivers,* J. Lewin (ed.), Allen and Unwin, London, pp. 1–53.

Whitney, M. and F. K. Cameron (1904) Investigations in soil fertility, *USDA Bureau of Soils Bulletin,* **23**: 6–21.

Widtsoe, J. A. (1914) *The principles of irrigation practice,* Macmillan, London, 496 pp.

Woo, M-K. and P. R. Waylen (1985) Macroscale estimation of evaporation from regional water balance, *Hydrol. Sci. J.,* **30**: 383–94.

Zabadal, T. J. (1974) A water potential threshold for the increase of abscisic acid in leaves, *Plant Physiol.,* **53**: 125–7.

Zumberge, J. H. and J. C. Ayers (1964) Hydrology of lakes and swamps, in *Handbook of applied hydrology,* V. T. Chow (ed.), sec. 23, McGraw-Hill, New York.

5. Soil water

5.1 Introduction and definitions

Most of the precipitation that reaches the ground surface is usually absorbed by the surface layers of the soil. The remainder, once any *depression storage* has been filled, will flow over the surface as *overland flow*, reaching the stream channels quite quickly. The water that *infiltrates* into the soil may subsequently be evaporated or it may move under gravity and *percolate* downwards to the groundwater zone or else flow laterally close to the surface as *interflow* or *throughflow*. The ability of the soil to absorb and retain moisture is crucial to the hydrology of an area. Thin or impermeable soils may characterize an area in which incoming rainfall runs quickly off the surface, possibly resulting in erosion, and little moisture is held in the soil to sustain plants and animals until the next rain. Deep, permeable soils, in contrast, can absorb and store large quantities of water, providing a moisture reserve through times of drought and helping to produce a more even pattern of river runoff.

Knowledge about the factors controlling water storage and movement in the soil is essential to an understanding of a wide range of processes, including the generation of runoff, recharge to underlying groundwater and water supply to plants, as well as the movement and accumulation of pollutants (Bear and Verruijt, 1987). Soil water is thus of interest to investigators in a number of disciplines in addition to the hydrologist, including agronomists, foresters, geomorphologists and civil engineers. Nevertheless, its position on the fringe of so many disciplines has led to it being the 'no man's land of hydrology' (Hackett, 1966). It is only in recent years that, with advances in field measurement techniques, a practical understanding has developed of the physical processes involved (Wellings and Bell, 1982).

Figure 5.1 shows in a simple diagrammatic cross-section of a river valley the four main zones into which subsurface water has been traditionally classified. Precipitation enters the soil zone at the ground surface and moves downwards to the water table which marks the upper surface of the zone of saturation. Immediately above the *water table* is the *capillary fringe* in which almost all the pores are full of water; between this and the soil zone is the intermediate zone, where the movement of water is mainly downwards. These zones vary between different areas of a catchment. On the valley flanks, water drains from the soil zone proper into the intermediate zone, and may or may not eventually reach the zone of saturation perhaps several hundred metres below. In the floodplain areas of the catchment, however, the capillary fringe often extends into the soil zone or even to the ground surface itself, depending on the depth of the water table and the height of the capillary fringe. In these areas, the upward flow of moisture from the underlying groundwater is likely to maintain a higher soil moisture content than on the valley flanks, where replenishment can come only from precipitation.

While this classification provides a convenient introductory picture, it must not obscure the fact that subsurface water is an essentially dynamic system. As well as varying spatially

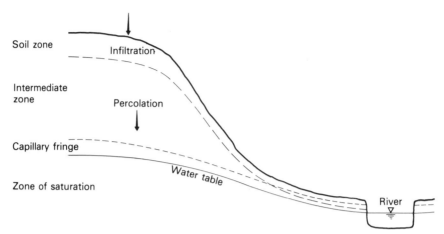

Figure 5.1 *Traditional classification of subsurface water.*

within a catchment, these zones may also vary over time with, for example, an increase in the saturated extent during periods of wetter weather. In the lower areas of the catchment seasonal fluctuations of the water table may, from time to time, bring the capillary fringe up into the soil zone. A number of different definitions have been adopted to try to distinguish between groundwater and soil water. These include, for example, limiting consideration of the latter to the surface metre or so, which contains the depth of plant roots (Shaw, 1988; Price, 1985). The most widely adopted definition of soil moisture is the subsurface water in the zone of aeration, i.e. the unsaturated soil and subsoil layers above the water table. This distinction between saturated and unsaturated conditions is an important one, which emphasizes the basic difference between the saturated layers, in which the pore spaces are almost completely filled with water and the pressure of water is equal to or greater than atmospheric pressure, and the zone of aeration, in which the pores are filled with both water and air and the pressure of water is less than atmospheric. However, in making such a distinction for the convenience of structuring this book, it must be emphasized that saturated conditions are really part of a continuum (Freeze and Cherry, 1979). Soil water conditions are continually in a state of flux. The zone of aeration is, in fact, a zone of transition in which water is absorbed, held or transmitted either downwards towards the water table or upwards towards the soil surface, from where it is evaporated. At times of prolonged or particularly intense rainfall part of the soil zone may become temporarily saturated but separated by unsaturated zones from the groundwater below. A perched water table may result from an impeding layer in the soil holding up the drainage of infiltrating water, or else the surface layers of the soil may be so slowly permeable as to result in saturated conditions. These often short-term and possibly localized areas of saturation may be very important in generating lateral flows to stream channels (see Section 7.4), but as they are not part of the general body of groundwater, and occur within the usually unsaturated surface soil layers, they are dealt with in this chapter.

In order to understand the behaviour of soil water it is necessary to first discuss the physical properties of the soil's solid constituents and their influence upon the storage and movement of soil water.

5.2 Physical properties of soils affecting soil water

Soil comprises the surface layers of fragmented and weathered rocks that have been disintegrated by physical and chemical processes, and may have been subject to further modification by the formation of secondary minerals (Brady, 1984; White, 1987). A vertical cross-section through the soil, i.e. the *soil profile*, normally comprises a number of layers or *horizons* having different physical characteristics. The nature of the soil profile depends upon a wide range of factors including the original parent material, the length of time of development and prevailing climate, as well as the vegetation and topography.

The soil system consists of three *phases* (Corey, 1977; Hillel, 1982): the solid particles comprising the porous medium, the gaseous phase of soil air and the liquid phase comprising the soil water (but more correctly termed the soil 'solution' as it always contains some dissolved substances). The soil water system is very complex, and this chapter provides an introduction to some of the most important phenomena of soil water theory and their practical consequences for the hydrologist. The subject is further complicated by the fact that soil properties may vary over short distances and do not even remain constant through time, due to factors such as swelling and shrinking of clays as well as compaction and disturbance by plants, animals and man. For these reasons soil physicists have often resorted to the study of 'idealized' media such as glass beads or bundles of capillary tubes. The challenge for the hydrologist, therefore, is to apply these often theoretical and laboratory-based concepts to field situations.

The amount of water that can be held in a given volume of soil and the rate of water movement through that soil depend upon both the soil *texture*, i.e. the size distribution of the mineral particles of the soil, and upon the soil *structure*, the aggregation of these particles. Water may occupy both interstructural voids and textural voids (between the particles). At high moisture contents water flow through the former may be dominant, but becomes rapidly less important as the soil becomes drier. In general, the coarser the particles, the larger will be the intervening voids and the easier it will be for water movement to take place. Thus sandy soils tend to be freely draining and permeable while clay soils are both slower to absorb water and slower to drain of water.

The finely divided clay material is the most important size fraction in determining the physical and chemical properties of the soil. The sand and silt fractions mainly comprise quartz and other primary minerals that have undergone little chemical alteration while the clays, in contrast, result from chemical weathering, forming secondary minerals with a great variety of properties (White, 1987; Wild, 1988). One difference is that the clay particles comprise platey sheets and have a much higher *specific surface*, i.e. the surface area per unit volume (Brady, 1984; Carter *et al.*, 1986). Most clays have negatively charged surfaces and are balanced externally by cations which are not part of the clay structure and which can be replaced or exchanged by other cations (see Section 8.5.2). Some types of clays have only weak bonds between adjacent sheets, and the 'internal' surfaces may also be available for taking part in reactions (Rose, 1966). This is important for the retention and release of nutrients and salts. Water can enter between these sheets causing them to separate and expand. Many clay soils swell on wetting and shrink and crack on drying, which may be important for the porosity and hydraulic properties of soils.

Soil structure results from the aggregation of the primary particles described above into the structural units, or *peds*, which are separated from one another by planes of weakness. The mechanics of soil structure formation and stability are very complex and depend on a number of factors. In the surface horizon of the soil the aggregates will change over time due to weather and soil tillage, but in deeper horizons they will be more constant. Plants are very important for the structure of surface horizons since their roots bind particles to help form stable aggregates. Grass swards, for example, are particularly effective due to the high density of roots near the surface (White, 1987). Soil structure is too complex for simple geometric characterization and is usually described qualitatively in terms of form: granular, blocky, prismatic, etc., and by the degree of development, whether structure-less or strongly aggregated (e.g. Hodgson, 1976). However, Bouma (1981) reported only poor relationships between such descriptions and hydraulic properties in structured clay soils due to the complexity of the soil pore pattern, and recommended instead the use of tracers to characterize preferential flow paths through the soil voids.

This broad description of the main soil properties provides a basis for subsequent sections, describing the processes governing the storage and movement of water in soils and the resulting patterns of soil moisture and flow rates occurring under field conditions.

5.3 Storage of soil water

If gravity were the only force acting on soil water then the soil would drain completely after each input of rain and the only water to be found would be below the water table (Terzaghi, 1942). In such a situation, plant growth would be restricted to those areas where rainfall occurred very frequently or to locations where the water table was near to the surface. In fact soils in natural conditions always contain some water, even at the end of long dry periods lasting many months or even years. This indicates that very powerful forces are holding moisture in the soil.

5.3.1 *Water retention forces*

The nature of the forces that act to retain water in the soil has prompted numerous investigations over the last century, but it is only comparatively recently that, building upon the pioneering work of Buckingham (1907), their true nature and influence on soil water retention has become understood. The main forces responsible for holding water in the soil are those of *capillarity*, *adsorption* and *osmosis*.

Capillary forces result from *surface tension* at the interface between the soil air and soil water. Molecules in the liquid are attracted more to each other than to the water vapour molecules in the air, resulting in a tendency for the liquid surface to contract. If the pressure is exactly the same on either side then the air–water interface would be flat, but pressure differences result in a curved interface, the pressure being greater on the inner, concave, side by an amount that is related to the degree of curvature (Baver *et al.*, 1972). At the interfaces in the soil pore space, the air will be at atmospheric pressure, but the water may be at a lower pressure. As water is withdrawn from the soil the pressure difference increases across the interfaces which becomes increasingly curved and can only be maintained in the smaller pores between the soil particles (Fig. 5.2). The force with which these small wedge-shaped films of water are held will vary, just as the capillary rise in a glass capillary tube varies with the radius of the tube, the curvature of the surface meniscus and the surface tension of the water. For a given viscosity and surface tension, water will be held more strongly in smaller pores than larger ones. Hence as the water content of a soil reduces, the larger pores empty at lower suctions than the smaller pores.

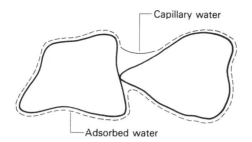

Figure 5.2 *Water is held by capillary forces between the soil particles and by adsorption as a thin film.*

In addition to capillary forces, soil water liquid or vapour can be adsorbed upon the surfaces of soil particles mainly due to electrostatic forces in which the polar water molecules are attached to the charged faces of the solids (Marshall and Holmes, 1979). Since the forces involved are only effective very close to the solid surface, only very thin films of water can be held in this way (Fig. 5.2). Nevertheless, if the total surface area of the particles (i.e. the specific surface) is large, and/or the charge per unit area is large, then the total amount of water adsorbed in a volume of soil may be considerable. The specific surface depends upon the size and shape of the particles. Its value increases as the grain size decreases and as the particles become less spherical and more flattened. Clay size particles and organic matter contribute most to the total surface area of a soil, with values of less than 1 m²/g for sand rising to over 800 m²/g for expanding layer clays such as montmorillonites. This helps to explain the very strong retention of water by clays during prolonged periods of drying.

Both the capillary and adsorption forces exert a *tension* or *suction* (negative pressure relative to atmospheric) on the soil water. Their respective importance depends upon soil texture and the water content of a given soil. Adsorptive forces are more important in clayey soils than sandy ones and as moisture content is reduced. In practice these forces cannot be easily measured separately and, since they are in a state of equilibrium with each other, one cannot be changed without affecting the other. It is therefore usual to deal with their combined effect, which, since they act to hold water in the soil matrix, is called *matric suction* (Marshall and Holmes, 1979).

A third force which acts to retain water in the soil results from osmotic pressure due to solutes in the soil water. Although much less important than the matric forces, it may be important in certain circumstances. It comes into play whenever there is a difference in solute concentration across a permeable membrane. This may be at a plant root surface, making water less available to plants, especially in saline soils, or across a diffusion barrier such as an air-filled pore, by allowing the movement of water vapour, but not the solute, across the pore from the more dilute to the more concentrated solution (Hillel, 1982). In the absence of such barriers soluble ions will diffuse throughout the soil solution by virtue of their kinetic energy, resulting in a uniform concentration (Baver *et al.*, 1972).

The total suction holding water in a soil is the sum of the matric and osmotic forces. The preceding discussion has shown how these retention forces vary with moisture content. In general, soils contain a wide range of pores of varying shapes and sizes. Those with large entry channels will empty at low suctions while those with narrow channels will empty at higher suctions. The relationship between soil moisture tension and moisture content is clearly of fundamental importance to an understanding of soil water behaviour.

5.3.2 *Soil moisture characteristics (retention curves)*

If a slowly increasing suction or tension is applied to a completely saturated soil, the air–water interface begins to fall below the soil surface and the matric forces of soil water retention come increasingly into effect. First, large pore channels will empty at low suctions, while narrow pores supporting air–water interfaces of much greater curvature will not empty until larger suctions are imposed.

The relationship between suction and the amount of water remaining in the soil can be determined experimentally in the laboratory using cores of soil (Klute, 1986b). The resulting function is known as the soil water or soil moisture *characteristic* (Childs, 1940). Examples of moisture characteristics obtained for different types of soil are shown in Fig. 5.3. The shape of the curve is related to the pore size distribution (Bouma, 1977). In general, sandy soils show a much more curved relationship than clayey soils since most of the pores are relatively large and once they are emptied there is little water remaining. Clayey soils, in contrast, have a wider distribution of pore sizes and consequently have a more uniform slope. The mechanism of water retention varies with suction. At very low suctions it depends primarily on capillary surface tension effects, and hence on the pore size distribution and soil structure. At higher suctions (lower moisture contents) water retention is increasingly due to adsorption, which is influenced more by the texture and the specific surface of the material. Due to the greater number of fine pores and the larger adsorption, clays tend to have a greater water content at a given tension than other soil types.

When an increasing suction is first applied to a saturated soil, little or no water may at first be released. A certain critical suction must be achieved before air can enter the largest

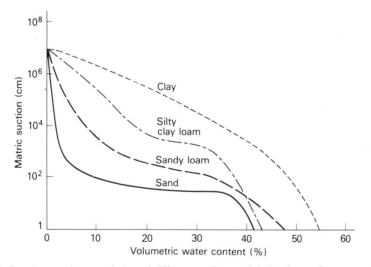

Figure 5.3 *Soil moisture characteristics of different soil materials (redrawn from an original diagram by Bouma, 1977).*

pores enabling them to drain. This is called the *bubbling* or *air entry pressure*. The critical suction will obviously be greater for fine-textured material such as clays than for coarse sands with larger maximum pore sizes. However, since the latter often have a more uniform pore size, their moisture characteristics may show the air entry phenomenon more distinctly than finer-textured soils.

The fact that the necessary tension to drain a pore depends upon its radius means that the slope of the moisture characteristic, the *specific* or *differential water capacity*, can be used to indicate the 'effective' pore size distribution of the soil. The volume of water withdrawn from the soil by increasing the suction is taken to represent the volume of pores having radii between the sizes corresponding to those two suctions. In making such an estimate it must be remembered that at high tensions adsorptive rather than surface tension forces may predominate and that, due to the often tortuous flow paths through the medium, not all pores of a given size will empty at the same time. A large water-filled pore may be surrounded by smaller pores, and cannot drain until these smaller pores drain first and air can pass through to the large pore (Childs, 1969). This phenomenon can lead to jumps in the water characteristic, especially at low tensions (Corey, 1977). Bouma (1977) noted that the correspondence between the 'effective' pore size distribution by this method and the size distribution obtained by micromorphological analysis of thin sections was better using the moisture characteristic obtained from a soil sample that was being wetted than from one that was being dried.

5.3.3 *Hysteresis*

One of the main limitations to the use of moisture characteristic curves is that the water content at a given suction depends not only on the value of that suction but also on the moisture 'history' of the soil. It will be greater for a soil that is being dried than for one that is being re-wetted (Fig. 5.4). This dependence on the previous state of the soil water leading up to the current equilibrium condition is called *hysteresis* (Haines, 1930). Pores

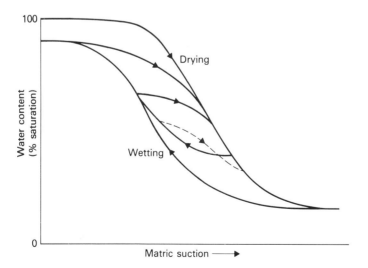

Figure 5.4 *Hysteresis in the moisture characteristic, showing the main wetting and drying boundary curves and the intermediate scanning curves (redrawn after Childs, 1969).*

empty at larger suctions than those at which they fill, and this difference is most pronounced at low suctions and in coarse-textured soils. Hysteresis has been attributed to a number of factors, the most important of which are the 'ink bottle' effect and the 'contact angle' effect (Baver *et al.*, 1972; Bear and Verruijt, 1987). Both of these are dependent on pore behaviour. The former results from the fact that a larger suction is necessary to enable air to enter the narrow pore neck, and hence drain the pore, than is necessary during wetting, which is controlled by the lower curvature of the air–water interface in the wider pore itself (Childs, 1969). The 'contact angle' effect results from the fact that the contact angle of fluid interfaces on the soil solids tends to be greater when the interface is advancing (i.e. wetting) than when it is receding (drying), so a given water content tends to be associated with a greater suction in drying than in wetting (Bear and Verruijt, 1987). The effect of entrapped air will decrease the water content of newly wetted soil, and the failure to attain true equilibrium in experimental conditions may create or accentuate the hysteresis phenomenon. The water content at zero suction may be only 80–90 per cent of the total porosity (Corey, 1977; Klute, 1986b), but may increase over time due to displacement by water flow and to the dissociation of the air into the water.

In fine-grained clay soils, wetting and drying may be accompanied by swelling and shrinkage. This leads to changes in pore sizes and the overall bulk density, and hence to a different volumetric water content at a given suction than if the matrix had remained stable and rigid. As water is withdrawn from the interstices between the plate-like particles, the particles move closer together, thus reducing the overall volume. The effect of this phenomenon is not yet completely understood, but Childs (1969) suggested that as the plates come together when water is withdrawn, they may reorientate themselves; on subsequent rewetting they may not necessarily return to their original alignments, resulting in a lower volumetric water content. This behaviour of clay soils is affected by the composition and concentration of the soil solution as well as the type of clay.

A number of methods for modelling hysteresis scanning curves were reviewed by Jaynes (1985) and may be applied to soils for which the boundary wetting and drying curves are already known. In practice, however, given the many problems associated with measuring the moisture characteristic accurately, the hysteresis phenomenon is usually ignored (Beese and van der Ploeg, 1976; Hillel, 1982). While it can be argued that a better indication of the pore size distribution is provided by the wetting curve (governed by the size of the pore entry channels), experimental difficulties with its measurement do much to offset its theoretical advantage, and the drying (retention) curve is almost always used (Childs, 1969; Hillel, 1982). For this reason some texts simply refer to the moisture characteristic as the moisture retention curve.

5.3.4 *Soil water 'constants'*

To facilitate comparisons between different soils a number of soil water 'constants' have been used, corresponding to particular values of matric suction. Clearly, since the moisture characteristic curve is a continuous function such points must be somewhat arbitrary and can have little intrinsic significance. Furthermore, under natural conditions rainfall and evaporation seldom, if ever, permit soil water contents to reach a state of equilibrium over the whole profile. Despite their arbitrary nature having been demonstrated as early as the 1920s (Gardner *et al.*, 1922), they have nevertheless been used widely in practical soil water problems, including applications to drainage, irrigation and watershed hydrology (van Bavel, 1949; White, 1987).

The 'constants' most frequently cited are the *wilting point* and the *field capacity*. The wilting point is defined as the minimum water content of the soil at which plants can extract water. Although this will vary between plant species and with their state of growth, the actual difference in the amount of water is quite small at such low water contents. Field capacity has been defined as the amount of water remaining in the soil after the downward movement under gravity has largely ceased (Veihmeyer and Hendrickson, 1931). There is no quantitative definition of 'largely ceased' and many soils continue draining for many weeks, but for field situations this has often been taken to be the moisture content 48 hours after a rainfall that thoroughly wetted the soil. In practice, permeable soils will drain more rapidly than less permeable soils and achieve a state of little further moisture change more quickly and at much lower suctions (Smedema and Rycroft, 1983). It is also assumed that no evaporation losses are occurring and that the water table is sufficiently deep so as to have no influence on the water content of the soil profile under consideration.

Despite such problems and limitations, soil water 'constants' are widely used and the difference in moisture content between field capacity and wilting point, for instance, provides a useful approximation to the available water capacity for plants growing in different soils (Fig. 5.5).

While these concepts are useful for making broad comparisons and generalizations, a fuller understanding of the behaviour of soil water requires consideration of the dynamic nature of the system. This needs to be based on the concepts and laws of soil physics.

5.3.5 *Soil water energy (potential)*

Soil water, like any other body, may possess both kinetic and potential energy. Due to the slow movement of water in soil the former can generally be considered negligible, whereas its potential energy, resulting from position or internal condition, is of major importance in determining the state and movement of soil water.

The concept of *soil water potential* expresses the total potential energy of soil water relative to that of water in a standard reference state. It has been defined (ISSS, 1976) as the amount of work that must be done per unit quantity of pure water in order to transport reversibly and isothermally an infinitesimal quantity of water from a pool of pure water at a specified elevation at atmospheric pressure to the soil water (at the point under consideration).

The total soil water potential (ϕ) at a given point comprises the sum of several components, of which only the gravitational potential (ψ_g), pressure potential (ψ_p) and osmotic potential (ψ_o) are usually considered:

$$\phi = \psi_g + \psi_p + \psi_o \tag{5.1}$$

Gravitational potential increases with elevation and, in the absence of strong retention forces or impeding factors, water will clearly drain downwards from higher to lower elevations. The pressure potential comprises the matric potential ($-ve$) above the water table or the piezometric pressure potential ($+ve$) below the water table. Some soil physicists prefer to distinguish between these two forms of pressure potential, although as Hillel (1982)

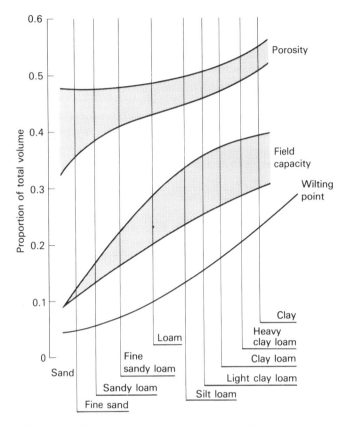

Figure 5.5 *General relationship showing the total porosity, field capacity and wilting point of various soils. The volumes quoted are illustrative only, but demonstrate the increase in the available water capacity to plants (between field capacity and wilting point) from sand to clay soils. (From an original diagram in Dunne et al., 1975. Blackwell Scientific Publications Ltd.)*

points out, there is an advantage in having a unified concept enabling a single continuous potential to describe the moisture profile. Osmotic potential results from solutes in the soil water, and like matric potential it acts to lower the total potential and to retain or draw water into the soil.

Water will move from a point where the total potential energy is *higher* to one where it is *lower*. It is thus not the absolute amount of potential energy that is important, but rather the relative levels in different regions of the soil. The difference in total potential between points depends upon both differences in retention forces and also differences in elevation. These component potentials may not necessarily act in the same direction and are not necessarily equally important in causing flow.

When calculating total potential it is conceptually easier to take the water table elevation as the reference level. Then both the gravitational potential and pressure potential are zero at this level. With height above the water table, gravitational potential becomes increasingly positive, while the matric potential is negative and variable depending upon gains and losses of water due to rainfall and evaporation. However, it is usually more practicable in studies of the unsaturated zone to use the ground level as the datum. Measurements may not be available at sufficient depth for the position of the water table to be accurately known, and even where such information is available its position will usually not be stationary over time. Since the same datum level applies in calculating the potential down the whole profile, different conventions do not in any way affect the relative values of the total energy profile.

An example of a total energy profile is given in Fig. 5.6, using the ground surface as the datum level. Osmotic pressure is ignored, as it is usually unimportant as a driving force for flow (Campbell, 1985). In this example matric suction increases with elevation above the water table and is much greater in the plant root zone near the ground surface due to drying by evaporation.

Flow will occur in the direction from higher to lower (i.e. more negative) total potential. The *zero flux plane* (ZFP) is given by the points of zero gradient on the total potential profile and divides the profile into zones of upward and downward flux. Figure 5.7 illustrates the annual cycle of water flux in a soil profile (Wellings and Bell, 1982). Taken together with measurements of water content variations, this information can be used to quantify both the amounts of deep percolation downwards to the groundwater and also the upward flux due to evaporation (Bell, 1987; Moser *et al.*, 1986).

Application of the concept of the total potential of soil water has revolutionized the study of soil water and has superseded some of the earlier arbitrary divisions of soil moisture, such as that of Briggs (1897) who referred to hygroscopic, capillary and gravitational water. While such a classification recognized that as drying of the soil proceeds the remaining water is held with increasing force, the forms of water so defined represent approximate equilibrium values only, and do not indicate real or abrupt changes in the physical condition of the soil water.

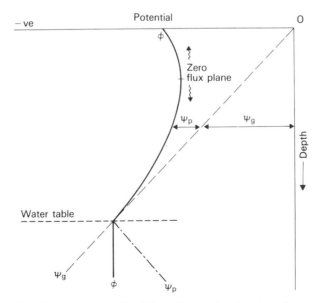

Figure 5.6 *Total soil water energy profile (ϕ) with depth, showing the gravitational (ψ_g) and pressure potential (ψ_p) components.*

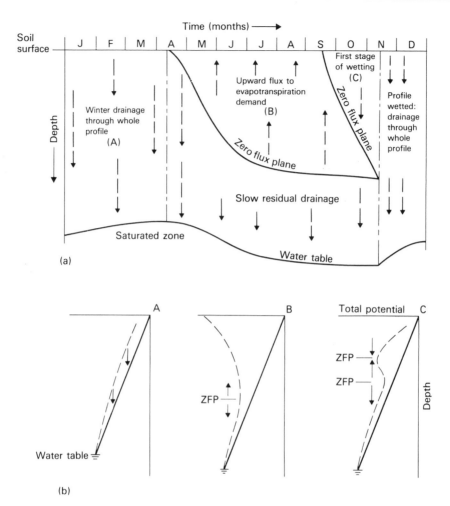

Figure 5.7 (a) *Annual cycle of water movement in a soil profile in a temperate climate (after an original diagram by Wellings and Bell, 1982).* (b) *Soil water potential profiles corresponding to times A, B and C.*

A note on terms and units

A great deal of confusion has arisen due to the use of three different systems of units to express potential energy (Baver *et al.*, 1972). These are:

(a) Energy per unit mass, ϕ. This is often taken as the fundamental expression.
(b) Energy per unit volume (i.e. pressure), P.
(c) Energy per unit weight (i.e. hydraulic head), H. This is often simpler and more convenient than the other systems of units, and is the one most often used for flows in the field.

However, although at first perplexing, the various expressions are in fact equivalent and can be translated directly into one another: i.e.

$$\phi = \frac{P}{\rho} = Hg$$

$$P = \rho\phi = H\rho g$$

$$H = \frac{P}{\rho g} = \frac{\phi}{g}$$

where ρ is the density of water and g is the acceleration due to gravity.

There are many different symbols in use, and even the Terminology Committee of the International Society of Soil Science found that it was 'a sheer impossible task to suggest a set of symbols which on the one hand corresponds with common usage and, on the other hand, is completely consistent' (ISSS, 1976). As Baver *et al.* (1972) point out, it is important to study carefully the definitions of symbols given in a particular text and the context in which they appear.

The most commonly used units of measurement for the different expressions of water potential are as follows. Pressure is generally measured in terms of bars (1 bar = 0.99 standard atmosphere) or pascals (1 newton/m²) while hydraulic head is measured in centimetres of water. Since the latter may introduce very large numbers for high suctions, it may be expressed as $\log_{10}$ (head), known as pF. To illustrate the use of these units, the water potential for a number of so-called soil water 'constants' (see Section 5.3.4) are given in Table 5.1. It may be noted that the terms tension, suction and pressure may be

Table 5.1 Typical values of water potential, expressed both in pressure and head units, for a number of widely used soil water 'constants' (approximate conversion assuming $g = 10$ m/s/s: 1 bar = 100 kPa = 1000 cm)

Soil water condition	Pressure		Head		Equivalent pore diameter (μm)
	bars	kPa	cm	pF	
Near saturation (only macropores drained)	0.001	0.1	1	0	3000
Aeration porosity	0.05	5	50	1.7	60
Field capacity	0.1	10	100	2	30
Wilting point	15	1 500	15 000	4.2	0.2
Air dry	220	22 000	220 000	5.4	0.01
Oven dry	10^4	10^6	10^7	7	—

used interchangeably. Tension or suction are simply negative pressures (i.e. pressures that are less than atmospheric). Thus, while suction or tension would be expressed as a positive quantity for unsaturated conditions (e.g. 80 cm water suction), the same value expressed as a pressure would be a negative amount (i.e. -80 cm water pressure).

5.3.6 *Measurement of soil water*

From the preceding discussions of soil water content and soil water suction it will be apparent that, in principle, the study of soil moisture behaviour may be approached by measurements of either soil water content or soil moisture suction, with conversion from one to the other by use of the moisture characteristic. Errors in the measurement of the two variables, especially with small samples, and problems of hysteresis mean, however, that where possible direct measurement of both variables is desirable (Hillel, 1982). There is a large number of techniques available and the following is a brief description of some of the most widely accepted methods for practical field situations.

Measurement of soil water content

The standard and most widely used technique for directly measuring the water content of soil is the *gravimetric* method. This involves taking a number of soil samples of known volume by coring or augering and determining their weight loss when dried in an oven. The methodology has been reviewed by Gardner (1986) and Reynolds (1970a, 1970b). Nevertheless, it is laborious, time consuming and prone to errors in sampling and repeated weighings. It does not distinguish between 'structural' and 'non-structural' water; after oven drying some clays may still contain appreciable amounts of adsorbed water (leading to underestimates of water content) while some organic matter may oxidize and decompose at temperatures as low as 50°C. For these reasons Gardner (1986) questioned the widespread and often uncritical acceptance of gravimetric values as 'correct'. The method is also destructive to the site and clearly is not suitable for a large number of repeated measurements over time.

To overcome some of these problems a number of indirect methods have been developed that, once calibrated against gravimetric samples, can be used to repeatedly make measurements more quickly, easily and with less disturbance (Schmugge *et al.*, 1980).

The *neutron probe* is the most commonly used indirect way of measuring soil water (Gardner and Kirkham, 1952; Holmes, 1956; Bell, 1987). A radioactive source of 'fast' (high-energy) neutrons is lowered into a borehole in the soil, and the number of 'fast' neutrons which are slowed or thermalized by collisions with hydrogen nuclei, mainly in the soil water, is measured by a detector. The effective volume of measurement varies inversely with the water content of the soil from about a 10 cm radius for wet soil to about 25 cm in dry soil. There is a fairly linear relationship between the detector count rate and the water content, but it varies from soil to soil. The readings are usually calibrated for a given soil against the gravimetric method, but due to uncertainties about the amount of water expelled by that method the neutron probe is normally used to measure differences rather than absolute moisture contents. Indeed, for many purposes the relative water content is more important than the absolute value (Marshall and Holmes, 1979).

The *capacitance probe*, which uses the dielectric constant of the soil as a measure of its moisture content, provides a non-radioactive method of measuring soil water in the field (Dean *et al.*, 1987; Bell *et al.*, 1987). As with the neutron probe, calibration with gravimetric samples is necessary for each soil type, and the method is best suited to water content changes rather than absolute values. Due to its smaller 'sphere of influence' the capacitance probe has the potential to make measurements close to the soil surface and to study the change in water content between different horizons of a soil profile. However, its small radial penetration means that local inhomogeneities are important, and the access tube must be installed very carefully in the ground since air gaps will affect the readings.

Measurement of soil water suction
Tensiometers are the most widely accepted technique for measuring matric suction, and probably date back to the 1920s (Childs, 1969). They comprise a liquid-filled porous cup connected to a pressure measuring device such as a mercury manometer or a pressure transducer. The cup is embedded in the soil *in situ*, and water can flow between the soil and cup until the pressure potential inside the cup equalizes with that of the soil water. The time taken for this equilibrium to be reached depends upon the flow rates through the cup and the surrounding soil and on the volume of water needing to be displaced to register a pressure change. Tensiometers can measure pressure heads below the water table, in which case they operate as piezometers, but are usually used to measure matric suction in the unsaturated zone. The lowest pressure that can be measured by this technique is about -800 cm (80 kPa) due to the effervescence of dissolved gases out of the water at the low pressure, making the system inoperative (Koorevaar *et al.*, 1983; Cassell and Klute, 1986). This range is adequate for many purposes, but in fine-textured soils quite a lot of water still remains available to plants at greater tensions, and in such circumstances *resistance blocks* may be useful. Two electrodes are embedded in a porous block buried in the soil and allowed to come into equilibrium with the soil water. The resistance varies with the water content and hence the matric potential of the surrounding soil. Wellings *et al.* (1985) described the use of gypsum resistance blocks for measuring potentials as low as $-15\,000$ cm (1500 kPa), and discussed methods of calibration, data processing and interpretation. Problems with the method include sensitivity to temperature and soil salinity, the need for blocks to be calibrated individually and gradual changes in block resistance over time.

The use of techniques such as the neutron probe give an indication of changes in water content over time, while measurements of soil moisture potential indicate the direction of moisture movement. Thus the combined use of these two measurements provides a powerful approach to study water fluxes in the unsaturated soil. This is now discussed in more detail.

5.4 Movement of soil water

It has been shown in earlier sections that soil water moves in response to a number of forces. Since gravity is not necessarily the dominant force, unsaturated flow may be in any direction. There is, however, a tendency for the main controlling forces to operate either from the ground surface (infiltration, evaporation) or from the bottom layers of the zone of aeration. This leads to the development of soil water potential gradients in the vertical direction, and the result is that vertical movement of water usually predominates.

The following sections deal with the general principles of flow in the unsaturated zone and with water movement in the vertical direction—either upwards or downwards. The factors limiting the rate at which water can infiltrate into a soil are then discussed, as this controls the partitioning of net precipitation into surface and subsurface flow paths. At the end of the chapter the roles of topography and soil layering are considered, looking at lateral soil water flow down slopes and the resulting spatially variable patterns of soil water content and movement that are observed in the field and are so important in many areas for runoff production.

5.4.1 *Principles of unsaturated flow*

It was shown in Section 5.3.5 that water in the soil will move from regions of high total potential to regions of lower potential. Expressing the potentials in units of head, the total potential (ϕ) is, for most practical purposes, taken as the sum of the matric potential (suction) (ψ) and the gravitational potential (z). If the ground surface is used as the datum level, the gravitational term, measured downwards as a depth, must be subtracted:

$$\phi = \psi - z \tag{5.2}$$

The way in which the gravitational and matric potentials combine to affect water movement in a soil is illustrated by reference to a simple example with two tensiometers at, say, 40 and 60 cm depths (Fig. 5.8). The ground surface is used as the datum level and the soil is initially unsaturated, so both matric suction and elevation terms of total potential are negative. At time A, the matric suction is the same at both depths, but due to the difference in elevation, water flow in the soil layer between these points will occur downwards, i.e. from higher to lower (more negative) potential. Over time the upper soil dries more quickly than that at greater depth until, at time B, the difference in matric suction balances the elevation difference. The total potentials are the same, hence there is no flow between these depths, and the zero flux plane will be at some intermediate point. If the upper soil continues to dry, its potential will become more negative than that of the deeper soil and upward flow will occur from 60 to 40 cm depth. After an input of rain, the soil becomes wetter, matric suctions reduce and, in this example, the water table rises to within 60 cm of the ground surface. This is indicated by the shaded area in the figure, and demonstrates the usefulness of the pressure potential for cases where both saturated and unsaturated conditions are considered.

It was first shown by Darcy (1856) for saturated conditions that the rate of flow of

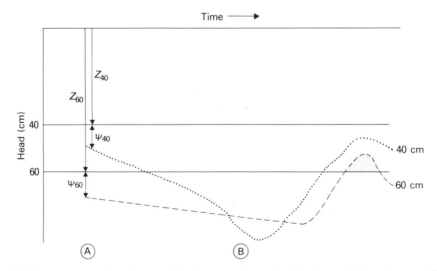

Figure 5.8 *The concept of total potential, ϕ, comprising elevation (gravitational potential), z, and pressure potential (matric suction), ψ, for two tensiometers at 40 and 60 cm depth below the ground-surface datum.*

water (flux) through a porous medium was proportional to the hydraulic gradient. Subsequent work, particularly by Richards (1931) and Childs and Collis-George (1950), showed that this relation is also applicable to the flow of soil water in unsaturated conditions. The Darcy equation may be simply expressed for unsaturated conditions as

$$v = -K(\theta)\nabla\phi \qquad (5.3)$$

where v is the macroscopic velocity of water, K is the hydraulic conductivity (and varies with the water content, θ) and $\nabla\phi$ is the gradient of the total or head. The minus sign indicates that water movement is in the direction of decreasing potential. The potential gradient and the resulting flow may be in any direction.

The flux calculation may be illustrated in the two simplest cases for:

(a) purely horizontal flow in the x direction, i.e. only a matric potential gradient ψ, and no gravity gradient:

$$v = -K(\theta)\frac{\delta\psi}{\delta x} \qquad (5.4)$$

and

(b) purely vertical flow downwards in the z direction, and with a matric potential gradient:

$$v = -K(\theta)\frac{\delta(\psi - z)}{\delta z}$$

$$= -K(\theta)\left(\frac{\delta\psi}{\delta z} - 1\right) \qquad (5.5)$$

In some situations information on moisture content gradients may be more readily available than hydraulic potentials. Childs and Collis-George (1950) introduced a function called the soil water *diffusivity*, D, which enables the flow equation to be transformed so that the flux is related to the gradient of water content (θ) rather than of potential. Like the hydraulic conductivity, diffusivity is also a function of water content, and the two are related:

$$D(\theta) = K(\theta)\frac{\delta\psi}{\delta\theta} \qquad (5.6)$$

The slope of the moisture characteristic $\delta\theta/\delta\psi$ is the specific water capacity $C(\theta)$, so it follows that

$$D(\theta) = \frac{K(\theta)}{C(\theta)} \qquad (5.7)$$

Then the relation between the flux and water content gradient is

$$v = \begin{cases} -D(\theta)\dfrac{\delta\theta}{\delta x} & \text{(horizontal flow)} & (5.8) \\[4mm] +K(\theta) - D(\theta)\dfrac{\delta\theta}{\delta z} & \text{(vertical flow)} & (5.9) \end{cases}$$

The theoretical unsaturated flow equations derived from the simple Darcy law are exceedingly complex. The introduction of the diffusivity term simplifies the mathematical treatment by rewriting the Darcy equation in a form analogous to equations of diffusion and heat conduction, for which solutions are readily available for a wide variety of boundary and initial conditions (Hillel, 1982). However, when using the diffusion terminology it must be borne in mind that the process being described is one of convective mass flow and not one of molecular diffusion. To emphasize this point and avoid confusion, it has been suggested that, by analogy with the hydraulic conductivity, the diffusivity, D, should be called the hydraulic diffusivity (Hillel, 1982).

There are a number of points that should be noted when using the Darcy equation, or equations derived from it. First, the equation describes flow at a 'macroscopic' level over the whole cross-section of flow, while in fact the flux is actually confined to the interstices and pore spaces. It is thus only applicable to cases in which the cross-section being considered is so much greater than the dimensions of its microstructure that it can reasonably be regarded as uniform. Second, the equation is restricted to steady-state situations, where the gradient and flux do not change, or change only slowly over time. Otherwise the movement of water alters the gradient and, indeed, the value of hydraulic conductivity. For normal field situations in which flow varies with space and time, Richards (1931) combined Darcy's equation with the continuity equation $(\delta\theta/\delta t = -\delta v/\delta z)$ to yield the important non-linear equation that bears his name:

$$\frac{\delta\theta}{\delta t} = \frac{\delta}{\delta z}\left[K(\theta)\left(\frac{\delta\psi}{\delta z} - 1\right)\right] \tag{5.10}$$

where t is time and flow is vertically downwards.

In conditions where the water content of the soil is low and liquid movement of water is negligible, the movement of water vapour in the soil may be quite important. This movement occurs as a result of differences in vapour pressure due to variations in soil moisture or soil temperature. In general the latter is the more important, and water vapour will move from warm soil to cold soil. It is a slow process in comparison with the mass flow of liquid water, and except for situations with very low soil water contents the contribution of vapour flow to the total fluid movement of water may be neglected (Childs, 1969).

Hydraulic conductivity

When a soil is saturated, its hydraulic conductivity depends mainly upon the geometry and distribution of the pore spaces. In addition to textural voids the presence of macropores such as interstructural cracks and root channels will greatly influence the hydraulic conductivity. As an illustration, Childs (1969) pointed out that a clay soil may have a textural porosity of, say, 1 per cent and a matric conductivity of 0.01 cm/h, but if the material is fissured with, for example, a structural porosity also of 1 per cent comprising 1 mm width cracks at 10 cm intervals, the saturated hydraulic conductivity of the soil could be of the order of 1000 cm/h. Thus while saturated hydraulic conductivity has been correlated with soil texture and also with descriptions of structure in some studies (McKeague et al., 1982; Rawls et al., 1982), such correlations can be hazardous and the resulting estimates of K may be seriously in error.

The relations between hydraulic conductivity and water content are such that whereas in saturated conditions K may be regarded as more or less constant for any given material, in unsaturated conditions K will vary considerably with soil water content and matric suction, as is shown in Fig. 5.9. It can be seen that hydraulic conductivity is largest at or near saturation, but that it decreases rapidly with reducing water content. In a saturated soil all the pore spaces are filled with water and therefore form an effective part of the water conducting system, since the movement of liquid moisture can take place only through existing films of water on and between the soil grains. By definition, however, in an unsaturated soil some of the pores are filled with air and therefore act as a non-conducting part of the system, reducing the effective cross-sectional area available for flow. It will be apparent that, the greater the decrease in soil water content, the greater will be the reduction in effectiveness of the conducting system and the smaller, therefore, the value of hydraulic conductivity.

Childs (1969) explained the validity of the Darcy equation to unsaturated materials (with appropriate values of K) as follows: in unsaturated soils some of the pores will be filled with air and so be ineffective for moisture flow. This is no different to those pores being filled with solid material and so the porous material could simply be regarded as a new saturated material with a different pore size distribution and a smaller value of K. The variation in hydraulic conductivity with moisture content will depend upon the pore size

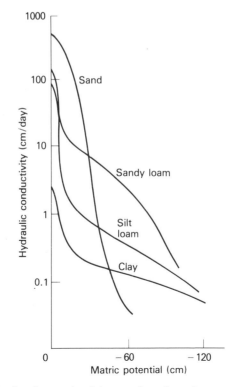

Figure 5.9 *Unsaturated hydraulic conductivity as a function of pressure potential (from an original diagram by Bouma, 1977).*

distribution of the soil. At high soil moisture contents, conductivity is broadly related to soil texture, and increases as the latter becomes coarser (Fig. 5.9). This relationship is to be expected because water will obviously be transmitted more easily through large water-filled pores than through small ones. At saturation coarser soils have a higher proportion of large pores through which water can flow easily, and K is larger than for clay soils. As the soil dries, the larger pores are the first to empty at low suctions, and K falls rapidly. As the suction increases and the moisture content decreases, the relationship between conductivity and texture is reversed so that, in dry conditions, clay soils are likely to have higher conductivities than loamy or sandy soils. Again, this relationship is to be expected, since finer-texture soils will have more water-filled pores at high suctions and, therefore, a larger cross-sectional area through which flow can take place, than will coarser soils in which only a small proportion of the pores contain water at high suctions. In shrinking soils, the increased suction that accompanies drying reduces the size of the pores that remain full of water, and this further helps to reduce the hydraulic conductivity.

Methods and associated problems of measuring soil hydraulic conductivity and hydraulic diffusivity in the laboratory and *in situ* are discussed in the texts edited by Burke *et al.* (1986) and Klute (1986a). A number of studies have shown that values obtained from small soil cores in the laboratory may be unreliable, and several times higher than for *in situ* determinations, due to short-circuiting of water through macrovoids that were continuous through the short lengths of the cores (Anderson and Bouma, 1973; Mensah-Bonsu and Lal, 1975; Lauren *et al.*, 1988).

Since the measurement of conductivity in unsaturated soils is difficult and expensive in the field a number of methods have been proposed to provide estimates based on other, more readily available soil properties. The simplest approach is that based on soil texture (e.g. Alexander and Skaggs, 1987). It is clear, however, that while texture is a major determinant, there are many other factors that may be important for a particular case. Better estimates of unsaturated conductivity may be expected where the moisture characteristic is available, and a large number of studies have been carried out to relate the two (e.g. Mualem, 1976). For satisfactory results from both groups of methods the predicted conductivities should be matched with measured values near saturation. Even so, these methods are not applicable to soils in which the hydraulic conductivity under saturated conditions is predominantly through macrovoids rather than through the soil matrix.

5.4.2 *Infiltration of water into soils*

The term *infiltration* is used to describe the process of water entry into the soil through the soil surface. The maximum rate at which water soaks into or is absorbed by the soil, its infiltration *capacity*, may in certain circumstances be very important in determining the partitioning of precipitation falling upon a catchment area. The relationship between rainfall intensity and infiltration capacity determines how much of the falling rain will flow over the ground surface, possibly directly into streams and rivers, and how much will enter the soil. Once in the soil it may move laterally as throughflow or interflow, or else be retained for some period of time before being either passed downwards as percolation or returned to the atmosphere by the processes of evaporation and transpiration. The reasons for variations of infiltration capacity in both time and space, and the methods by which it may be determined, therefore merit careful consideration.

The terms infiltration *capacity* and infiltration *rate* are often used interchangeably in the literature. This can lead to some confusion since infiltration may be limited by the rate of supply of water (rainfall or irrigation) or by the capacity of the soil to absorb water. Thus in this text the terms infiltration capacity and rate will be distinguished, the latter term being used to indicate that infiltration is proceeding at less than the infiltration capacity. When this occurs, all the falling rain not held as surface storage will infiltrate into the soil so that there will be a direct relationship between the rate of infiltration and the intensity of rainfall. When, however, rainfall intensity exceeds the infiltration capacity, this relationship breaks down and may, indeed, be replaced by an inverse relationship between infiltration and rainfall intensity. This is normally the case when an increase in rainfall intensity is reflected in an increase in the compacting force as the drops strike the ground surface.

Much of the early, pioneering, field research on infiltration (Horton, 1933, 1939) was conducted in semi-arid areas where rainfall intensity was often greater than the ability of the soil to absorb it, resulting in widespread ponding and overland flow (see Chapter 7). So influential and widely accepted was this concept, that the subsequent period of hydrology has been called the 'era of infiltration' (Cook, 1946). From the late 1960s, however, it became generally recognized that this model was not appropriate for well-vegetated temperate areas where widespread overland flow is uncommon (Hewlett and Hibbert, 1967; Betson and Marius, 1969; Dunne, 1978). Comparisons of measured infiltration capacities and long-term rainfall statistics confirmed that many of the soils in such areas can absorb the rainfall of all but the most intense storms (Rubin, 1968; Freeze, 1972; Kirkby 1969). Nevertheless, a knowledge of the infiltration capacity of a soil may be important in many parts of the world as well as in cases of very large water inputs such as extreme rainstorms or artificial irrigation.

Infiltration capacity

It is generally observed that the infiltration capacity of a soil decreases rapidly over time during a storm (Fig. 5.10). The infiltration capacity will be determined by a wide range of factors. Horton (1933) considered that it was probably controlled more by processes operating at the soil surface than by flow processes within the soil profile, and he laid some stress upon the limitations imposed on infiltration by surface and surface cover conditions. He thus implicitly distinguished between infiltration and percolation, the latter term being used to describe the downward flow of water through the zone of aeration towards the water table, the former being restricted to the entry of water through the surface layers of the soil. Clearly, however, these two processes are closely related. In some circumstances infiltration will be limited by soil surface and surface cover conditions and in other circumstances it will be dependent upon the rate of downward movement of water through the soil profile.

Soil surface conditions may impose an upper limit to the rate at which water can be absorbed, despite the fact that the capacity of the lower soil layers to receive and to store additional infiltrating water remains unfilled. In general the infiltration capacity is reduced by surface compaction, the washing of fine particles into surface pores and by frost, and increases with the depth of standing water on the surface, the number of cracks and

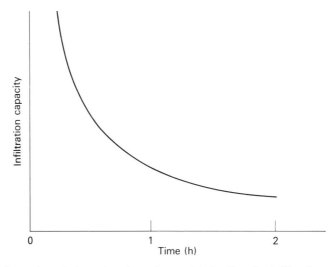

Figure 5.10 *Schematic relation showing the rapid decline in infiltration capacity generally observed.*

fissures at the surface and the ground slope. Cultivation techniques may either increase or decrease infiltration capacity. The nature of the vegetation cover also has an important influence on the infiltration process (US Soil Conservation Service, 1972). Vegetation tends to increase the infiltration capacity of a soil by retarding surface water movement, by reducing raindrop compaction and by improving soil structure (Section 5.2). Infiltration is generally higher beneath forest than beneath grass, due to the presence of ground litter. Grain crops tend to be intermediate between row crops and grass in their effects on infiltration (Musgrave, 1955). Snow may have a similar effect to ground litter, and urban surfaces considerably reduce the infiltration capacity over large areas.

Water cannot continue to be absorbed by the soil surface at a given rate unless the underlying soil profile can conduct the infiltrated water away at a corresponding rate. The ability of a given soil to conduct water depends upon its properties, including texture and structure, and is greater for permeable, coarse-textured, soils than for slowly permeable clays (Musgrave and Holtan, 1964). Other factors include soil stratification and the initial soil moisture profile.

Soil water movement during infiltration
The process of infiltration under field conditions may be quite complex due to non-uniformity of the initial moisture content and soil properties, hysteresis, changes of various soil and boundary conditions over time and the existence of two- or three-dimensional flows. For this reason, much of the work on infiltration theory has been applied to homogeneous soils with a uniform initial water content and assuming ponded water at the upper boundary of the soil.

Some of the classic experimental work on the entry of water into the soil in this situation was performed by Bodman and Colman (1943), who found that the wetted portion of a column of soil into which infiltration is taking place comprises a number of zones, which are illustrated in Fig. 5.11. The saturated zone, as its name implies, is a shallow saturated layer, a centimetre or so in thickness, at the ground surface. Immediately below this is another shallow zone, only a few centimetres in thickness, in which the water content decreases very rapidly from top to bottom and which is, therefore, known as the transition zone. Below this, again, is the transmission zone, through which water from the upper two zones is transmitted, with little or no change in moisture content, to the underlying wetting zone which, like the transition zone, has fairly steep moisture gradients and where the water content also changes appreciably with time. Finally, at the base of the wetting zone is the sharply defined wetting front, which is characterized by very steep moisture gradients and which marks the limit between the wetted soil above and the dry soil below. Provided that the supply of water to the soil surface from rainfall or irrigation continues, the wetting front advances steadily downwards into the unwetted soil as a result of the passage of water through the transmission zone.

It follows, therefore, that variations in the duration of surface ponded infiltration will result in variations in the depth of wetted soil rather than in continued increases in the water content of the surface layers. Indeed, during infiltration and percolation, the only part of the soil profile in which the moisture content changes significantly with time is the

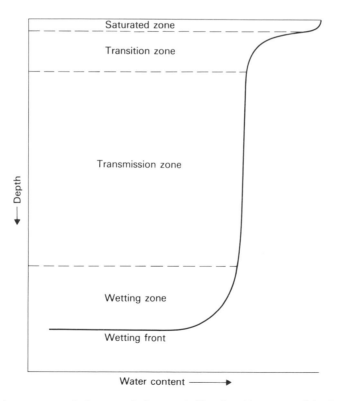

Figure 5.11 *Moisture zones during ponded water infiltration (from an original diagram by G. B. Bodman and E. A. Colman,* Proceedings of Soil Science Society of America, **8**, *pp. 116–22, 1943, and adapted by permission of the publisher).*

wetting zone and the wetting front. The saturated zone, of course, remains saturated, the moisture gradient in the transition zone remains fairly constant, and so, too, does the water content and suction in the transmission zone. Thus, with continuing infiltration, the transmission zone becomes longer and the wetting zone and the wetting front move farther downwards into the soil. It may be noted that the saturation and transition zones described by Bodman and Colman may have been an experimental artefact due to looseness, structural instability or swelling of the soil at the surface (Hillel, 1982). Later investigators have generally found that the water content of the surface soil is below complete saturation due to air entrapment and that in field soils about 8–20 per cent of the pore space is commonly occupied by air when the soil is at maximum saturation (Corey, 1977).

A similar moisture profile during sustained surface flooding, but without the saturated and transition zones, was derived from theoretical considerations by Philip (1964). Figure 5.12 shows typical soil water profiles computed by Philip (1964) as a function of the time since the start of infiltration. The soil has a clay loam texture (Constantz, 1987). The increase in profile water over time consists mainly of an extension of the nearly saturated

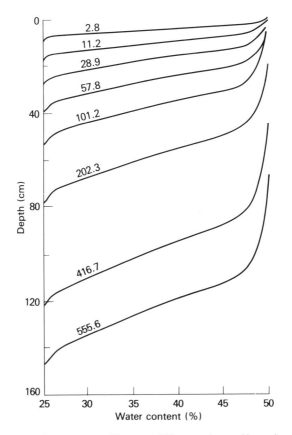

Figure 5.12 *Computed soil water profiles at different times (hours) during ponded water infiltration into a clay loam soil (from an original diagram by Philip, 1964).*

transmission zone. The sharp change in moisture content in the wetting zone is a consequence of the dependence of hydraulic conductivity on moisture content. Thus, from Darcy's law it is clear that a steep hydraulic gradient is necessary in this zone to achieve a flux equal to that in the (near-saturation) transmission zone.

From the preceding discussion it can be seen that in the early stages of infiltration into a uniformly dry soil the matric suction gradient in the surface layer will be very steep, and is likely to be the most important factor determining the amount of infiltration and downward movement of water. Thus, the initial rates of movement are likely to be high, and this early rapid penetration of the moisture profile is clearly illustrated in Fig. 5.12. As the wetting proceeds the transmission zone lengthens and the gravitational gradient becomes relatively more important. The rates of infiltration and downward movement decrease until the infiltration tends to settle down to a steady, gravity-controlled rate which approximates the saturated hydraulic conductivity (Hillel, 1982).

For vegetated, humid, areas a more realistic situation is one in which the soil infiltration capacity exceeds the rainfall intensity so that no surface ponding occurs, and the rate of infiltration will be equal to the rate of water supply to the soil surface. The development of moisture profiles in such circumstances was described by Rubin (1966); profiles for two steady rainfall intensities are shown in Fig. 5.13. The surface soil does not become saturated and the moisture content increases until it reaches a value at which the hydraulic conductivity becomes equal to the rainfall rate. A wave of moisture percolates downwards, wetting the soil to this moisture content. Increasing the rainfall intensity results in a higher water content throughout the profile, and hence a larger conductivity. As in the ponded infiltration case, the moisture content at the wetting front remains steep. Thus, although reached in different ways, the soil moisture profiles developed by both ponded infiltration and rainfall-limited infiltration are similar in shape (Childs, 1969).

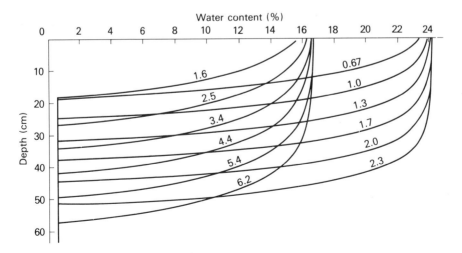

Figure 5.13 *Computed soil water profiles at different times (hours) during non-ponding infiltration for constant rate rainfall intensities of 12.7 and 47.0 mm/h (after an original diagram by Rubin, 1966).*

Time variations in infiltration capacity

Numerous field studies have demonstrated that the infiltration capacity decreases with time through a period of rain until, after a couple of hours or less, a more or less constant value is reached. Following the work of Horton (1939), this decrease was widely attributed in the literature to factors operating at the soil surface, including swelling of clays, raindrop impact and inwashing of fine material (e.g. Foster, 1948). Such mechanical effects do occur, particularly in bare soils and certain clays, but, as noted by Gardner (1967), there was an apparent reluctance to accept the importance of flow processes within the soil in limiting surface infiltration. The work by Bodman and Colman (1943), however, performed in the absence of compaction or inwashing, demonstrated how a decreasing infiltration capacity results directly from the reduction in the water potential gradient. According to Hillel (1982), the decrease in infiltration capacity through time is primarily due to the reduction in the matric suction gradient.

Figure 5.10 shows a typical curve of declining infiltration capacity over time for a deep soil of relatively uniform texture. The capacity is initially high but falls very rapidly, generally within the first hour of rainfall, to a comparatively constant value which closely approximates the saturated hydraulic conductivity. Where, however, the soil profile is complicated by the existence of a relatively impermeable layer at some distance below the surface, the curve of infiltration capacity versus time may show a further sudden reduction of infiltration capacity, reflecting the fact that, when the available storage in the surface soil horizons has been filled, infiltration is then governed by the rate at which water can pass through the layer of lower saturated hydraulic conductivity.

A number of formulae have been proposed to describe the variation in infiltration capacity for a uniform soil with ponded water on the surface and to account for the initial rapid decrease leading to an asymptotic approach to a constant value. One of the earliest attempts was that by Green and Ampt (1911) in a classic study of the flow of air and water through soils. Starting with the Hagen–Poiseuille equation for flow through capillary tubes, they made the assumption that the wetting front was a sharply defined surface that separated completely saturated soil above from uniformly unsaturated and unaffected soil below (cf. Figs 5.11 and 5.12). For a soil with water ponded on the surface to a depth H_o and with the wetting front at a depth L, the rate of infiltration is given as

$$f = \frac{K(H_o + L - H_f)}{L} \tag{5.11}$$

where H_f is the capillary suction at the wetting front and K is the *effective* hydraulic conductivity above the wetting front (less than in saturated conditions due to air entrapment). Despite the assumptions of a 'step-like' form to the profile soil moisture distribution and of a piston type of flow in capillary tubes, the equation was satisfactorily applied by its authors and by a number of subsequent workers (Whisler and Bouwer, 1970; Mein and Larson, 1973). Since it is a physically based equation the parameters can be evaluated experimentally from infiltration data (Brakensiek and Onstad, 1977) or estimated, less reliably, from basic soil property measurements (Rawls *et al.*, 1983).

Philip (1957) showed that when water is applied to a soil with a uniform initial moisture content, the solution of the Richards soil moisture equation can be expressed as an infinite power series, which for vertical infiltration and short times may be approximated as

$$F = St^{1/2} + At \tag{5.12}$$

where *F* is the accumulated infiltration and *t* is the time since surface ponding began. The coefficients of the individual terms in the series are functions of both the soil water diffusivity and the initial and surface moisture contents of the soil. The first term represents the slow filling-up of soil pores from the surface downwards due to moisture gradients, and decreases in importance over time as water enters the soil. The second term represents conductivity flow under gravity through a continuous network of large pores (Kirkby, 1969). The model has proved useful in predicting the initial decline in infiltration capacity, but the equation becomes unreliable for longer-term rates (Gardner, 1967). Philip called the parameter *S* the *sorptivity*, as it is essentially a measure of the soil's ability to absorb water by matric forces.

The preceding discussion of the infiltration process has, for simplicity, assumed a uniform initial moisture content. In practice this state is rarely, if ever, found in the field due to continued moisture movement between storms.

5.4.3 *Soil water redistribution following infiltration*

The downward movement of water under the influence of gravity and matric forces often continues long after infiltration at the surface ceases. During this *redistribution* the transmission zone which existed during infiltration becomes a draining zone as water moves from the infiltration-wetted upper layers of the soil to deeper, drier, layers. This process is important since it controls the quantity of water retained in the plant root zone, the available air-filled porosity for subsequent storage of water in the next storm and the recharge to the groundwater resource.

The redistribution process inevitably slows with time as the hydraulic conductivity in the former wetted zone reduces with the decreasing water content and also because the suction gradients become weaker as the moisture content becomes more uniform. The wetting front continues to move down the profile, but its advance becomes progressively slower and less distinct. After a few days the water content changes only very slowly and the soil is said to be at 'field capacity' (Section 5.3.4). Figure 5.14 demonstrates these points and shows successive soil water profiles in a fine sandy loam soil during redistribution, in the absence of evaporation or water table influences. Since redistribution is a continuing process there is no distinct or unique point when field capacity is reached. Rather than specifying an arbitrary period of time after infiltration ceases or a particular value of moisture tension, it may be preferable to make repeated measurements of soil moisture and to designate field capacity, as when the rate of change becomes less than a predetermined value.

If uninterrupted, redistribution would continue until ultimately a state of equilibrium was reached in which gravity and retention forces were in balance, producing equal total soil water potentials at all depths. There would be no hydraulic gradient, and hence no flow. In field situations this rarely occurs because of rainfall and evaporation. Under natural conditions evaporation will normally take place from the soil and plants after the cessation of rainfall or irrigation, and will result in a drying-out of the surface layers. In this way, a suction gradient is created which encourages the upward movement of water from the draining zone and which, therefore, assists in further reducing the downward movement of water over time.

An interesting comparison of the progress of moisture distribution is provided in Fig.

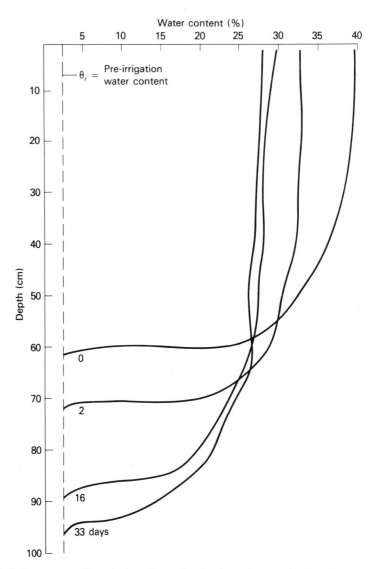

Figure 5.14 *Soil water profiles during the redistribution of water in the absence of evaporation following the irrigation of a fine sandy loam soil. (From an original diagram by W. R. Gardner, D. Hillel and Y. Benyamini,* Water Resources Research, **6**, *p. 857, Fig. 3. © 1970 American Geophysical Union.)*

5.15 for three situations: (a) redistribution without evaporation, (b) simultaneous evaporation and redistribution and (c) evaporation only. The curves of the simultaneous processes resemble those for redistribution alone, except for the obvious evaporation effect in the surface layers. In particular, the lower portions of the curves indicate that evaporation had little effect on the shape and rate of advance of the wetting front. Gardner *et al.* (1970), in fact, estimated that evaporation reduced drainage by only about 10 per

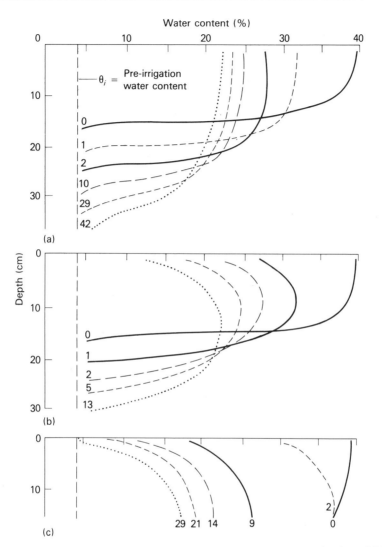

Figure 5.15 *Successive soil water profiles in soil columns following an irrigation of 50 mm. Values indicate the time in days since irrigation ended. (a) Redistribution with no evaporation, (b) redistribution and evaporation, (c) evaporation only. (From an original diagram by W. R. Gardner, D. Hillel and Y. Benyamini,* Water Resources Research, **6**, *p. 1149, Fig. 2. © 1970 American Geophysical Union.)*

cent, although redistribution greatly detracted from evaporation, reducing it by about 75 per cent. The Figure also shows that the upper part of the profile, which was initially wetted during infiltration, drains monotonically, though increasingly slowly, while the immediate sublayer at first becomes wetter before eventually beginning to drain. Clearly, then, in this part of the soil profile, hysteresis will severely complicate attempts to measure or estimate the process of moisture redistribution. Hysteresis presents a more general problem, however, in the sense that during redistribution, as has been shown, the upper part of the profile is drying through drainage and evaporation, while the lower part is becoming wetter. The relation between water content and suction will, therefore, be different at different depths depending on the history of wetting and drying that takes place at each point in the soil (Section 5.5.1).

The post-infiltration redistribution of water is still not well understood, and clearly can involve a number of simultaneous processes. Most of the studies have consequently dealt in idealized conditions and have assumed deep profiles unaffected by the water table. However, in many soils the depth to the saturated zone may be fairly small and may exert a considerable influence on the distribution of water in a soil profile.

5.4.4 *Upward movement of soil water from the water table*

Earlier consideration of the retention forces of matric suction (Section 5.3.1) indicated that water would rise in capillary tubes to a height above a free water surface that was largely governed by the radius of the tube. The hydrologist is interested in both the *extent* of this rise (since it affects soil moisture profiles) and, more importantly, the *rate* of this capillary flow (which will determine the ability of the soil to supply water to evaporation at the soil surface or by crops).

For a certain height above the water table capillarity keeps the soil pores full of water. This zone is called the capillary fringe or tension saturated zone. Its thickness corresponds to the air entry value (Childs, 1969), which is the tension necessary for air to enter the largest pores. In general, this will be greater for clays than for sandy soils but since soils contain a range of sizes of both particles and pore spaces, the height of rise will vary spatially within a given soil. In some areas where the top of the capillary fringe is close to the ground surface it may play an important role in generating rapid groundwater flow to streams (Section 7.4.2), since only a small addition of water will reduce the suction to zero causing a large and sudden rise in the water table (Sklash and Farvolden, 1979; Gillham, 1984).

Capillary rise also extends above the tension saturated zone by means of movement through films of water in the irregularly shaped and variously sized interparticle spaces of the unsaturated zone. The speed and direction of this movement will be largely determined by the unsaturated hydraulic conductivity and the combined gradients of matric suction and gravity. Evaporation and transpiration create a suction gradient encouraging the upward movement of water towards the soil surface or root zone. Unless the upward rate of movement can keep pace with these losses, progressively deeper layers of the soil will lose water until the rate of evaporation becomes limited.

The maximum rate of capillary rise is related to the water table depth and the soil texture (Gardner, 1958; Wind, 1961; Bloemen, 1980). Figure 5.16 compares the maximum rates of steady capillary flow at different heights in a coarse-textured and a fine-textured soil. It can be seen that, particularly in the case of the coarse-textured soil, the curves become almost horizontal at the higher suctions, indicating that the maximum rate of capillary flow is dependent on height above the water table rather than on the suction imposed at the soil surface. The fact that the soil profile can so limit the rate of evaporation loss, through the medium of hydraulic conductivity, was described by Hillel (1982) as a remarkable and useful feature of the unsaturated flow system. Thus, for the soil in this example, with a

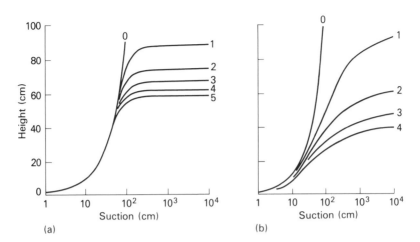

Figure 5.16 *Maximum capillary flow rate (mm/day) at different heights above the water table, related to the matric suction at the ground surface for (a) coarse-textured and (b) fine-textured soils (from original diagrams by Wind, 1961).*

water table depth of 60 cm, the maximum capillary flow is about 5 mm per day, whereas, with a depth of 90 cm, the rate declines to about 1 mm per day. In the fine-textured soil the horizontal sections of the curves are not well developed, although the influence of water table depth can still be clearly seen.

It can be deduced from Fig. 5.16 that, with an appropriate suction gradient, very small rates of capillary flow may occur at a considerable height above the water table, and, in fact, laboratory experiments have shown that measurable movement may occur over vertical distances of at least 7 metres (Gardner and Fireman, 1958). Especially in arid and semi-arid areas, therefore, where rates of evaporation are high, such movements may lead to damaging accumulations of salt at the ground surface, even where the water table occurs at a considerable depth (Wind, 1961). Since capillary rise may occur over a large vertical range of the unsaturated zone it is usually necessary to consider the effect of the different horizons in the soil profile upon the rate of capillary movement (Bloemen, 1980). It was noted in Section 5.4.1 under 'Hydraulic conductivity' that in wet conditions coarser-textured soils have a higher conductivity than clays, while in dry conditions the reverse is true. Thus the highest rates of capillary rise occur in soils where the texture becomes progressively finer with height above the water table (Wind, 1961).

5.5 Soil water behaviour under field conditions

The preceding sections of this chapter have discussed the basic properties of soil moisture retention, storage and movement by reference to a number of studies which have, of necessity, dealt mainly with idealized soil systems. Such studies have often been carried out on homogeneous soils; with uniform initial moisture conditions and with a controlled environment, e.g. usually plants, evaporation losses and water table influences are absent and 'rainfall' is at a constant rate. It is only since the end of the 1960s, with the development of reliable field instruments, that field studies of water movement under natural conditions became widespread. Prior to that most studies of moisture movement in the unsaturated zone were theoretical or laboratory based (Wellings and Bell, 1982). Similarly, due to the considerable complexities of natural conditions, in which initial and boundary conditions are usually not constant, mathematical descriptions have almost always assumed considerably simplified conditions (see reviews of models in Freeze, 1969; De Jong, 1981).

The soil profile in the field is a very complex and heterogeneous system in which none of these restrictions and simplifications need apply. Much still needs to be learnt about the behaviour of soil water under natural conditions. The following sections, which are based both on more realistic laboratory experiments and on actual field situations, can provide only a brief insight into some of the phenomena that may need to be considered in dealing with the soil water physics of a particular field site. Many of these phenomena are interrelated and have important implications for the transport of solutes (see Section 8.5.3).

5.5.1 *Soil water hysteresis*

As previously noted in Section 5.3.3 it is known that there is hysteresis in the relation between matric suction and the soil water content. This depends both upon whether the soil is currently being wetted or dried as well as on the previous pattern of moisture changes. Although $\psi(\theta)$ hysteresis has generally been ignored in soil water studies, this practice has been criticized following work that emphasized its importance under field conditions (Royer and Vachaud, 1975; Beese and van der Ploeg, 1976). It has generally been found that in comparison with the $\psi(\theta)$ curves there is much less hysteresis for unsaturated hydraulic conductivity, $K(\theta)$.

Under natural conditions the existence of a uniform water content throughout the profile is virtually impossible, even for a homogeneous soil. Immediately after precipitation the soil surface layers will be wetter than those deeper down, while subsequent downward infiltration and redistribution, together with evaporation, will tend to dry out the surface layers so that the moisture content increases with depth. This pattern will be repeated for successive storms, their effects becoming superimposed, so that the distribution of soil water in the profile is both non-uniform and continually changing. As a result, the water content changes at different depths in the soil will follow different scanning curves (Fig. 5.17). The situation will obviously be much more complicated for soils that are not homogeneous.

5.5.2 *Soil heterogeneity*

Depending upon the hydrological problem under study, variation in soil properties may be viewed at a number of different scales. These include areal variations across a catchment or between catchments, differences down a soil profile and differences within a small volume of soil in a single soil horizon.

Spatial variability

Soil properties tend to vary continuously over the earth's surface, with very few sharp changes. Thus the identification and classification of soil units by pedologists is often

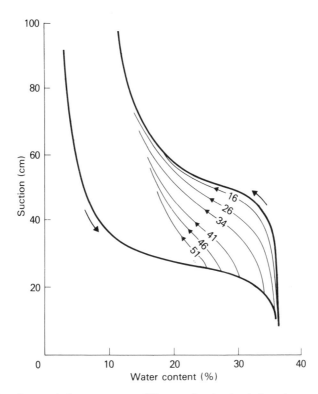

Figure 5.17 *Soil moisture drying curves at different depths (cm) in a homogeneous soil profile (redrawn after diagrams by Vachaud and Thony, 1971).*

based on subjectively and arbitrarily chosen boundaries in the continuum of soil change (Cruikshank, 1972). These soil units are broad classifications, often intended primarily for agricultural use, and based on a variety of features which can be readily and unambiguously noted by surveyors in the field. Consequently, physical properties that are of interest to the hydrologist may play only a minor role in the classification (Warrick and Nielsen, 1980). Nevertheless, in the absence of direct measurements, these soil units may be satisfactory 'carriers' of certain types of basic data (Bouma, 1986). Broad groupings of soil types have been widely used for certain hydrological purposes (e.g. Musgrave and Holtan, 1964; Farquharson *et al.*, 1978).

The spatial variation of soil physical properties in a 'uniform' 150 ha field belonging to the same Panoche soil series was studied by Nielsen *et al.* (1973). They found very large variations in hydraulic conductivity and soil water diffusivity, which both had log-normal frequency distributions. Texture, bulk density and water content exhibited much less variation and were normally distributed. Geostatistical techniques such as kriging may be used to interpolate between measured values, and the *similar media* concept has been applied by using a scaling coefficient to describe heterogeneity from site to site (Warrick and Nielsen, 1980). Hopmans (1987) provided a comparison of different scaling methods for the hydraulic properties of the soil in a research catchment in the eastern Netherlands.

Price and Bauer (1984) found that small textural changes within a sandy soil gave rise to large differences in water content, with zones of preferential retainment (finer texture) and preferential percolation (coarser texture) over a distance of only a few metres.

Soil layering
Natural soil forming processes lead to layering of the soil profile which may greatly affect the movement of water in a soil profile. The stratification of a soil profile often results in a considerable variation of hydraulic conductivity and porosity with depth. In many humid areas, for example, leaching of minerals and fine particles from the surface layers and their accumulation at greater depths often results in a marked decrease in the number of large pores in the zone of accumulation. In extreme cases the formation of an iron pan or hard pan may result in near-zero permeability and consequently in the waterlogging of the surface layers. Although the surface soil tends generally to have a higher saturated hydraulic conductivity than the subsoil, this is not always the case, and in fact any one soil horizon may limit the overall transmissibility of the complete profile. It has been found that, in general, layering reduces the infiltration capacity of a profile.

Where a coarse layer of higher saturated hydraulic conductivity overlies a finer-textured layer the infiltration capacity is initially controlled by the coarse layer, but once the wetting front extends into the finer layer it is the latter that controls the rate of water movement. If infiltration is prolonged a perched water table may develop in the coarse soil, just above its boundary with the impeding finer layer. This saturated zone may then lead to subsurface lateral flow (e.g. Whipkey, 1965; Calver *et al.*, 1972).

In the opposite case, where fine material overlies coarse, the infiltration capacity is again controlled initially by the upper layer but, surprisingly, may also slow down when the wetting front reaches the underlying coarser material. This effect was observed by Miller and Gardner (1962) and results from the fact that the soil moisture suction at the wetting front may be too high to allow water to enter the larger pores of the coarse material. Continued infiltration raises the moisture content in the upper layer until the matric potential has reduced sufficiently for water to penetrate the coarser layer. Clearly, then, a layer of sand in fine-textured soil may actually impede water movement through the profile rather than increase it (Brady, 1984; Hillel, 1982). Corey (1977) described the case in which water flows downwards through a sequence of layers.

The pattern of soil water content and pressure potential during a constant, supply-limited, infiltration was studied by Vachaud *et al.* (1973) for a layered soil comprising fine sand over coarse sand (Fig. 5.18). The saturated conductivity of both layers was considerably greater than the rate of application of water. A number of points can be seen from this figure. Since water transfer occurs between the two layers, the soil water potential must be continuous across the boundary because any pressure discontinuity would imply an infinite hydraulic gradient (Corey, 1977). Similarly, the flux must be equal across the boundary. The sudden jump in moisture content, in contrast, results from the difference in the moisture characteristic of the two materials. As infiltration continues, the wetting front moves down and the water content of each layer increases until its hydraulic conductivity becomes sufficient to carry the flow (Section 5.4.2 under 'Time variations in infiltration capacity').

In climatic conditions where evaporation exceeds precipitation the net upward movement of water and solutes may result in surface deposition and the formation of a

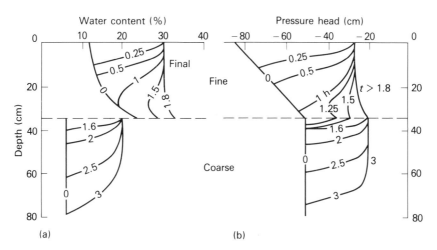

Figure 5.18 *Change in (a) water content and (b) suction profiles with time (hours) during a constant rate of water application for a layered soil (redrawn after diagrams by Vachaud et al., 1973).*

crusted or indurated surface layer. Such a layer may also result from compaction by raindrops (Tackett and Pearson, 1965) or from the breakdown of soil aggregates during wetting (Hillel, 1960). Even a thin surface crust can considerably impede infiltration and must, therefore, be taken into account in infiltration studies. Analyses of crust effects on infiltration were presented by Edwards and Larson (1969) and by Hillel and Gardner (1969, 1970).

In high latitudes, and in some high altitude areas, frozen soils introduce a further complication to the entry of water into the soil profile. Some of the work in these conditions was reviewed by Gray and Norum (1968) and by Gray *et al.* (1970) who emphasized that one of the important factors governing the rate of infiltration into frozen soils is the soil moisture content at the time of freezing, since that determines the number and size of the ice-free pores. The wetter the soil is when freezing takes place the greater the number of ice-blocked pores and the lower the rate of infiltration, so that, when a saturated soil freezes, its intake rate will be virtually zero.

Macropores
While traditional concepts of infiltration have been based on uniform media, there is a growing awareness that in natural soils water does not always move in the manner predicted from Darcy's equation. In certain circumstances, water movement may be dominated by flow through a few large openings or voids (Bouma, 1977; Quisenberry and Phillips, 1976). These large pores may comprise structural cracks and fissures or may be derived from biological activity including earthworms and other burrowing creatures, as well as paths taken by plant rooting systems. A detailed review of a large number of studies relating to the role of macropores on the flow of water through soils was given by Beven and Germann (1982).

The effect of such macropores (which may only comprise 1 or 2 per cent of the bulk soil

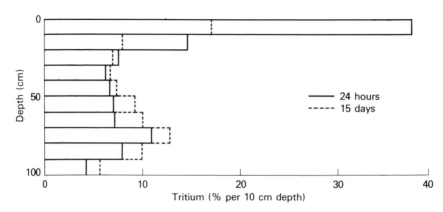

Figure 5.19 *Distribution of tritium down the soil profile at 24 hours and 15 days after application at the ground surface (reproduced from G. Blake, E. Schlichting and U. Zimmermann,* Proceedings of Soil Science Society of America, **37**, *1973, pp. 669–72 by permission of the Soil Science Society of America, Inc.).*

volume) upon flows will be particularly marked in fine-textured clay soils which have very low matrix hydraulic conductivities. From the earlier discussion of water retention it will be evident that these macropores will only conduct water when the capillary potential is sufficiently low. Beven and Germann (1982) suggested an arbitrary potential of 0.1 kPa, which is equivalent to pores larger than about 3 mm in diameter. When the soil is drier than this potential, which will be the majority of the time for most soils, rapid water flow through macropores will not take place. Thus the role and importance of macropores in infiltration is largely confined to periods of heavy rainfall or artificial irrigation when the rate of water supply exceeds the infiltration capacity of the adjacent soil peds. Under such circumstances water will not move down the profile as a well-defined wetting front. Blake *et al.* (1973) applied 50 mm of tritiated water to the surface of a dry clay soil to demonstrate that some of the water flowed down cracks to reach the subsoil (Fig. 5.19). Tritium concentrations were much higher on the faces of the cracks than within the soil peds.

Thus, in such soils, rather than the classic picture of infiltration proceeding with a well-defined wetting front as water moves slowly through the soil peds by means of matrix flow, a certain proportion of the water moves quickly down cracks and root channels. By allowing infiltration directly from the surface to the subsoil, macropores alter the pattern of wetting of the soil profile. Hodnett and Bell (1986), working on a swelling clay soil in central India, found that the monsoon rains led to initial saturation at about 1.6 m depth rather than from the surface downwards (Fig. 5.20). This rapid, preferential movement of water down the profile may have a number of consequences. The macropores will obviously considerably increase the infiltration capacity of soils and help to reduce the incidence of overland flow. Water flowing down these fissures will bypass, and so fail to wet, the soil matrix. This may mean that less water is available in the surface layers for plants. Macropore flow may allow deep percolation and recharge to groundwater, even when the overlying soil is dry, but since this flow does not pass through the natural

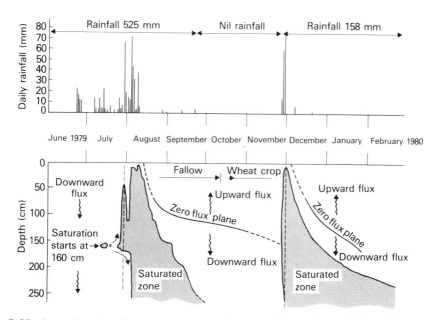

Figure 5.20 *Annual cycle of water movement in a swelling clay soil in India, showing the development of saturated conditions starting at 1.6 m depth due to infiltration down shrinkage cracks. (From an original diagram by Hodnett and Bell, 1986. Blackwell Scientific Publications Ltd.)*

filtration and purification of the soil matrix it may lead to contamination of groundwater supplies. Hodnett and Bell (1986) found that recharge to an aquifer in the Betwa catchment in India was confined to areas where the clay soil was shallow and fully penetrated by well-developed shrinkage cracks. The continuity of macropores is very important for flow processes: while many macropores have closed bottom ends and consequently fill up with water, others may be connected to large air-filled cavities such as animal burrows or artificial drains. Robinson and Beven (1983) found that flows from artificial drains in a clay soil, which exhibited seasonal shrink–swell, were much more responsive to storms in summer when the ground was dry and cracked than in winter when the topsoil was close to saturation. Macropores may also be important in generating lateral flows during heavy rains in certain soils (Whipkey, 1965) (see also Section 7.4.2).

In the presence of macropores, flow theory will no longer be adequately described by equations based on Darcy's law, since the assumptions of homogeneous soil hydraulic properties and a well-defined hydraulic gradient will no longer apply (Beven and Germann, 1982). The importance of such flows will be likely to vary over time for a given soil, with macropore flows being dominant during, and shortly after, rainfall and matrix flows dominant during the subsequent period of redistribution. De Vries and Chow (1978) observed very irregular hydraulic gradients during infiltration into a forest soil, and found that they then became more regular over time once infiltration had ceased.

A great deal of further work will be necessary in order to understand the behaviour of flows in macropores and to determine the circumstances (soils, rainfall rates, etc.) under which their role needs to be considered.

5.5.3 *Topography*

Up to now the discussion of soil water behaviour has dealt predominantly with water in a single, vertical, soil profile. On level ground soil water behaviour may be dominated by soil properties but on steep ground it may be well correlated to the topography. Ground slope will clearly influence the magnitude, speed and direction of *lateral* flows in the soil, promoting spatial variations in soil water contents across a catchment and leading to a gradient of increasing moisture content downslope (Hewlett and Hibbert, 1967). In fact, soil scientists have long recognized a 'hydrologic sequence' of progressively deteriorating drainage condition downslope (Cruikshank, 1972; White, 1987). Both topographic and soil maps may therefore provide hydrologists with a useful starting point in studying an area (e.g. Bouma, 1986).

Figure 5.21 shows, in a generalized form, the changing pattern of soil water tension and flow patterns in a uniform soil on a hillslope during the course of a storm:

(a) Initially the soil water state is close to complete gravity drainage. Tension increases

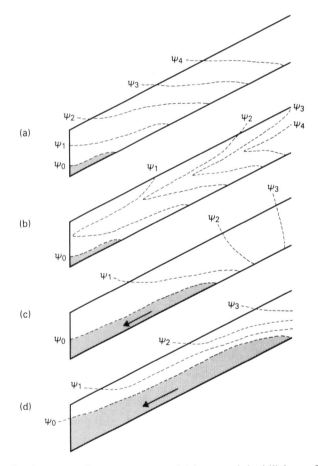

Figure 5.21 *Generalized pattern of pressure potential in a straight hillslope. See text for details (redrawn after an original diagram by Weyman, 1973).*

with elevation, approximately balancing the gravity potential, except near the base of the slope, where there is some saturated lateral flow.

(b) Once rain begins, the surface layers of the soil become wetted and tension is reduced.

(c) Percolation wets deeper layers, and the saturated layer begins to grow as it is fed by unsaturated vertical and lateral flows.

(d) After the rain, drainage of water from upper to lower layers continues, with drying of the upper layer and further expansion of the saturated conditions.

The extent of this saturated zone at the base of the slope is important in consideration of mechanisms of storm runoff generation (Section 7.4). With continued drainage of soil water after the storm, the soil water pattern reverts to that of the initial state. Such studies of two-dimensional soil water movement down slopes do not consider the influence of contour curvature on water movement. Anderson and Burt (1978) monitored the pressure potential at a grid of points; Fig. 5.22 shows the changing spatial patterns, and areas of saturation, for one storm.

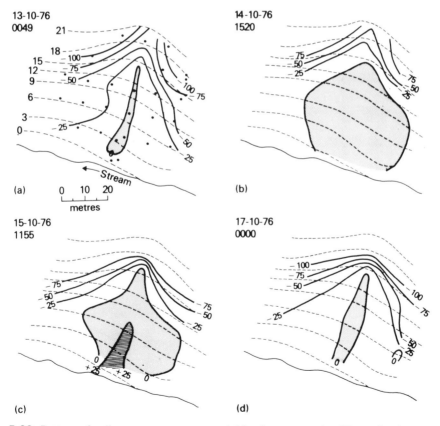

Figure 5.22 *Pattern of soil water pressure potential (cm), measured at 60 cm depth, over a hillside for one storm: (a) prior to rainfall, (b) near end of rain, (c) at stream hydrograph peak, (d) two days after storm. (From an original diagram by Anderson and Burt, 1978, by permission of John Wiley and Sons Ltd.)*

Saturation of the soil is also often observed close to stream channels. This is not simply due to the water in the channel maintaining the water table at a similar level: water can only flow out from a soil if the soil water pressure exceeds atmospheric pressure (Richards, 1950). This applies to soil water entering field drains or large macropores, as well as seepage faces such as channel banks and collector trenches for throughflow measurements (Freeze and Cherry, 1979). This 'boundary effect' elevation of the water table is especially marked at times of baseflow between storms (Troendle, 1985). The zero pressure outflow boundary condition has been known since at least the 1930s (Richards, 1950) and yet appeared to be ignored in some much later research (cf. Anderson and Burt, 1977).

In areas of steeply sloping terrain there is often a good relation between the spatial pattern of soil water content and receiving areas such as valley bottom lands and areas of converging flows (Kirkby and Chorley, 1967). Areas of shallow soils are also easily saturated (Betson and Marius, 1969). In the more widespread areas of gentle topography, soil water patterns are more difficult to predict. The natural vegetation and soil properties may provide useful indicators of the extent of saturated conditions (Dunne et al., 1975). Remote sensing techniques (Schmugge et al., 1980) provide a means of studying spatial variations over large areas. In a particular catchment, other factors may be dominant in controlling the areal pattern of soil moisture including, for example, water being forced up to the surface layers of the soil as springs (Price, 1985). Another, more general, effect on soil moisture is the impact of human activities.

5.5.4 *Effects of Man*

Man's activities can alter soil conditions in a large number of ways, ranging from irrigation schemes, which considerably increase the amount of water entering the soil, to the construction of large impermeable surfaces in urban areas, which prevent water from infiltrating into the soil beneath. These effects, which are described in engineering and agricultural texts, are outside the scope of this chapter. Accordingly, only a brief account of a few examples is given below.

Agricultural practices have the most widespread effect on soil water conditions. Irrigation and artificial drainage are used throughout the world as a means to increase crop production (Framji et al., 1982). Agricultural drainage schemes comprise open ditches or subsurface pipes (Smedema and Rycroft, 1983; Farr and Henderson, 1986). These are deeper and closer together than the natural stream channels, so increasing the hydraulic gradient in the soil and lowering the water table more rapidly between storms than would otherwise occur. A detailed account of the distribution and purpose of recent field drainage in England and Wales, the most intensively drained part of Europe, was given in Robinson and Armstrong (1988).

Tillage and cultivation operations may also alter the movement and distribution of soil water. Ploughing increases the pore spaces in the upper soil (Kuipers and van Ouwerkerk, 1963) and may encourage lateral flow in the topsoil, with less downward flow into the subsoil (Goss et al., 1978). It has been shown by tracer studies that ploughing disrupts the vertical continuity with pores in the soil below (Quisenberry and Phillips, 1976; Douglas et al., 1980). Infiltrating water was found to penetrate to greater depths on land that had not been ploughed.

A change in agricultural land use from grassland to arable cropping may also affect interception and evaporation losses, especially if the arable farming leaves the soil bare at times of the year. Heavy rainfall on land with little vegetation cover may lead to crusting and sealing of the soil surface, reducing infiltration. Forestry may have a large effect on interception and evaporation losses, causing soils under trees to be much drier than under other types of vegetation (see Chapters 3 and 4). In areas where the natural water table is close to the ground surface, groundwater abstraction may lower the water table, causing significant drying of the soil and a reduction in plant growth (van der Kloet and Lumadjeng, 1987). The most extreme case of man's influence on soil water conditions, however, is perhaps found in areas of steep topography, where deforestation and bad farming practices lead to accelerated erosion and may, in severe cases, ultimately result in the complete destruction of the soil.

References

Alexander, L. and R. W. Skaggs (1987) Predicting unsaturated hydraulic conductivity from soil texture. *Journal of Irrigation Drainage Engineering*, **113**: 184–97.

Anderson, J. L. and J. Bouma (1973) Relationships between saturated hydraulic conductivity and morphometric data of an argillic horizon. *Proc. SSSA*, **37**: 408–13.

Anderson, M. G. and T. P. Burt (1977) A laboratory model to predict the soil moisture conditions on a draining slope. *J. Hydrol.*, **33**: 383–90.

Anderson, M. G. and T. P. Burt (1978) The role of topography in controlling throughflow generation, *Proc. Earth Surf.*, **3**: 331–4.

Baver, L. D., W. H. Gardner and W. R. Gardner (1972) *Soil physics*, 4th edn, John Wiley and Sons, New York, 498 pp.

Bear, J. and A. Verruijt (1987) *Modelling groundwater flow and pollution*, D. Reidel Publishing Co., Boston, 414 pp.

Beese, F. and R. R. van der Ploeg (1976) Influence of hysteresis on moisture flow in an undisturbed soil monolith, *Proc. SSSA*, **40**: 480–4.

Bell, J. P. (1987) Neutron probe practice, *Report 19*, 3rd edn, Institute of Hydrology, Wallingford, 51 pp.

Bell, J. P., T. J. Dean and M. G. Hodnett (1987) Soil moisture measurement by an improved capacitance technique. II. Field techniques, evaluation and calibration. *J. Hydrol.*, **93**: 79–90.

Betson, R. P. and J. B. Marius (1969) Source areas of storm runoff, *WRR*, **5**: 574–82.

Beven, K. J. and P. Germann (1982) Macropores and water flow in soils, *WRR*, **18**: 1311–25.

Blake, G., E. Schlichting and U. Zimmermann (1973) Water recharge in a soil with shrinkage cracks. *Proc. SSSA*, **37**: 669–72.

Bloemen, G. W. (1980) Calculation of steady state capillary rise from the groundwater table in multi-layered soil profiles, *Zeitschrift für Pflanzenernährung und Bodenkunde*, **143**: 701–19.

Bodman, G. B. and E. A. Colman (1943) Moisture and energy conditions during downward entry of water into soils. *Proc. SSSA*, **8**: 116–22.

Bouma, J. (1977) Soil survey and the study of water in the unsaturated soil, *Soil Survey Paper 13*, Netherlands Soil Survey Institute, Wageningen, 106 pp.

Bouma, J. (1981) Soil morphology and preferential flow along macropores, *Agric. Water Manag.*, **8**: 235–50.

Bouma, J. (1986) Using soil survey information to characterize the soil-water state, *J. Soil Sci.*, **37**: 1–7.

Brady, N. C. (1984) *The nature and properties of soils*, 9th edn, Macmillan Publishing Co., New York, 560 pp.

Brakensiek, D. L. and C. A. Onstad (1977) Parameter estimation of the Green and Ampt infiltration equation, *WRR*, **13**(6): 1009–12.

Briggs, L. J. (1897) The mechanics of soil moisture, *Bureau of Soils Bulletin 10*, USDA.

Buckingham, E. (1907) Studies on the movement of soil moisture, *Bureau of Soils Bulletin 38*, USDA, 61 pp.

Burke, W., D. Gabriels and J. Bouma (eds) (1986) *Soil structure assessment*, A. A. Balkema, *Rotterdam*, 92 pp.

Calver, A., M. J. Kirkby and D. R. Weyman (1972) Modelling hillslope and channel flows, in *Spatial analysis in geomorphology*, R. J. Chorley (ed.), Methuen and Co. Ltd, London, pp. 197–218.

Campbell, G. S. (1985) *Soil physics with BASIC*, Developments in Soil Science 14, Elsevier, Amsterdam, 150 pp.

Carter, D. L., M. M. Mortland and W. D. Kemper (1986) Specific surface, Ch. 16 in *Methods of soil analysis, I. Physical and mineralogical methods*, A. Klute (ed.), 2nd edn, ASA/SSSA, Madison, Wis., pp. 413–23.

Cassell, D. K. and A. Klute (1986) Water potential, tensiometry, Ch. 23 in *Methods of soil analysis, I. Physical and mineralogical methods*, A. Klute (ed.), 2nd edn, ASA/SSSA, Madison, Wis., pp. 563–96.

Childs, E. C. (1940) The use of soil moisture characteristics in soil studies, *Soil Sci.*, **50**: 239–52.

Childs, E. C. (1969) *An introduction to the physical basis of soil water phenomena*, John Wiley and Sons Ltd., London, 493 pp.

Childs, E. C. and N. Collis-George (1950) The permeability of porous materials. *Proc. Roy. Soc.*, **A, 201**: 392–405.

Constantz, J. (1987) R. E. Moore and Yolo Light Clay, in *History of geophysics*, R. Landa and S. Ince (eds), vol. 3, *History of hydrology*, American Geophysical Union, pp. 99–101.

Cook, H. L. (1946) The infiltration approach to the calculation of surface runoff, *Trans. AGU*, **27**: 726–47.

Corey, A. T. (1977) Mechanics of heterogeneous fluids in porous media, in *Water Resources Publications*, Fort Collins, Colo., 259 pp.

Cruikshank, J. G. (1972) *Soil geography*, David and Charles, Newton Abbott, 256 pp.

Darcy, H. (1856) *Les fontaines publiques de la ville de Dijon*, V. Dalmont, Paris.

Dean, T. J., J. P. Bell and A. J. B. Baty (1987) Soil moisture measurement by an improved capacitance technique. I. Sensor design and performance, *J. Hydrol.*, **93**: 67–78.

De Jong, R. (1981) Soil water models, a review, *Contribution 123*, Land Resource Research Institute, Ottawa, Canada, 39 pp.

De Vries, J. and T. L. Chow (1978) Hydrologic behaviour of a forested mountain soil in coastal British Columbia, *WRR*, **14**: 935–42.

Douglas, J. T., M. J. Goss and D. Hill (1980) Measurements of pore characteristics in a clay soil under ploughing and direct drilling, including the use of a radioactive tracer (114 ce) technique, *Soil and Tillage Research*, **1**: 11–18.

Dunne, T. (1978) Field studies of hillslope flow processes, Ch. 7 in *Hillslope hydrology*, M. J. Kirkby (ed.), John Wiley and Sons, New York, pp. 227–93.

Dunne, T., T. R. Moore and C. H. Taylor (1975) Recognition and prediction of runoff-producing zones in humid regions, *Hydrol. Sci. Bull.*, **20**: 305–26.

Edwards, W. M. and W. E. Larson (1969) Infiltration of water into soils as influenced by surface seal development, *Trans. ASAE*, **12**: 463, 465, 470.

Farquharson, F. A. K., D. Mackney, M. D. Newson and A. J. Thomasson (1978) Estimation of runoff potential of river catchments from soil survey, *Soil Survey Special Survey 11*, Harpenden, 29 pp.

Farr, E. and W. C. Henderson (1986) *Land drainage*, Longman, London, 251 pp.

Foster, E. E. (1948) *Rainfall and runoff*, Macmillan, New York, pp. 487.

Framji, K. K., B. C. Garg and S. D. L. Luthra (eds) (1982) *Irrigation and drainage in the world, a global review*, 3rd edn, International Commission on Irrigation and Drainage, New Delhi.

Freeze, R. A. (1969) The mechanism of natural groundwater recharge and discharge. 1, One dimensional, vertical, unsteady, unsaturated flow above a recharging or discharging groundwater flow system, *WRR*, **5**: 153–71.

Freeze, R. A. (1972) Role of subsurface flow in generating surface runoff. 2, Upstream source areas, *WRR*, **8**: 1272–83.

Freeze, R. A. and J. A. Cherry (1979) *Groundwater*, Prentice-Hall, Englewood Cliffs, N. J., 604 pp.

Gardner, W., O. W. Israelsen, N. E. Edlefsen and H. Clyde (1922). The capillary potential function and its relation to irrigation practice, *Physics Review*, **20**: 196.

Gardner, W. H. (1986) Water content, Ch. 21 in *Methods of soil analysis, I. Physical and mineralogical methods*, A. Klute (ed.), ASA/SSSA, Madison, Wis., pp. 493–544.

Gardner, W. R. (1958) Some steady state solutions of the unsaturated moisture flow equation with application to evaporation from a water table, *Soil Sci.*, **85**: 228–32.

Gardner, W. R. (1967) Development of modern infiltration theory and application in hydrology, *Trans. ASAE*, **10**: 379–81, 390.

Gardner, W. R. and M. Fireman (1958) Laboratory studies of evaporation from soil columns in the presence of a water table, *Soil Sci.*, **85**: 244–9.

Gardner, W. R. and D. Kirkham (1952) Determination of soil moisture by neutron scattering, *Soil Sci.*, **73**: 391–401.

Gardner, W. R., D. Hillel and Y. Benyamini (1970) Post irrigation movement of soil water, 2. Simultaneous redistribution and evaporation, *WRR*, **6**: 1148–53.

Gillham, R. W. (1984) The capillary fringe and its effect on water-table response, *J. Hydrol.*, **67**: 307–24.

Goss, M. J., K. R. Howse and W. Harris (1978) Effects of cultivation on soil water retention and water use by cereals in clay soils, *J. Soil Sci.*, **29**: 475–88.

Gray, D. M. and D. I. Norum (1968) The effect of soil moisture on infiltration as related to runoff and recharge, *Soil Moisture, Proceedings of Hydrology Symposium No. 6*, National Research Council of Canada, Ottawa, pp. 133–53.

Gray, D. M., D. I. Norum and J. M. Wigham (1970) Infiltration and the physics of flow of water through porous media, Sec. V in *Handbook on the principles of hydrology*, D. M. Gray (ed.), National Research Council of Canada, Ottawa.

Green, W. H. and G. A. Ampt (1911) Studies in soil physics, part I. The flow of air and water through soils, *J. Agric. Sci.*, **4**: 1–24.

Hackett, O. M. (1966) Groundwater research in the United States, *Circular 527*, US Geological Survey, 8 pp.

Haines, W. B. (1930) Studies in the physical properties of soils. V. The hysteresis effect in capillary properties, and the modes of moisture distribution associated therewith, *J. Agric. Sci.*, **20**: 97–116.

Hewlett, J. D. and A. R. Hibbert (1967) Factors affecting the response of small watersheds to precipitation in humid areas, in *Forest hydrology*, W. E. Sopper and H. W. Lull (eds), Pergamon Press, New York, pp. 275–90.

Hillel, D. (1960) Crust formation in loessial soils, *Trans. Int. Soil Sci. Congr., Madison*, **7**: 330–9.

Hillel, D. (1982) *Introduction to soil physics*, Academic Press, New York, 364 pp.

Hillel, D. and W. R. Gardner (1969) Steady infiltration into crust-topped profiles, *Soil Sci.*, **108**: 137–42.

Hillel, D. and W. R. Gardner (1970) Transient infiltration into crust-topped profiles. *Soil Sci.*, **109**: 69–76.

Hodgson, J. M. (ed.) (1976) Soil survey field handbook, *Soil Survey Technical Monograph 5*, Soil Survey of England and Wales, Harpenden, 99 pp.

Hodnett, M. G. and J. P. Bell (1986) Soil moisture investigations of groundwater recharge through black cotton soils, in Madhya Pradesh, India, *Hydrol. Sci. J.*, **31**: 361–81.

Holmes, J. W. (1956) Calibration and field use of the neutron scattering method of measuring soil water content, *Australian Journal of Applied Science*, **7**: 45–58.

Hopmans, J. W. (1987) A comparison of various methods to scale soil hydraulic properties, *J. Hydrol.*, **93**: 241–56.

Horton, R. E. (1933) The role of infiltration in the hydrological cycle, *Trans. AGU*, **14**: 446–60.

Horton, R. E. (1939) Analysis of runoff plot experiments with varying infiltration capacity, *Trans. AGU*, **20**: 693–711.

ISSS (1976) Soil physics terminology. Report of the Terminology Committee (Chairman G. H. Bolt) of Commission I (Soil Physics), *International Soil Science Society Bulletin*, **49**: 26–36.

Jaynes, D. B. (1985) Comparison of soil water hysteresis models, *J. Hydrol.*, **75**: 287–99.

Kirkby, M. J. (1969) Infiltration, throughflow and overland flow, in *Water, earth and man*, R. J. Chorley (ed.), Methuen and Co., London, pp. 213–27.

Kirkby, M. J. and R. J. Chorley (1967) Throughflow, overland flow and erosion, *Int. Assoc. Hydrol. Sci. Bull.*, **12**: 5–21.

Klute, A. (ed.) (1986a) *Methods of soil analysis, I. Physical and mineralogical methods*, 2nd edn, Agronomy 9(1), American Society of Agronomy/Soil Science Society of America, Madison, Wis., 1188 pp.

Klute, A. (1986b) Water retention, laboratory methods, Ch. 26 in *Methods of soil analysis, I. Physical and mineralogical methods*, A. Klute (ed.), 2nd edn, ASA/SSSA, Madison, Wis., pp. 635–62.

Koorevaar, P., G. Menelik and C. Dirksen (1983) *Elements of soil physics*, Developments in Soil Science 13, Elsevier, Amsterdam.

Kuipers, H. and C. van Ouwerkerk (1963) Total pore-space estimations in freshly ploughed soil, *Neth. J. Agric. Sci.*, **11**: 45–53.

Lauren, J. G., R. J. Wagenet, J. Bouma and J. H. M. Wosten (1988) Variability of saturated hydraulic conductivity in a Glossaquic Hapludalf with macropores, *Soil Sci.*, **145**(1): 20–8.

McKeague, J. A., C. Wang and G. C. Topp (1982) Estimating saturated hydraulic conductivity from soil morphology, *Proc. SSSA*, **46**: 1239–44.

Marshall, T. J. and J. W. Holmes (1979) *Soil physics*, Cambridge University Press, Cambridge, 345 pp.

Mein, R. G. and C. L. Larson (1973) Modelling infiltration during a steady rain, *WRR*, **9**(2): 384–94.

Mensah-Bonsu and R. Lal (1975) Field and laboratory determination of hydraulic conductivity of an Alfisol in Nigeria, *5th Latin American Congress of Soil Science*, Medellin, Colombia. Quoted in R. Lai and D. J. Greenland (eds) (1979) *Soil physical properties and crop production in the tropics*, John Wiley and Sons, New York, 551 pp.

Miller, E. E. and W. H. Gardner (1962) Water infiltration into stratified soil, *Proc. SSSA*, **26**: 115–18.

Moser, H., W. Rauert, G. Morgenschweis and H. Zojer (1986) Study of groundwater and soil moisture movement by applying nuclear, physical and chemical methods, in *Technical documents in hydrology*, UNESCO, Paris, 104 pp.

Mualem, Y. (1976) A new model for predicting the hydraulic conductivity of unsaturated porous media, *WRR*, **12**(3): 512–22.

Musgrave, G. W. (1955) How much of the rain enters the soil?, in *Water*, USDA Yearbook of Agriculture 1955, pp. 151–9.

Musgrave, G. W. and H. N. Holtan (1964) Infiltration, in *Handbook of applied hydrology*, V. T. Chow (ed.), McGraw-Hill, New York, pp. 12.1–12.30.

Nielsen, D. R., J. W. Biggar and K. T. Erh (1973) Spatial variability of field measured soil water properties, *Hilgardia*, **42**: 215–60.

Philip, J. R. (1957) The theory of infiltration, 4. Sorptivity and algebraic infiltration equations, *Soil Sci.*, **84**: 257–64.

Philip, J. R. (1964) The gain, transfer and loss of soil water, in *Water resources use and management*, Melbourne University Press, pp. 257–75.

Price, A. G. and B. O. Bauer (1984) Small-scale heterogeneity and soil moisture variability in the unsaturated zone, *J. Hydrol.*, **70**: 277–93.

Price, M. (1985) *Introducing groundwater*, George Allen and Unwin, London, 195 pp.

Quisenberry, V. L. and R. E. Phillips (1976) Percolation of surface-applied water in the field, *Proc. SSSA*, **40**: 384–9.

Rawls, W. J., D. L. Brakensiek and K. E. Saxton (1982) Estimation of soil water properties, *Trans. ASAE*, **25**: 1316–20.

Rawls, W. J., D. L. Brakensiek and N. Miller (1983) Green–Ampt infiltration parameters from soils data, *Journal of Hydraulic Engineering*, **109**: 62–70.

Reynolds, S. G. (1970a) The gravimetric method of soil moisture determination, Part I. A study of equipment, and methodological problems, *J. Hydrol.*, **11**: 258–73.

Reynolds, S. G. (1970b) The gravimetric method of soil moisture determination, Part II. Typical required sample sizes and methods of reducing variability, *J. Hydrol.*, **11**: 274–87.

Richards, L. A. (1931) Capillary conduction of liquids through porous mediums, *Physics*, **1**: 318–33.

Richards, L. A. (1950) Laws of soil moisture, *Trans. AGU*, **31**: 750–6.

Robinson, M. and A. C. Armstrong (1988) The extent of agricultural field drainage in England and Wales, 1971–80, *Trans. IBG*, **13**: 19–28.

Robinson, M. and K. J. Beven (1983) The effect of mole drainage on the hydrological response of a swelling clay soil, *J. Hydrol.*, **64**: 205–23.

Rose, C. W. (1966) *Agricultural physics*, Pergamon Press Ltd, London, 226 pp.

Royer, J. M. and G. Vachaud (1975) Field determination of hysteresis in soil water characteristics, *Proc. SSSA*, **39**: 221–3.

Rubin, J. (1966) Theory of rainfall uptake by soils initially drier than their field capacity and its applications, *WRR*, **2**: 739–94.

Rubin, J. (1968) Numerical analysis of ponded rainfall, *Int. Assoc. Sci. Hydrol. Publ.*, **82**: 440–51.

Schmugge, T. J., T. J. Jackson and H. L. McKim (1980) Survey of methods for soil moisture determination, *WRR*, **16**: 961–79.

Shaw, E. M. (1988) *Hydrology in practice*, 2nd edn, Van Nostrand Reinhold (UK) Co. Ltd, 539 pp.

Sklash, M. G. and R. N. Farvolden (1979) The role of groundwater in storm runoff, *J. Hydrol.*, **43**: 45–65.

Smedema, L. K. and D. W. Rycroft (1983) *Land drainage*, Batsford Academic and Educational Ltd, London, 376 pp.

Tackett, J. L. and R. W. Pearson (1965) Some characteristics of soil crusts formed by simulated rainfall, *Soil Sci.*, **99**: 407–13.

Terzaghi, K. (1942) Soil moisture and capillary phenomena in soils, in *Hydrology*, O. E. Meiner (ed.), McGraw-Hill, New York, pp. 331–63.

Troendle, C. A. (1985) Variable source area models, in *Hydrological forecasting*, M. G. Anderson and T. P. Burt (eds), John Wiley and Sons, pp. 347–403.

US Soil Conservation Service (1972) *Hydrology*, National Engineering Handbook Section 4, Washington, D.C.

Vachaud, G. and J. L. Thony (1971) Hysteresis during infiltration and redistribution in a soil column at different initial water contents, *WRR*, **7**: 111–27.

Vachaud, G., M. Vauclin, D. Khanji and M. Wakil (1973) Effects of air pressure on water flow in an unsaturated stratified vertical column of sand, *WRR*. **9**: 160–73.

van Bavel, C. H. M. (1969) The three-phase domain in hydrology, in *Proceedings of Wageningen Symposium on Water in the Unsaturated Zone*, vol. 1, UNESCO/IASH, pp. 23–32.

van der Kloet, P. and H. S. Lumadjeng (1987) The development of an economic objective function for decision making in a water resource control problem, in *Decision making in water resources*, A. J. Carlsen (ed.), Proceedings of UNESCO Symposium, pp. 221–37.

Veihmeyer, F. J. and A. H. Hendrickson (1931) The moisture equivalent as a measure of the field capacity of soils, *Soil Sci.*, **32**: 181–94.

Warrick, A. W. and D. R. Nielsen (1980) Spatial variability of soil physical properties in the field, Ch. 13 in *Applications of soil physics*, D. Hillel (ed.), Academic Press, New York, pp. 319–44.

Wellings, S. R. and J. P. Bell (1982) Physical controls of water movement in the unsaturated zone, *QJEG*, **15**: 235–41.

Wellings, S. R., J. P. Bell and R. J. Raynor (1985) The use of gypsum resistance blocks for measuring soil water potential in the field, *Inst. Hydrology Rept.*, 92, Wallingford, 32 pp.

Weyman, D. R. (1973) Measurements of downslope flow of water in a soil, *J. Hydrol.*, **20**: 267–88.

Whipkey, R. Z. (1965) Subsurface stormflow from forested slopes, *Int. Assoc. Hydrol. Bull. Sci.*, **10**: 74–85.

Whisler, F. D. and H. Bouwer (1970) Comparison of methods for calculating vertical infiltration and drainage in soils, *J. Hydrol.*, **10**: 1–19.

White, R. E. (1987) *Introduction to the principles and practices of soil science*, 2nd edn, Blackwell Scientific Publications, 244 pp.

Wild, A. (ed.) (1988) *Russell's soil conditions and plant growth*, 11th edn, Longman/Wiley, New York, 991 pp.

Wind, G. P. (1961) Capillary rise and some applications of the theory of moisture movement in unsaturated soils, in *TNO Committee for Hydrological Research. Proceedings and Information No. 5*, Wageningen, pp. 186–99.

6. Groundwater

6.1 Introduction

Groundwater may be defined as the subsurface water in soils and rocks that are fully saturated. The following discussion of groundwater therefore completes the survey of subsurface water begun in the previous chapter on soil water. As was noted in that chapter, but is worth reemphasizing here, the distinction between saturated and unsaturated conditions is an artificial one, made largely for convenience in structuring the material in this book. There is an essential unity between all types of subsurface water and indeed also between surface water and subsurface water.

Groundwater is the earth's largest accessible store of fresh water and, excluding ice sheets and glaciers, has been estimated to account for 94 per cent of all fresh water (see Table 1.1). Half of this water is held within 800 m of the ground surface (Price, 1985). The magnitude of this invisible store can perhaps be better appreciated when it is considered that it is equivalent to an average depth of about 60 m of water over the earth's land surfaces, although of course its distribution is in fact quite variable. Heath (1983) quoted values for the total pore space (occupied by water, gas and petroleum) for the United States ranging from 3 m under the Piedmont plateau to about 2500 m under the Mississippi delta. The role of groundwater as a vast regulator in the hydrological cycle can be seen from the large residence time, averaging about 300 years, although with considerable site-to-site variation. Groundwater sustains streamflow during periods of dry weather and is a major source of water in many arid regions. Due to its long residence time, areas that currently have an arid climate with little opportunity for percolation to depth, may nevertheless have significant groundwater reserves which are the result of recharge in past, more pluvial, times. There are, for example, enormous groundwater reserves, equivalent to over 5×10^5 km³ of water, under the Sahara.

Although there has been much controversy in the past regarding the source of groundwater, it is now clear that almost all of it is *meteoric*, i.e. composed of precipitated atmospheric moisture that has percolated downwards through the zone of aeration. In addition, small quantities of groundwater may comprise *connate* water, which originated as sea water trapped in some rocks at the time of their deposition.

6.2 Geological background

A geological formation comprising layers of rock or unconsolidated deposits that contain sufficient saturated material to yield significant quantities of water is known as an *aquifer* (Lohman, 1972). Other formations that are much less permeable and can only transmit water at much lower rates than the adjacent aquifers are commonly known as *aquitards* (Freeze and Cherry, 1979; Price, 1985). The terms are deliberately imprecise and indicate relative properties. Thus a silt bed would represent an aquitard in a stratigraphic sequence comprising alternate sand and silt layers, while if interlayered with much less permeable clay beds the silt could be considered an aquifer. Most of the major aquifers are composed of *sedimentary* deposits formed from the erosion and deposition of other rocks. In contrast, *igneous* and *metamorphic* rocks, formed under conditions of high temperatures and pressures, generally have few interconnected pore spaces and consequently mostly have only low water-bearing capacities.

The lower limit of groundwater represents a zone in which the interstices are so few and so small that further downward movement is virtually impossible. The boundary is thus frequently formed by a stratum of very dense rock, such as clay, slate or granite, or by the upper surface of the parent rock where the groundwater body occurs only in a surface deposit of weathered material. Alternatively, the gradual compression of strata with depth, as a result of the increasing weight of the overlying rocks, gives rise eventually to a zone in which the interstices have been so reduced in both size and number that downward movement is effectively limited. The depth at which this occurs will naturally depend on the nature of the water-bearing rock, so that the lower limit of the zone of saturation would be shallower where a dense granite outcropped at the ground surface than where a porous sandstone extended to great depths. Nevertheless, the number of interstices tends generally to decrease with depth, and below about 10 km all rocks may, for practical purposes, be considered to be impermeable (Price, 1985).

The form of an aquifer is determined by the geological conditions. A common distinction, depending on the presence or absence of an overlying aquitard or confining layer, is made between confined and unconfined aquifers.

6.3 Confined and unconfined aquifers

The upper boundary of the zone of saturation varies according to whether the groundwater is confined or unconfined (see Fig. 6.1). In the case of *unconfined*

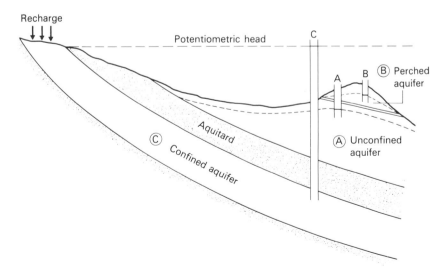

Figure 6.1 *Diagrammatic relationship between unconfined aquifer (A), perched aquifer (B) and confined aquifer (C). Note that the potentiometric levels may be different.*

groundwater, this boundary is normally known as the water table, which is defined as the level where the porewater pressure is equal to atmospheric pressure (Lohman, 1972).

It is well known that the water table tends to follow the contours of the overlying ground surface, although in a more subdued form. Assuming a similar amount of infiltration from rainfall over both high and low ground, the amplitude of relief of the water table depends largely upon the texture of the material comprising the zone of saturation. In the case of very open-textured rock, such as well-jointed limestone, groundwater will tend to move through the interstices at such a rate that it is able rapidly to find its own level, and thus form a more or less horizontal surface. On the other hand, with a fine-textured rock, groundwater movement will be so slow that water will still be draining towards the valleys from beneath the higher ground when additional infiltration from subsequent precipitation occurs, so that its height is built up under the latter areas. This tendency is magnified by the fact that precipitation normally increases with relief.

Perched groundwater represents a special case of an unconfined aquifer in which the underlying impermeable or semi-permeable bed is not continuous over a very large area and is situated at some height above the main groundwater body. The most common occurrence of perched aquifers is probably where an impermeable bed intersects the side of a valley (Price, 1985). In many areas the first unconfined groundwater encountered in drilling a borehole is of this perched type and constitutes a more or less isolated body of water whose position is controlled by structure or stratigraphy (Davis and De Wiest, 1966). In addition, as is shown in discussion of throughflow in Chapters 5 and 7, water percolating through the zone of aeration after heavy rainfall may also be regarded as a temporary perched water body.

In the case of *confined groundwater* (see Fig. 6.1), the upper boundary of the water body is formed by an overlying less permeable bed. The distinction between unconfined and confined groundwater is often made because of hydraulic differences between the flow of water under pressure and the flow of free, unconfined groundwater. Hydrologically, however, the two form part of a single, unified system. Thus, most confined aquifers have an unconfined area through which recharge to the groundwater occurs by means of infiltration and percolation, and in which a water table, as defined above, represents the upper surface of the zone of saturation. Furthermore, the confining impermeable bed rarely forms an absolute barrier to groundwater movement so that there is normally some interchange, and therefore a degree of hydraulic continuity, between the confined groundwater below the confining bed and the unconfined groundwater above it. Indeed, attention has already been drawn to the relative sense in which terms such as aquifer and aquitard must be used and to the fact that a rock forming an aquitard in one situation may form an aquifer in another.

Since the water table in the unconfined groundwater area, through which recharge takes place, is situated at a higher elevation than the confined area of the aquifer, it will be apparent that the groundwater in the latter area is under a pressure equivalent to the difference in hydrostatic level between the two. If, then, the pressure is released locally, as by sinking a well into the confined aquifer, the water level will theoretically rise in the well to the height of the hydrostatic head, i.e. the water table, in the recharge area minus the height equivalent of any energy losses resulting from friction between the moving groundwater and the solid matrix of the aquifer between the point of recharge and the point of withdrawal. The imaginary surface to which water rises in wells tapping confined aquifers is called the *potentiometric surface* (Lohman, 1972; Freeze and Cherry, 1979). This term has now replaced earlier names, such as piezometric surface, and can be applied to both confined and unconfined aquifers (in the latter case the water table is the potentiometric surface).

An alternative name sometimes used in the literature for a confined aquifer is *artesian* (Todd, 1980). The term has, however, been used in a number of different ways by different authors with meanings including any well that penetrates a confined aquifer, the confined aquifer itself or any well producing freely flowing water up to the ground surface. Some of the classic and best-known free-flowing 'artesian' conditions are found in areas of gently folded sedimentary strata such as the type area in the province of Artois in northern France, the London Basin in England, or the great artesian basins of east-central Australia and the Great Plains of the USA. Early wells in these last two basins encountered water with sufficient pressure to flow upwards, more than 45 m into the air although, of course, the pressure head subsequently diminished rather rapidly (Davis and De Wiest, 1966). Artesian conditions have also been found in fissured and fractured crystalline rocks, particularly where they are overlain by relatively impervious superficial deposits. Natural artesian springs may also result from faulting in areas of folded sedimentary rocks. Artesian conditions do not, however, require an overlying confining bed and can occur in steep areas as a result of topographic controls (Section 6.5.4).

Groundwater systems may be extremely complicated in structure and the simple distinction between unconfined, confined and perched groundwater tends to overemphasize the difference between the three types of groundwater body. It is often difficult to apply in practice even in simple geological conditions. In areas of complex geology the terms become almost meaningless. Despite these disadvantages, however, the terms have been widely adopted and are used in this chapter as a convenient basis for discussion.

6.4 Groundwater storage

Aquifers are commonly regarded as both reservoirs for groundwater storage and as pipelines for groundwater movement, although as Nace (1969) observed, due to the often slow nature of the flow, the pipeline analogy is rather less apt. Estimates of over 20 000–30 000 years have, for example, been made for the age of water in some aquifers in England and Libya (Downing *et al.*, 1977; Wright *et al.*, 1982), while some groundwater in central Australia may be 1.4 million years old (Habermehl, 1985).

Consideration will now be given to the main features of groundwater storage, and particularly to the aquifer characteristics affecting it, to its role in the groundwater balance and to aspects of storage changes in unconfined and confined aquifers.

6.4.1 *Porosity*

The amount of groundwater stored in a saturated material depends upon its *porosity*. This is normally expressed as the percentage of the total volume of a rock or soil which is represented by its *interstices*, or voids. While most interstices are small intergranular spaces some are cavernous. A knowledge of the nature of these interstices is clearly essential to an understanding of the storage and movement of groundwater, and consequently several methods have been proposed to classify them. The most frequently used classification is based upon their mode of origin, and considers original and secondary interstices (Todd, 1980; Heath, 1983). *Original interstices*, as the name implies, were created at the time of origin of the rock in which they occur; thus in sedimentary rocks they coincide with the intergranular spaces, while in igneous rocks, where they normally result from the cooling of molten magma, they may range in size from minute intercrystalline spaces to large caverns. *Secondary interstices* result from the subsequent actions of geological, climatic or biotic factors upon the original rock. Faults and joints, enlarged perhaps by weathering and solution, are the most common. Such interstices are often found in old, hard, crystalline rocks which have virtually no intergranular porosity, and over large areas of Africa, northern North America, northern Europe and India, for example, form the main channels for the storage and movement of groundwater (UNESCO, 1984). The problem with this type of 'genetic' classification of the interstices is that the original intergranular spaces are often later modified by processes including cementation and solution. A very similar, but perhaps more appropriate, classification is, therefore, that between the primary porosity due to intergranular spaces in the soil or rock matrix (Fig. 6.2a,b,c,d) and secondary porosity due to processes such as solution along joints and bedding planes (Fig. 6.2e) or to jointing and fracturing (Fig. 6.2f).

Confusion sometimes arises in the case of, say, a well-jointed rock between the porosity of the solid rock matrix (which may be very low) and the porosity of the whole stratum or formation that it comprises (which may be relatively high). It is important to realize that all interstices are involved in the concept of porosity, so that joints, bedding planes and fractures, including those greatly enlarged by solution and weathering, must be included as part of the total interstitial volume. Sometimes porous media contain voids that are not interconnected to other voids and so can play no part in the storage and movement of water (McWhorter and Sunada, 1977). They are consequently of no importance in hydrology and are not considered further.

In analyses of aquifer systems it is common to assume that the aquifer is *homogeneous*

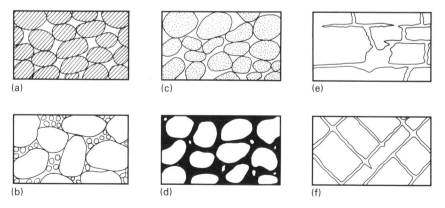

Figure 6.2 *Types of interstices:* (*a*) *between relatively impermeable well-sorted particles,* (*b*) *between relatively impermeable poorly sorted particles,* (*c*) *between permeable particles,* (*d*) *between grains that have been partially cemented,* (*e*) *formed by solution along joints and bedding planes in carbonate rocks and* (*f*) *formed by fractures in crystalline rocks* (*from an original diagram by Meinzer, 1923, with acknowledgements to the US Geological Survey*).

and *isotropic*; that is that certain properties, such as porosity, have the same values in different parts of the aquifer (homogeneity) and in different directions from the same point (isotropy). The very nature of primary and secondary geological processes means, however, that even apparently uniform deposits may have a preferred orientation of particles or fractures (*anisotropy*), and the stratification in most sediments often imparts a marked *heterogeneity* (Downing and Jones, 1985).

The porosity of a medium will depend upon a number of factors, including the shape, arrangement and degree of sorting of the constituent particles, and the extent to which modifications arising from solution, cementation, compaction and faulting have occurred. For example, poorly sorted material (with a large range of particle sizes) will have a low porosity since the interstices between the larger fragments will be filled with smaller particles and the porosity correspondingly reduced in comparison with material composed of uniformly sized grains. The combined effect of these various factors is illustrated in Fig. 6.3, which shows the normal range of porosity values for a number of different types of material. It can be seen that, in general, rocks such as sandstone, shale and limestone have lower porosities than soils and other unconsolidated deposits. Initially it may seem strange that clay, which so often forms a barrier to water movement, has a very high porosity, while good aquifers, such as sandstone, have low to medium porosities. Further consideration, however, reveals that although porosity determines how much water a saturated medium can hold, by no means all of this water will be readily available for movement in the hydrological cycle. The proportion of the groundwater that is potentially 'mobile' will depend partly on how well the interstices are interconnected and partly on the size of the interstices and, therefore, by implication on the forces by which the water is retained in the medium.

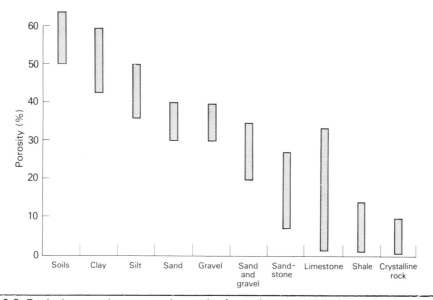

Figure 6.3 *Typical ranges in measured porosity for various materials (compiled from data from various sources).*

6.4.2 *Specific yield and specific retention*

While the porosity is very important in determining the maximum amount of water that an aquifer can hold when it is saturated it is also important to know that only a part of this is available to supply, for example, a well or spring. The *specific yield* may be defined as the volume of water that can freely drain from a saturated rock or soil under the influence of gravity, and it is normally expressed as a percentage of the total volume of the aquifer (*not* just the pore space). It can be measured by a variety of methods, but well pumping tests generally give the most reliable results (Todd, 1980). The remaining volume of water (also usually expressed as a percentage of the total aquifer volume), which is retained by surface tension forces as films around individual grains and in capillary openings, is known as the *specific retention*. This term is analogous to 'field capacity' which is used when referring to soil water (Section 5.3.4), and is similarly imprecise in the sense that there is no fixed water content at which gravity drainage ceases. An extreme example of the difference between the porosity and specific yield of an aquifer is given by chalk. This is a very fine-grained limestone in which the matrix pore sizes are typically less than 1 μm and only a very small part of the pore water can drain freely under gravity (Price *et al.*, 1976). At a site in southern England, Wellings (1984) found that the porosity was about 30 per cent, but the specific yield was only about 1 per cent. A much larger proportion of the water was available to plants, and in the upper layers suctions in excess of 1000 kPa were recorded.

The relationship between porosity, specific yield and retention for different types of unconsolidated material is shown in Fig. 6.4. It is clear that, as the texture of the material becomes coarser, and by implication the importance of the larger interstices increases, both the specific retention and the total porosity decrease. Although clay has a high total porosity, the available water in terms of the specific yield is very small.

These physical characteristics control the ability of an aquifer to store and retain water,

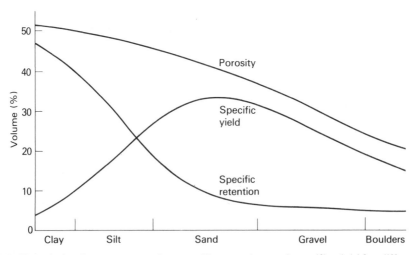

Figure 6.4 *The relation between porosity, specific retention and specific yield for different types of unconsolidated material, showing typical values which may differ significantly from values at a particular site (from an original diagram by Eckis, 1934).*

but the amount of water actually in an aquifer will also depend upon the balance between water inflows and outflows.

6.4.3 *The groundwater balance*

If the recharge into a groundwater system exactly equals the outflow from that system the amount of water in storage will remain constant; if recharge exceeds discharge, storage will increase while, if discharge is greater than recharge, storage will decrease. This may be conveniently expressed in the form of a simple water-balance equation:

$$\Delta S = Q_r - Q_d \tag{6.1}$$

where ΔS is the change in groundwater storage, Q_r is the recharge to groundwater and Q_d is the discharge from groundwater. A knowledge of the main components of this equation is essential in assessing the 'safe yield' that can be pumped from an aquifer, so as not to reduce the water table to an undesirable level.

Recharge

As mentioned earlier, it is now generally accepted that virtually all groundwater is derived, directly or indirectly, from precipitation. The main components of this groundwater recharge are: (a) infiltration of part of the total precipitation at the ground surface; (b) seepage through the bed and banks of surface water bodies such as lakes, rivers and even the oceans; (c) groundwater leakage and inflow from adjacent aquitards and aquifers; and (d) artificial recharge from irrigation, spreading operations, injection wells and leakage from water supply pipelines. These are now briefly discussed in turn.

(a) In general terms, the proportion of precipitation infiltrating to the water table depends upon a large number of factors, including the characteristics of the precipitation, vegetation, topography, soils and geology. Infiltration is discussed in detail in Section 5.4.2. Recharge to the groundwater must traverse the unsaturated zone; some water may be intercepted by field drains or by layers of lower permeability leading to lateral throughflow to streams and some water may be used by plants.

Valuable theoretical work by Freeze (1969) based on numerical simulation of infiltration into homogeneous isotropic soils indicated that recharge to the water table is likely to be greater for conditions with long-duration low-intensity rainfall, moist soils and a shallow water table, and for soils with a high hydraulic conductivity and/or a low specific water capacity. Subsequent comparisons between laboratory data and field observations in the Canadian prairies (Freeze and Banner, 1970) showed, however, that infiltration to the water table was isolated both in space and time; it normally resulted only from very intense rainfall and favourable conditions of water table depth, soil type and antecedent soil moisture, and spring snowmelt was the major source of groundwater recharge on the prairies, particularly in topographic depressions. Their work demonstrated that attempts to estimate groundwater recharge simply from a knowledge of the texture and saturated hydraulic conductivity of the overlying soil are likely to fail. It is necessary to know the depth to the water table and to have detailed information on the functional relationships between matric potential, hydraulic conductivity and water content. Very small differences in these properties can result in large differences in the hydrological response of similar field soils to the same infiltration event (Freeze and Banner, 1970).

Investigating the mechanism of recharge by infiltration, Horton and Hawkins (1965) described a process of displacement whereby the water that is added to the water table during rainfall is not 'new' rainfall but previously stored rainfall that has been displaced downwards by successive bouts of infiltration. This process, which was referred to as *translatory flow* by Hewlett and Hibbert (1967) and is also called *piston flow* (see also Section 7.4), undoubtedly helps to explain the often rapid response of water tables to precipitation, even in low permeability material, especially when the soil water content is at field capacity or wetter. One result of translatory flow is that even deep water table levels may respond rapidly to precipitation even though rates of downward percolation are very low. The movement of water through the unsaturated zone to the water table has been studied using tracers (Section 8.5.4). Instead of a slow and uniform percolation of water down to the saturated layers it is known that in many soils there are preferential flow paths down cracks, fissures and decayed root channels (Section 5.5.2 under 'Macropores'). These may be important in enabling pollutants such as fertilisers, pesticides and bacteria to bypass the filtering and purifying medium of the soil and be directly transmitted down to the groundwater.

(b) Where groundwater occurs in direct contact with surface water bodies such as lakes, ponds and streams, there will normally be a movement of water between the two water bodies. Either flow will take place from the lake or stream to the groundwater body, in which case it is known as *influent seepage*, or the reverse movement, *effluent seepage*, will occur, in which case groundwater seeps into and adds to the volume and flow of the surface water body. Stephenson (1971) and Winter (1976) studied the interaction between lakes and groundwater and showed the importance of topography and geology in determining the rate and direction of groundwater movements.

The seepage relationship between surface and undergound water is seldom static and will also be subject to changes with the changing levels of, say, a stream and the adjacent water table so that, in a matter of a few hours, influent seepage may supersede effluent seepage and then, in turn, be replaced once more by the latter. Thus, in Fig. 6.5a, the normal humid situation of effluent seepage is represented by water table profile A. Occasionally, during periods of heavy rainfall or after snowmelt, the water levels in the stream channels rise above those of the adjacent water table (profile B), resulting in groundwater recharge through influent seepage. During the summer, withdrawals of groundwater by evaporation may induce groundwater recharge, as shown in profile C.

In arid and semi-arid areas, where evaporation losses normally far exceed precipitation, direct recharge from widespread infiltration of precipitation is of limited importance. Lloyd (1986) considered that indirect methods of local recharge are the major form of the (limited) current recharge in desert areas. He argued that channel bed transmission losses during floods is the main form of recharge, together with localized ponding in depressions and fissure bypass flow in

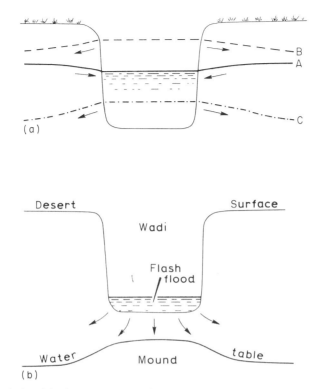

Figure 6.5 *The relationship between groundwater and surface water:* (*a*) *humid area,* (*b*) *arid area. See text for explanation.*

high-intensity storms. The few-perennial or semi-permanent streams that do exist are allogenic, and almost without exception lose water by influent seepage throughout the 'desert' sections of their courses. The rate and amount of groundwater recharge resulting from such seepage depends partly upon channel characteristics, such as the length of the wetted perimeter and the permeability of the channel bed, which may be reduced by fine silt deposits, and partly upon water characteristics such as temperature, with seepage increasing as temperature increases. Other desert streams are ephemeral, flowing only in flash floods after heavy rainfall, and in these cases the total flow may be completely absorbed by evaporation and influent seepage within a short distance. Influent seepage from ephemeral streams may lead to the development of groundwater mounds beneath surface channels and depressions, as shown in Fig. 6.5b, the mounds often having a markedly lower salt content than the groundwater farther away from the seepage areas (Bergstrom and Aten, 1965). Groundwater mounds, of course, may also be formed where a shallow water table occurs in humid areas, as a result of the concentration of runoff in surface depressions.

(c) It has already been noted that confining beds rarely form an absolute barrier to water movement, so that there is almost always some slow drainage of water from adjacent aquifers through the intervening aquitards. It is, however, very difficult to quantify the amount of water involved in this type of movement, and the estimates that are normally made of this component of the groundwater-balance equation may often be considerably in error. However, since the rate of movement of water through an aquitard is likely to be relatively very slow, the importance of this component is probably comparatively small, except in specific local conditions. Particularly in arid areas, recharge may also result from groundwater inflow from higher level groundwater basins within a complex system.

(d) In terms of its total contribution to the groundwater balance, artificial recharge is virtually insignificant; in some areas of water shortage, however, it has become increasingly important in recent years. Worldwide, nearly 250 million hectares are now irrigated for agriculture, predominantly in China, India, the USA and the USSR (Kharchenko and Maddock, 1982). In many irrigated areas inadvertent artificial recharge results from the leakage of irrigation canals or from the deep percolation of water applied in excess to the irrigated crops. Nace (1969) estimated that annual recharge from these sources is equivalent to between a few centimetres and a metre of precipitation. For example, in the south-western part of the the Indus valley in India, the water table has risen by about 0.5 m per year since the start of the century as a result of percolation from the irrigated land and leakage from the system of irrigation canals (Kharchenko and Maddock, 1982). As the water table approaches the ground surface it may bring dissolved salts with it and so create problems of salinity as well as of waterlogging. In some areas leakage from mains water supply pipes may be significant. In England and Wales it has been estimated that about 20–30 per cent of the water in the public water supply system is lost to leakage. This is equivalent to just over half of the total groundwater abstracted for all purposes (Price and Reed, 1989), and the resulting groundwater recharge is concentrated in urban areas. A number of major cities including New York, Paris, London and Cairo are experiencing rising groundwater levels due to a decline in local abstractions and possibly to leakages from the water mains. This rise has resulted in problems of flooding of basements, tunnels and subways (Hurst and Wilkinson, 1986). Conversely, urban areas have large impermeable surfaces which act to reduce the natural infiltration into the soil below.

Discharge
The main components of groundwater discharge are (a) evaporation, particularly in low-lying areas where the water table is close to the ground surface; (b) natural discharge by means of spring flow and effluent seepage into surface water bodies; (c) groundwater leakage and outflow through aquitards and into adjacent aquifers; and (d) artificial abstraction.

(a) The role of evaporation in groundwater discharge is relatively complicated and it is necessary to consider not only its indirect and direct effects upon variations of groundwater level but also factors such as the effect of water table depth on the rate of evaporation losses.

   The main indirect effect of evaporation concerns the reduction of soil water content and the consequent effectiveness of rainfall, which have been discussed above. Short-term direct effects are evidenced in the diurnal fluctuation of shallow water tables which has been observed on numerous occasions during periods when evaporation rates are high (Godwin, 1931; Troxell, 1936). In valley bottom areas, particularly, losses of groundwater by evaporation during the hottest part of the day may exceed the rate at which groundwater inflow from surrounding higher areas takes place and, as a result, the water table falls. During the evening and at night, when the plant stomata close, the evaporation rate is much reduced, and it will be exceeded by groundwater inflow, so that the water table level will recover. This regular diurnal rhythm will be maintained for much of the summer, although it will be interrupted by periods of rainfall and reduced evaporation (see Fig. 6.6).

Provided that the night-time recovery of the water table level equals the daytime drawdown, the long-term effects of evaporation may not be readily apparent. In the situation illustrated in Fig. 6.6, however, the daily drawdown exceeded the subsequent recovery for much of the time, with the result that, over a period of about two weeks, the average water table level fell significantly. From the discussions of soil water movement in Chapter 5, it will be apparent that this process cannot be continued indefinitely, because eventually the water table will drop to such a level that the capillary rise of water will be unable to satisfy the demands of the transpiring vegetation at the soil surface. The long-term effect of evaporation in these circumstances would be an exponential reduction of groundwater levels similar to that shown by the composite curves in Fig. 4.10.

(b) Effluent seepage, either directly into surface water bodies or as seepages and springs at the ground surface, is normally the major form of groundwater discharge and occurs where, for a variety of reasons, the upper surface of the zone of saturation intersects the ground surface. This seepage is very important for sustaining the baseflow in rivers during dry weather periods. In humid areas where rainfall exceeds evaporation losses, the normal relationship between surface water and groundwater is that depicted by profile A in Fig. 6.5a, where the water table slopes gently towards the stream, thereby facilitating the continuous discharge of groundwater into the surface channel. In this situation, the rate of seepage will obviously depend mainly on the hydraulic gradient which will be represented by the

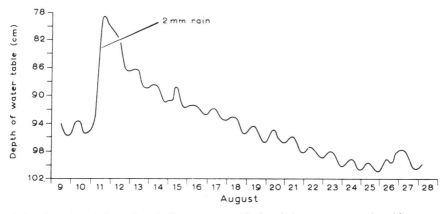

Figure 6.6 *Diurnal variation of a shallow water table level due to evaporation (from an original diagram by Meyboom, 1965).*

difference in elevation between the stream surface and the adjacent water table. Seasonal modifications, represented by profiles B and C, which result in influent seepage, have already been discussed in Section 6.4.3 in relation to groundwater recharge.

Although major streams and other water bodies are normally also major sinks for groundwater discharge, there are many instances where broadly distributed seepages occur over the lower valley slopes as well (Hubbert, 1940; Toth, 1962, 1963). Springs are often distinguished from seepage areas on the basis that a spring is a concentrated discharge of groundwater which appears as a definite flow of water at the surface, whereas a seepage area indicates a slower movement of groundwater to the surface (Todd, 1980), although for the purposes of this discussion the difference is a matter of degree only.

The discharge of most springs is variable to a greater or lesser extent; those that flow throughout the year are classed as *perennial*, while those that flow for only part of the year are normally referred to as *intermittent* springs. The variations in flow are directly related to variations in groundwater storage, and for many aquifers this relationship may be represented by a straight line on a log-log plot. It follows, therefore, that spring flow from thick porous aquifers having a very large storage capacity will tend to be relatively constant throughout the year, since the volume of storage change will probably represent a comparatively small proportion of the total storage volume of the aquifer. Similarly, 'artesian' spring flow from confined aquifers will not vary greatly from one season to another. In the case of thin aquifers, however, particularly veneers of superficial deposits such as glacial gravels, changes of storage in relation to total capacity are likely to be large. Therefore outflow tends to be very variable through the year, sometimes occurring only for short periods after rainfall in the case of, say, a spring at the foot of a scree slope.

(c) Groundwater leakage through aquitards and between adjacent aquifers was briefly discussed in relation to groundwater recharge. Clearly, a movement of water that represents a source of recharge to one part of a groundwater system must necessarily represent a discharge from another part of the system.

(d) Groundwater is abstracted for a wide range of water supply purposes—industrial, agricultural and domestic. Its generally good quality makes it particularly suitable for drinking water. Groundwater supplies over 90 per cent of the drinking water in West Germany, Austria and Denmark, and over 70 per cent in Belgium, Switzerland, Hungary and the USSR (Margat, 1985). Groundwater abstraction has had a profound effect on storage and storage changes in some areas. If excessive abstraction takes place from a large number of wells over a prolonged period, it will result in a gradual lowering of the potentiometric surface over a wide area, and this has, in fact, happened in many major groundwater supply areas. In Britain, for example, the eastern part of the London Basin has had substantial groundwater abstraction beginning from about the 1820s, causing a steady decline in the groundwater levels although, more recently, there has been a rise in levels due to a decline in extractions and, more controversially, probably an increase in leakage inputs from the mains water supply network (Marsh and Davies, 1984) (Fig. 6.7). In other areas groundwater abstraction has resulted in the incursion of inferior

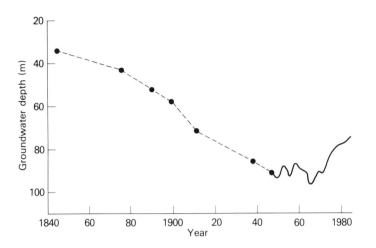

Figure 6.7 *Water table depth under Trafalgar Square, London, 1845–1986, showing the steady decline to the 1960s due to groundwater abstraction and the more recent partial recovery (data from British Geological Survey, national groundwater level archive, Institute of Hydrology, 1988).*

quality, saline, groundwater from adjacent aquifers or by direct seepage from the oceans along coastal areas (Section 6.5.8).

In Britain, groundwater supplies about 30 per cent of the public water supplies, although this figure hides regional variations, ranging from 70 per cent in parts of south and east England to under 10 per cent in Wales and Scotland. The principal aquifers are the Cretaceous chalk in south and east England and the Permo-Triassic sandstones of the Midlands (Fig. 6.8).

Storage coefficient
Now that the components of storage change have been described in largely qualitative terms, it remains to discuss the mechanism of storage change in both unconfined and confined conditions. The *coefficient of storage*, or *storativity*, of an aquifer is defined as the volume of water that an aquifer releases from, or takes into, storage per unit surface area of aquifer per unit change in head. It is, therefore, the ratio of a volume of water to a volume of aquifer (Lohman, 1972). This is illustrated in Fig. 6.9 in which the volume of water released from storage in the aquifer prism divided by the product of the prism's cross-sectional area and the decline in head results in a dimensionless number, i.e. the coefficient of storage. In unconfined conditions (Fig. 6.9b) the coefficient of storage approximates the specific yield (Section 6.4.2), provided that gravity drainage is complete, and it normally ranges from about 0.01 to 0.3 (Heath, 1983). In confined conditions (Fig. 6.9a), where no dewatering of the aquifer occurs, the volume of water released for a unit decline of the piezometric surface may be attributed to compression of the granular structure of the aquifer and to the expansion of the water itself. Although definite limits are difficult to establish, the storage coefficients of confined aquifers may range from about 0.00001 to 0.001 (Heath, 1983).

Clearly, there are significant differences between the mechanism of storage changes in

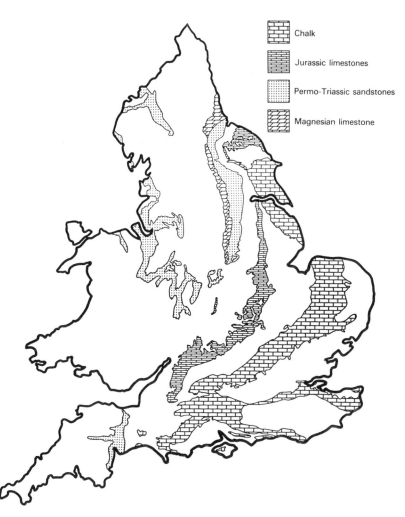

Chalk

Jurassic limestones

Permo-Triassic sandstones

Magnesian limestone

Figure 6.8 *Principal aquifers in England and Wales (based on an original map by IH, 1988).*

confined and unconfined conditions, and some of the more important of these differences will now be discussed.

Storage changes in unconfined aquifers

It will be apparent from the foregoing discussions that storage changes in unconfined conditions are relatively uncomplicated and may be directly reflected in variations of groundwater level. When recharge exceeds discharge, water table levels will rise, and when discharge exceeds recharge, they will fall. It is worth noting, though, that when the water table falls, much of the water actually comes from above the original water table level, as drainage from the capillary zone. Since, in most natural circumstances, recharge to and discharge from the same groundwater system occur simultaneously, groundwater level fluctuations, in fact, reflect the net change of storage resulting from the interaction of

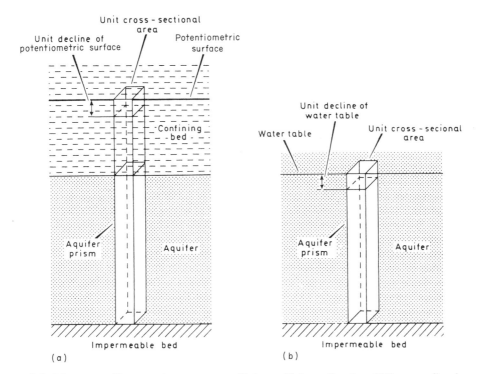

Figure 6.9 *Diagram to illustrate the storage coefficient of (a) confined and (b) unconfined aquifers (from an original diagram by Todd, 1980).*

these two components. The study and interpretation of water table fluctuations thus forms an integral part of the study of groundwater storage.

In many areas water table fluctuations tend to follow a fairly rhythmic seasonal pattern. In the British Isles, for example, high water levels occur during the winter months and low levels during the summer months, and it is on the basis of this broad differentiation that the hydrological or water year has been delimited as beginning on the first day of October and ending on the last day of September. The water table levels illustrated in Fig. 6.10 refer to a chalk aquifer well in southern England and may be considered as generally typical. In some areas, considerable attention has been paid to the study of groundwater regimes, and to the extent to which the normal climatically determined regime may be modified by artificial factors such as irrigation and drainage (Brown *et al.*, 1972; Kharchenko and Maddock, 1982).

Seasonal fluctuations, reflecting as they do seasonal changes in storage, are normally of considerable interest to the hydrologist, in the sense that they represent large-scale contrasts in water availability for both natural and artificial uses. Short-term fluctuations, usually on a much smaller scale, may also be of relevance, however, in particular circumstances. Thus, in many coastal areas regular short-term fluctuations of water table level are associated with tidal movements since, if the sea level varies with a simple harmonic motion, a train of sinusoidal waves is propagated inland from the submarine outcrop of an aquifer (Todd, 1980). Investigations of this phenomenon have shown that

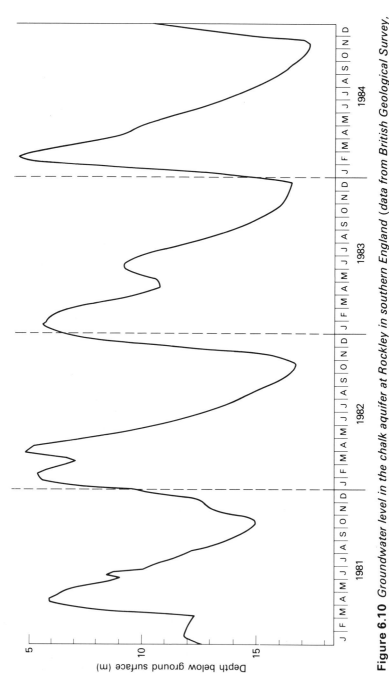

Figure 6.10 *Groundwater level in the chalk aquifer at Rockley in southern England (data from British Geological Survey, national groundwater level archive, Institute of Hydrology, 1988).*

the groundwater fluctuations are of decreased amplitude and lag behind the tidal fluctuations, the extent of the reduction and lag depending largely on distance from the sea and the ease with which groundwater can move through the aquifer.

Long-term irregular or secular fluctuations of water table level are also commonly experienced and are often of considerable hydrological significance. These are mainly associated with secular variations of rainfall (Section 2.5.3). Numerous investigations of this phenomenon have been made, especially in the United States, since the classic analyses by Jacob (1944) of water levels and precipitation on Long Island.

The relationship between storage and water level is, however, complicated by the fact that the latter responds to factors other than storage changes, for as Meinzer (1932) noted, 'The water level in a well is sensitive to every force that acts upon the body of water with which the well communicates'. Although this is particularly true in the case of confined groundwater, in some circumstances it may also apply to unconfined groundwater. Thus, in the late nineteenth century, experiments by Baldwin Latham in the North Downs near Croydon indicated that spring flow increased with a sudden fall of barometric pressure, presumably because of a slight increase in water table level, and spring flow decreased with a rise in barometric pressure (Slichter, 1902). The actual magnitude of water table variations must, however, have been very small.

Water level change maps may be used to calculate changes in the saturated volume of an unconfined aquifer and, if the storage coefficient is known, the actual change in groundwater storage. Such maps are constructed by plotting the change of water levels over a period of time at a number of wells. Over short periods, data from the same wells can normally be used but for longer periods initial and terminal water level data may be from different sets of wells. The maps of present and past water levels may be superimposed and the water level changes at contour intersections transferred to a third map on which lines of equal water level change can be drawn (Davis and De Wiest, 1966). This procedure is illustrated in Fig. 6.11 which also shows that water level change maps may be useful in delimiting the local effects of recharge or discharge, which are often difficult to detect simply from a comparison of successive water table maps. Thus in Fig. 6.11 the effects of recharge, which show only as a deflection of the water table contours, are well defined by the curved isopleths in the water level change map.

Storage changes in confined aquifers

In confined aquifers the relationships between changes in potentiometric level and changes of groundwater storage are greatly complicated by the compressibility and elasticity of the aquifers, whereas under water table conditions aquifer compressibility is virtually negligible with respect to gravity drainage of the interstices. Before the classic work of Meinzer (1928) on the Dakota sandstone of North Dakota, it was generally assumed that confined aquifers were rigid incompressible bodies and that changes in pore water pressure were not accompanied by changes in pore volume. Meinzer studied the long-period excess of discharge over natural recharge due to pumping from the sandstone. He found that the difference could not be accounted for by the compressibility of the water alone and that, therefore, the pore space of the sandstone was presumably reduced to the extent of the unexplained excess volume of water. Although later fieldwork discovered significant inflows to the aquifer that were unknown to Meinzer, the concept

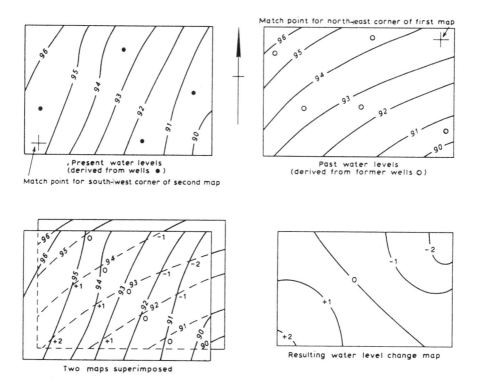

Figure 6.11 *Diagram to illustrate the construction of a water level change map by superimposing water level contour maps at different dates (from an original diagram by Davis and De Wiest, 1966).*

that aquifers are subject to compression and elasticity is now an established fact (Price, 1985). Meinzer's pioneering work was subsequently further developed by Theis (1935) and by Jacob (1940), who also noted that water may be derived from the compression of adjacent and included clay beds in interbedded aquifers. Subsequent valuable contributions have been made by Lohman (1961) on elastic compression and Poland (1961) on permanent compression.

Within a confined aquifer there are *intergranular pressures* in the solid phase at the points of contact between individual particles, resulting from the weight of the overlying deposits, and there are *hydrostatic pressures* due to the weight of the contained pore water. The former are referred to as the effective stress and the latter as neutral stress, and their sum represents the total vertical stress acting on a horizontal plane at any depth. The neutral stresses act equally on all sides of the solid particles of the matrix and although they may compress each grain, they have little effect upon the overall porosity of the aquifer. In practice, all measurable effects such as compression and expansion result from changes in the effective stress; i.e. in order for an aquifer to undergo compression, there must be an increase in the grain-to-grain pressures within the matrix and, conversely, in order for it to expand, there must be a decrease in the grain-to-grain pressure (Domenico, 1972). Thus, when water is pumped from a confined aquifer, lowering the potentiometric head, there is no change in the total stress, so that the reduction in pore water pressure must result in a proportional increase in that part of the total load carried by the grain structure. Owing to the reduced fluid pressure the water expands to the extent permitted by its elasticity, while compression of the aquifer results in reduced porosity and the consequent release of water from the saturated volume. If pumping ceases and the pressure is restored, the aquifer grains gradually return to their original position and the water is compressed. If the aquifer is perfectly elastic and the water levels in the recharge and discharge areas do not change, the original potentiometric head will ultimately be restored (Walton, 1970). Some aquifers and aquitards may not, however, be perfectly elastic, and especially in those with a high clay content there may well be some residual (permanent) compaction.

It will be clear from the foregoing and from earlier discussion of the storage coefficient that comparatively small yields of water from a confined aquifer may be accompanied by large variations in the potentiometric level compared with the corresponding water table fluctuation in an unconfined aquifer. It will also be evident that any variation of loading on a confined aquifer may result in fluctuations of the potentiometric surface. Such variations of loading may result from factors including barometric changes (Jacob, 1940; Parker and Stringfield, 1950; Clarke, 1967) and variable tidal and gravitational loading (Roberts, 1883; Robinson, 1939), and, in certain circumstances, from the occurrence of an earthquake (Scott and Render, 1964) or a nuclear explosion (Ineson, 1962); in each case they may provide valuable information about the elastic and storage properties of the aquifers concerned.

With a continuing excess of discharge over recharge to a confined aquifer the ever-increasing intergranular pressures and the resulting compression of the aquifer may ultimately result in the subsidence of the overlying ground surface. Substantial subsidence resulting from reductions of groundwater storage have been recorded in many areas including California (Poland and Davis, 1969), Venice (Carbognin and Gatto, 1986), Bankok (Bergado et al., 1986), Mexico City (Figueroa Vega, 1977) and Shanghai (Guangxiao and Yiaoqi, 1986).

The amount of subsidence at any location depends upon the subsurface lithology, the thickness of the compressible units and their storage characteristics, and upon the magnitude and duration of the decline in head. Almost all the main areas suffering from subsidence due to groundwater extraction are underlain by deposits of young, poorly consolidated material of high porosity, and much of the subsidence occurs due to compression of the clayey aquitards (Poland, 1984). Subsidence is normally inelastic and permanent, and in the worst affected areas may amount to between 10 and 5 per cent of the observed decline in potentiometric level (Walton, 1970).

Just as water table maps may be used to throw light on storage changes in unconfined aquifers, so potentiometric maps may be used in the case of confined aquifers, although since the potentiometric surface is determined both by water pressure and the elevation of the aquifer, it is misleading and erroneous to imply that there is one potentiometric surface for a given aquifer. Normally, potentiometric surfaces in confined aquifers are much smoother than the water tables in unconfined aquifers since local changes in head will be more rapidly propagated as a recharge or discharge wave (Davis and De Wiest, 1966). Used with caution, therefore, potentiometric change maps may provide valuable information about storage changes.

6.5 Groundwater movement

Groundwater taking an active part in the hydrological cycle is in more or less continuous motion from areas of recharge to areas of discharge in response to the
prevailing hydraulic gradient. Hydrological interest is largely concerned with the speed
and direction of groundwater movement, and it is mainly upon these two factors that
attention will be concentrated in the following pages. It is commonly remarked that not
only is the rate of movement of groundwater very slow, compared with that of surface
water, but also that it is very variable. Buchan (1965), for example, quoted rates of
groundwater movement through permeable strata ranging from as low as a fraction of a
millimetre per day, in some fine-grained pervious rocks, to as much as 5500 metres per day
through fissured chalk in Hertfordshire, England.

The direction of groundwater movement is similarly variable since, like surface water,
groundwater tends to follow the line of least resistance. Other things being equal,
therefore, flow tends to be concentrated in areas where the interstices are larger and better
connected, and the hydrologist's problem is to locate such areas, often from rather scanty
geological information. Theoretical analyses commonly assume ideal and greatly simplified conditions, and the results from them may be difficult to apply in field conditions.
These often assume that the medium under consideration is homogeneous and isotropic,
and generally attempt to define more or less complete and independent flow systems, e.g.
bounded by impermeable beds. In most real situations, however, flow systems are
bounded by semi-permeable rather than by completely impermeable beds, so that very
complex and widespread flow systems develop. It is therefore encouraging to note that
there has been a considerable growth of interest in the definition and analysis of regional
groundwater flow systems and a corresponding decline in the trend towards increasingly
sophisticated, idealized solutions to problems in the field of well hydraulics. Even so,
simplifying assumptions are normally essential, and while these may be reasonable for
many real situations, it should be emphasized that important 'untypical' groundwater flow
systems are found in, say, limestone (Legrand and Stringfield, 1973; Price, 1987) and
volcanic rocks (UNESCO, 1984).

The direction and rate of groundwater movement in a porous medium may be calculated
from the prevailing hydraulic gradient and the hydraulic conductivity of the water-bearing
material by the use of the Darcy equation.

6.5.1 *Darcy's law*

The main features and limitations of the Darcy equation were discussed in Section 5.4.1 in
relation to unsaturated flow, and so the present discussion will therefore be brief. Most
groundwater movement takes place in small interstices so that the resistance to flow
imposed by the material of the aquifer itself may be considerable, with the consequence
that the flow is *laminar*, i.e. with successive fluid particles following the same path or
streamline and not mixing with particles in adjacent streamlines. As the velocity of flow
increases, especially in material having large pores, the occurrence of turbulent eddies
dissipates kinetic energy and means that the hydraulic gradient becomes less effective in
inducing flow. In very large interstices, such as those found in many limestone and
volcanic areas, groundwater flow is almost identical to the turbulent flow of surface water.

The law that expresses the relationship between capillary or laminar flow and the hydraulic gradient was stated by Poiseuille (1846), and is actually a special case of Darcy's law (Hubbert, 1956). Later, Darcy (1856) confirmed the application of this law to the movement of groundwater through natural materials and, for hydrologists, the law has since become associated with his name.

Darcy's law for saturated flow may be written as

$$v = -K\left(\frac{\delta h}{\delta l}\right) \tag{6.2}$$

where v is the macroscopic velocity of the groundwater, K is the saturated hydraulic conductivity and $\delta h/\delta l$ is the hydraulic gradient comprising the change in hydraulic head (h) with distance along the direction of flow (l). The negative sign is to indicate that flow is in the direction of decreasing head. The total head is defined as

$$h = \psi + z \tag{6.3}$$

where ψ is the pressure head and z is the elevation head above a selected datum (Fig. 6.12). Head may be converted to potential (ϕ) by applying the gravitational constant (i.e. $\phi = gh$). The reader should consult Section 5.3.5 for a discussion of the potential energy of water and the units of measurement used.

Over the years there has been some confusion in the terminology relating to groundwater flow in general and to Darcy's law in particular. The *hydraulic conductivity*, K, in the Darcy equation refers to the characteristics of both the porous medium and the fluid. This is virtually synonymous with the earlier term coefficient of permeability. It should not be confused with the *intrinsic or specific permeability*, usually denoted as k, which depends only upon the characteristics of the porous medium itself.

As was emphasized in the discussion of unsaturated flow, the Darcy equation yields only a macroscopic velocity value through the cross-sectional area of solid matrix and interstices. Clearly, flow velocities through the interstices alone will be higher than the macroscopic value, and since the interstices themselves vary in shape, width and direction, the actual velocity in the soil or rock is highly variable. Furthermore, due to the

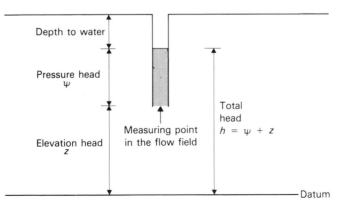

Figure 6.12 *Diagram showing the elevation head (z), pressure head (ψ) and total head for a point in a flow field (from an original diagram by Domenico, 1972).*

tortuous nature of the flow path of a water particle around and between the grains in an aquifer, the actual distance travelled exceeds the apparent distance given by the measured length of the porous medium in the average direction of flow.

Another factor complicating the field application of the Darcy equation is that hydraulic conductivity is often markedly anisotropic, particularly in fractured and jointed rock. Anisotropic versions of the Darcy equation have been presented by a number of investigators (Freeze and Cherry, 1979).

Extreme flow velocities may result in deviations from Darcy's law. The effect of turbulence in modifying the relationship between the hydraulic gradient and high rates of flow has already been mentioned. At the other extreme, some investigators have claimed that in clay soils, with small pores and low hydraulic gradients, the very low flow rates are less than proportional to the hydraulic gradient (Miller and Low, 1963; Swartzendruber, 1962). A possible explanation is that much of the water in such material is strongly held by adsorptive forces and may be more rigid and less mobile than ordinary water (Hillel, 1982).

Apart from minor deviations of this sort, however, the Darcy law can be successfully applied to virtually all normal cases of groundwater flow and is equally applicable to both confined and unconfined conditions. An understanding of most groundwater problems demands information not only about the velocity of water movement but also about the velocity of head transmission which is usually many hundreds of times faster. The velocity of head transmission is proportional to the square root of the hydraulic diffusivity (a), which is defined as

$$a = \frac{Kb}{S} \tag{6.4}$$

where K is the hydraulic conductivity, b is the saturated thickness of the aquifer and S is the coefficient of storage (Section 6.4.3 under 'Storage coefficient'). For confined conditions, this ratio therefore depends not only on the conductivity of the aquifer material but also on its elastic properties.

By itself, the Darcy law suffices to describe only steady flow conditions, so that for most field applications it must be combined with the mass-conservation law to obtain the general flow equation or, for saturated conditions, the Laplace equation. A direct solution of the latter equation for groundwater flow conditions is generally not possible so that it is necessary to resort to various approximate or indirect methods of analysis, some of which will be referred to later in Section 6.5.4.

6.5.2 *Factors affecting hydraulic conductivity*

Fundamental to the application of Darcy's law is a knowledge of the hydraulic conductivity of the saturated medium. The factors affecting hydraulic conductivity may be conveniently grouped into those pertaining to the water-bearing material itself and those pertaining to the groundwater as a fluid.

One of the most important of the aquifer properties concerns the geometry of the pore spaces through which groundwater movement occurs. Although methods have been evolved to determine the pore size distribution of a porous medium directly (Childs and Collis-George, 1950), the difficulties in so doing have, for the most part, resulted in indirect approaches whereby the pore space geometry is related to factors such as grain size distribution on the not always very sound assumption that there is a definable relationship between these properties and pore size distribution.

A second aquifer property relates to the geometry of the rock particles themselves, particularly in respect to their surface roughness, which may have an important effect on the speed of groundwater flow.

Finally, an important influence is exerted by secondary geological processes such as faulting and folding, which may increase or decrease groundwater movement, depending on the lithology of the beds involved (Ineson, 1956), secondary deposition, which will tend to reduce the effective size of the interstices and, therefore, also the flow of water, and secondary solution in, for example, limestone, to which reference has already been made. Figure 6.13 illustrates the effect of factors such as these on groundwater movement in the chalk of East Anglia; the areas of high transmissivity (the product of hydraulic conductivity and the saturated thickness of the aquifer) tend to be related to topographic valleys, which in turn are associated with fold or fault structures, or with increased fissuring, and the map suggests that these relationships may continue beneath relatively impermeable overlying deposits. In the London Basin, the effects of folding are even more

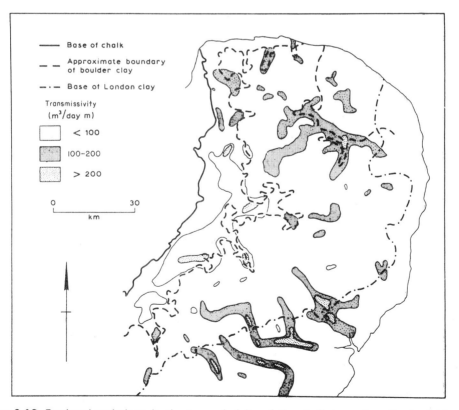

Figure 6.13 *Regional variations in the transmissivity of the chalk in eastern England (from an original diagram by Ineson, 1963).*

pronounced, and here synclinal areas in the chalk are associated with low permeability as a result of compaction, while anticlinal areas frequently tend to encourage high rates of groundwater flow (Ineson, 1963).

Numerous attempts have been made to represent porous media by idealized theoretical models that are amenable to mathematical treatment. In a comprehensive review, Scheidegger (1960) noted that since most natural porous media are extremely disordered, they are unlikely to be adequately represented by such idealized models and suggested that a statistically based model might be preferable.

The effects of fluid characteristics, such as density and viscosity, on hydraulic conductivity tend to be rather less important than the effects of the aquifer characteristics. Certainly, in relation to normal conditions of groundwater flow, the physical properties of the groundwater are likely to be influenced only by changes in temperature and salinity. Temperature, by inversely affecting the viscosity, has a direct influence on the speed of groundwater flow. Since, however, most groundwater is characterized by relatively constant temperatures, this factor is unlikely to be important except in special circumstances.

Variations of salinity are also unlikely to be significant in normal groundwater conditions. Where saline infiltration occurs, however, hydraulic conductivities may be affected both by changes in the ionic concentrations of the groundwater and also by the chemical effect of the saline water on the aquifer material itself, particularly where this is of a clayey nature (Ineson, 1956). Increasing salinity will increase water density and so may affect hydraulic heads and gradients.

Pumping tests of wells are the most reliable and widely used methods of determining the storage coefficient and transmissivity of an aquifer. Well flow equations have been developed for steady and unsteady flows, and most solutions can be reduced to convenient graphical or mathematical form for practical application. These methods are described in numerous groundwater texts (e.g. Todd, 1980; Freeze and Cherry, 1979; Davis and De Wiest, 1966) and will not be discussed further here.

6.5.3 *Flow nets*

Since it is impossible to observe groundwater flow directly, it is frequently convenient to make use of the relationship between flow and the hydraulic or potential gradient and thereby to examine two-dimensional groundwater flow indirectly by reference to the subsurface distribution of groundwater potential, which can be quite easily measured. Lines joining points of equal potential (ϕ) are known as *equipotentials*, and the potentiometric surface of an aquifer (confined or unconfined) above a datum plane may be contoured at regular increments of ϕ by a family of such lines (see Fig. 6.14). In accordance with Darcy's law, the maximum gradient potential, in which direction the water is driven, is perpendicular to the equipotential lines (Childs, 1969). The *streamlines* labelled '*S*' in Fig. 6.14 depict the direction of the force on the moving water and represent the paths followed by particles of water. Since at any one point the flow can only have one direction, it follows that streamlines never intersect. The network of meshes formed by the two families of equipotentials and streamlines is known as a *flow net* (Freeze and Cherry, 1979).

Flow nets not only show the direction of groundwater movement but can also be used

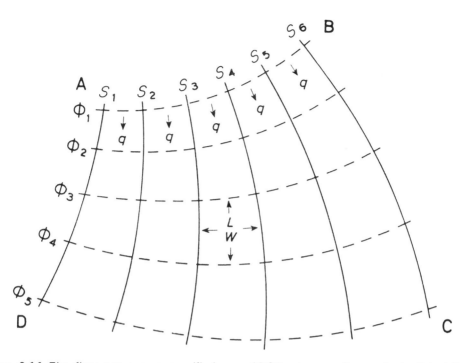

Figure 6.14 *The flow net over a specified area ABCD, showng the equipotentials* (ϕ) *and streamlines* (S). *See text for explanation (from an original diagram by Childs, 1969).*

to estimate the rate of flow, either by graphical construction or mathematically. The zone between any pair of neighbouring streamlines is known as a *streamtube*, and at every cross-section of a streamtube the total rate of flow (q) remains the same. If adjacent equipotentials differ by the same increment of potential ($\Delta\phi = \phi_1 - \phi_2 = \phi_2 - \phi_3$, etc.) and the streamlines are chosen to be evenly spaced so as to give the same rates of flow in all streamtubes, one can apply Darcy's law to any of the elements of the flow net having width W and length L so that

$$q = K\Delta\phi\left(\frac{W}{L}\right) \qquad (6.5)$$

As Childs (1969) observed, with a porous medium having uniform hydraulic conductivity, the ratio of W to L, i.e. the shape of each of the rectangular elements of the flow net, is the same, since q and $\Delta\phi$ are the same for all.

Nearly all groundwater systems comprise both aquifers and aquitards, and involve flow through aquifers and across confining beds (Heath, 1983). Hydraulic conductivities in aquifers are usually several orders of magnitude greater than in confining beds and so offer least resistance to flow, the result being that head loss (and hence hydraulic gradient) becomes much less in aquifers. As Fig. 6.15 indicates, streamlines are refracted at the boundary between media of different permeabilities, in the direction that produces the shortest flow path through the confining bed. Streamline refraction permits the

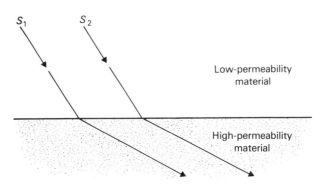

Figure 6.15 *Refraction of streamlines across a boundary between two materials of differing permeability (from an original diagram by Todd, 1980).*

conservation of fluid mass when flow takes place across the permeability boundary. Thus, in accordance with Darcy's law, other factors being equal, the higher the permeability, the smaller the area required to pass a given volume of water in a given time (Domenico, 1972). The streamlines are thus widely spaced in the low-permeability material and closely spaced in the high-permeability material.

The use of flow nets, which are commonly constructed in either a vertical plane or a horizontal plane (i.e. the potentiometric map), probably originated with Forchheimer in the nineteenth century (Maxey, 1969) but was revived again with important modifications by Hubbert (1940) and more recently, in connection with computer models of groundwater flow, by Toth (1962). With the growing interest in regional flow patterns the use of flow nets has received a new impetus.

6.5.4 *Two-dimensional groundwater flow*

Although virtually all groundwater flow is three dimensional, in practice it is usually both necessary and reasonable to consider flow as taking place in two dimensions rather than three. This simplification is ideally suited to the use of flow nets and also to the use of electrical analogues and parallel-plate 'Hele-Shaw' models of groundwater flow.

In the horizontal plane, two-dimensional flow is represented by the potentiometric map. A simple example for unconfined groundwater flow in a non-homogeneous aquifer is shown in Fig. 6.16 where the equipotentials ($\phi_1 - \phi_9$) represent the water table levels indicated in metres. Since the change in potential between adjacent pairs of equipotentials is equal and the hydraulic gradient varies inversely with the distance between equipotentials, then if inflow for any section is just balanced by outflow, the relative steepness of the hydraulic gradient reflects the hydraulic conductivity, as indicated in Darcy's law. Thus in Fig. 6.16 the hydraulic conductivity is lowest in the north-west of the area and increases towards the east and south, as is evidenced by the wider spacing of the equipotentials.

If Fig. 6.16 presented potential distribution in a homogeneous, rather than a non-homogeneous, aquifer, the variable spacing of the equipotentials would reflect a variation

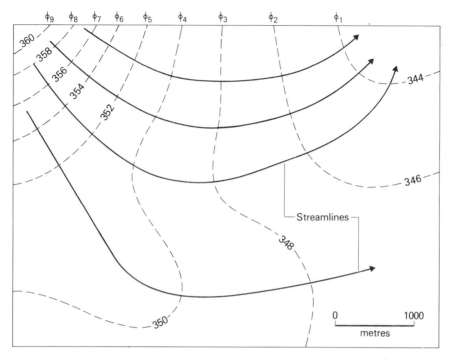

Figure 6.16 *Water table contour map showing the direction of groundwater flow.*

in groundwater flow—in this case a decrease of flow in the direction of flow, as represented by the decrease in hydraulic gradient towards the east and south.

Because of the readier availability of the appropriate data, groundwater flow is normally considered as a two-dimensional problem in the vertical, rather than the horizontal, plane, and in this case it is information on the rate of change of hydraulic head or potential with depth that facilitates the definition of the vertical groundwater flow pattern. If groundwater potential increases with depth, flow will be upwards (Fig. 6.17a) and if it decreases with depth, flow will be downwards (Fig. 6.17b). Under normal, or hydrostatic, pressure there is no change of groundwater potential with depth and therefore no groundwater movement in the vertical plane.

However, a zero change of potential with depth is also characteristic of situations where the direction of unconfined groundwater flow is approximately horizontal, and where as a result the equipotentials are approximately vertical and are labelled by the height at which they intersect the water table and where the potential gradient is simply the slope of the water table immediately above the given point (Childs, 1969). This might be the situation, for example, where an extensive thin permeable bed rests on an underlying horizontal impermeable bed and is an approximation proposed by Dupuit (1863) and elaborated by Forchheimer (1914) and subsequently known as the Dupuit–Forchheimer approximation. This is an empirical approximation to the actual flow field and ignores vertical flows, but in practice the results often compare favourably with more complex methods, and it is widely used in engineering calculations (Scheidegger, 1974).

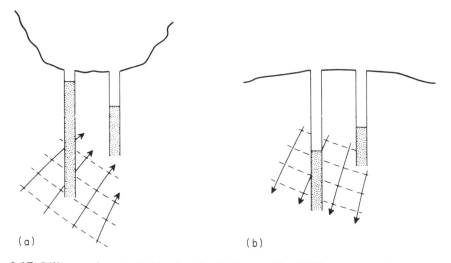

Figure 6.17 *Difference in water table elevation in* lined *wells of different depths in the presence of vertical groundwater flows: (a) potential increases with depth and (b) potential decreases with depth (from an original diagram by Domenico, 1972).*

6.5.5 *Unconfined groundwater flow—classical models*

Increasing emphasis on a hydrodynamic approach to the solution of groundwater problems (including the application of flow net techniques, which has been a notable feature of groundwater hydrology, has stemmed largely from the classic work of Hubbert (1940), who presented the groundwater system as a dynamic mechanism subject to all the requirements of the conservation equation within the context of accepted hydrodynamic principles (Maxey, 1969). Figure 6.18 illustrates the essential features of Hubbert's presentation for a homogeneous, isotropic material. Equipotentials are shown as broken lines and the value of hydraulic head for each line is equal to the elevation of its intersection with the water table. Streamlines, indicating the groundwater flow paths,

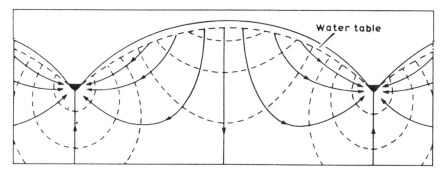

Figure 6.18 *Approximate groundwater flow pattern in a uniformly permeable unconfined aquifer. (From an original diagram by M. K. Hubbert,* Journal of Geology, **48**, *University of Chicago, 1940. Used with permission of the University of Chicago Press.)*

connect the source areas in which recharge is dominant with the sinks in which discharge is dominant. Under closed conditions, groundwater flow would ultimately result in the complete drainage of water from the topographic highs and the production of a flat surface of minimum potential energy (the hydrostatic condition) (Domenico, 1972). This tendency, however, is counteracted by continuous replenishment from precipitation. The result of this continuous movement and renewal is the flow pattern shown whose upper surface, the water table, is a subdued replica of the topography. The source areas are the topographic highs and in this diagram the sinks are shown as streams, and each groundwater flow cell is bounded by the lines of vertical flow beneath the groundwater divides and the sinks or by widely distributed impermeable beds or both.

Toth (1962) suggested that, whereas major streams may also be major groundwater sinks, as in the Hubbert model just discussed, for valleys having only low-order streams groundwater discharge is not concentrated at the stream but is broadly distributed on the down-gradient side of a mid-line between the valley bottom and the groundwater divide, as shown in Fig. 6.19. As can be seen, the mid-line is an approximately vertical equipotential about which the flow pattern is broadly symmetrical, giving a central area of lateral flow where groundwater potential does not vary with depth, an up-gradient area of downward flow or recharge where groundwater potential decreases with depth and the down-gradient area of upward flow or discharge where groundwater potential increases with depth, as in Fig. 6.17. Toth's work stimulated other contributions and in particular the development of a similar model for arid and semi-arid areas having the same three major components but differing somewhat as a result of physiographic and topographic conditions (Maxey, 1968; Maxey and Farvolden, 1965). As Domenico (1972) pointed

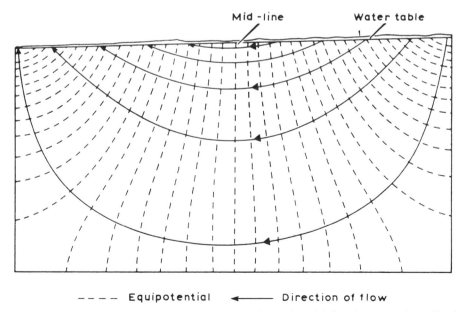

Figure 6.19 *Groundwater flow pattern symmetrical about the mid-line between the valley bottom and the groundwater divide. (From an original diagram by J. Toth,* Journal of Geophysical Research, **67**, *pp. 4380, Fig. 3.* © *1962 American Geophysical Union.)*

out, however, the concepts of Toth and subsequent investigators are by no means new but were first introduced by Meinzer (1917) to describe the flow system in Big Smokey Valley, Nevada.

The unconfined groundwater models discussed above all assume that the porous medium is hydrologically isotropic and homogeneous. In many areas this may not be an unreasonable assumption, for, as Maxey (1969) observed, even in the Great Basin area in which he worked, the rocks have been broken up to such an extent by tectonic stresses that the whole mountain mass may be regarded as a single hydrological unit, just as homogeneous and isotropic on a large scale as is a permeameter filled with sand on a small scale.

6.5.6 *Confined groundwater flow*

At the other extreme from the homogeneous isotropic unconfined groundwater situation is a field situation where, with alternating beds of markedly different lithology and permeability, groundwater is confined beneath an impermeable layer and the potentiometric surface of the flow field is completely independent of surface topography and of the configuration of the water table in the upper, unconfined groundwater body. Domenico (1972) suggested, however, that what is commonly found in actual conditions is neither a completely confined system nor a completely unconfined system, but rather a system of flow that possesses distinct characteristics of both extremes. For example, as has already been pointed out (Section 6.3), confining beds rarely form an absolute barrier to water movement so that there is normally some degree of hydraulic continuity, which suggests that the potential distribution with depth in a confined groundwater body is

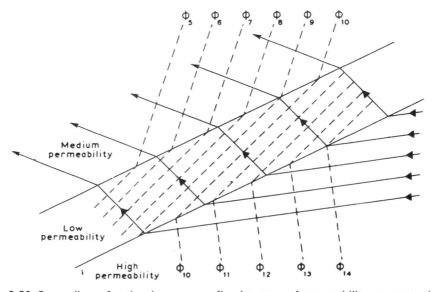

Figure 6.20 *Streamline refraction in an unconfined system of permeability contrasts.* (*From an original diagram by M. K. Hubbert,* Journal of Geology, **48**, *University of Chicago, 1940. Used with permission of the University of Chicago Press.*)

partly affected by the potential distribution of the overlying water table. On the other hand, in an apparently unconfined situation, the flow field may possess characteristics of confinement whenever flow is refracted on emerging from a low-permeability bed, so that it proceeds almost tangentially to the lower surface of that bed. This is clearly illustrated in Fig. 6.20, which represents an unconfined system with hydraulic continuity in the vertical direction. Streamlines in the high-permeability bed are almost tangential to the lower surface of the low-permeability bed. It will be observed that there is very little difference in groundwater potential along an imaginary vertical line passing through the high-permeability bed, but if this line is extended upwards, it crosses several equipotentials in the low-permeability bed and relatively few in the medium-permeability material. Thus, if a well was drilled along this line, there would be a large increase in potentiometric head when it first entered the high-permeability bed because static levels can establish themselves more rapidly here than elsewhere in the system. This increase in head is often attributed to confinement of water under pressure, although in reality it results from the movement of water through the low-permeability bed, which suggests that the conditions implied by the term confinement will arise when a unit of low permeability overlies a unit of high permeability (Domenico, 1972). The so-called artesian condition is an extreme case of this (Price, 1985).

6.5.7 *Regional groundwater flow*

Recognition of the blurred distinction between confined and unconfined groundwater flow has facilitated the definition and study of regional groundwater flow systems, which have been defined as large groundwater flow systems that encompass one or more topographic basins. This work owes much to the earlier contributions of Meinzer (1917) and especially of Hubbert (1940) who was concerned with applying his theory of steady-state groundwater flow to large-scale regional problems. However, as Freeze and Witherspoon (1966) pointed out, this work was overshadowed for a quarter of a century up to 1960 by work on the application of mathematical solutions of transient flow to well and wellfield hydraulics. During and since the 1960s, however, attention has once again been turned to the regional situation, with the groundwater basin as the unit of hydrological study. Especially prominent in this field were a group of Canadian hydrologists including J. Toth, P. Meyboom and R. A. Freeze.

Toth (1962, 1963) significantly extended Hubbert's work by introducing the concept that groundwater flow patterns can be obtained mathematically as solutions to formal boundary value problems. He assumed two-dimensional vertical flow in a homogeneous isotropic medium bounded below by a horizontal bed and above by the water table, which is a subdued replica of the topography. The lateral flow boundaries are the major groundwater divides. Toth (1963) defined a flow system as a set of flow lines (streamlines) in which any two flow lines adjacent at one point remain adjacent through the whole region and which can be intercepted anywhere by an uninterrupted surface across which flow takes place in one direction only. Toth (1963) recognized the three broad categories of flow system illustrated in Fig. 6.21. These are the *local* system, which has its recharge area at a topographic high and its discharge area at an adjacent topographic low, and is therefore identical to the classic Hubbert model shown in Fig. 6.18; an *intermediate* system having one or more topographic highs and lows located

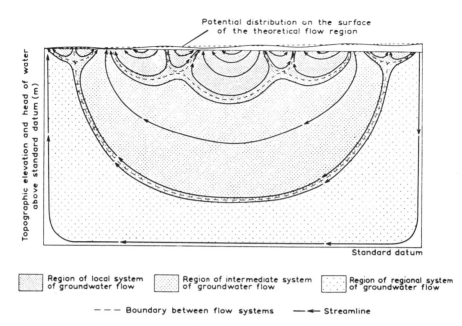

Figure 6.21 *Theoretical flow pattern and boundaries between local, intermediate and regional flow systems. (From an original diagram by J. Toth,* Journal of Geophysical Research, **68**, *p. 4807, Fig. 3.* © *1963 American Geophysical Union.)*

between its recharge and discharge areas; and a *regional* system, where the recharge area is at the main topographic high and the discharge area at the lowest part of the basin. Even in basins underlain by uniform material, topography can create complex systems of groundwater flow. Toth (1963) suggested that if local relief variations are negligible but there is a general topographic slope, only regional systems will develop. With increasingly pronounced local relief, deeper local systems will develop and extensive unconfined regional systems are unlikely to develop across the valleys of large rivers or pronounced watersheds. Under extended flat areas, which will be characterized by local waterlogging and groundwater mineralization from concentration of salts, neither regional nor local systems will develop.

Meyboom (1963, 1967a) described a general model of groundwater flow in a prairie environment which he designated the 'prairie profile' (Fig. 6.22). By definition this consists of a central topographic high bounded on both sides by areas of major natural discharge. Geologically the profile consists of two layers of different permeability, the upper layer having the lower permeability, with a steady flow of groundwater towards the discharge areas. The ratio of permeabilities is such that the groundwater flow is essentially downwards through the low-permeability material and lateral and upwards through the more permeable underlying layer.

Since most of the natural discharge occurs by means of evaporation, Meyboom examined areas in which this component of the groundwater balance appeared to be important. He concentrated particularly on the occurrence of willow rings, areas of saline soil which occur where there is a net upward movement of mineralized groundwater, as in

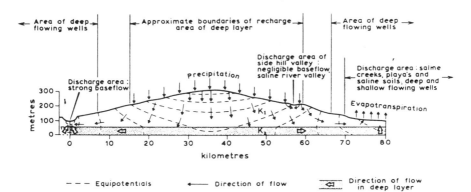

Figure 6.22 *Flow patterns and areas of recharge and discharge in the prairie profile. (From an original diagram by Meyboom, 1963. Reproduced by permission of Information Canada.)*

the major areas of regional groundwater discharge, and the occurrence of lakes and bogs and their relationship to groundwater flow (Meyboom, 1966, 1967a, 1967b).

The work of Toth and Meyboom was itself expanded and generalized in a major contribution by Freeze and Witherspoon (1966, 1967, 1968) which introduced a more versatile mathematical modelling technique based on numerical solutions. Their model is capable of determining steady-state flow patterns in a three-dimensional, non-homogeneous anisotropic groundwater basin, with any water table configuration given knowledge of the dimensions of the basin, the water table configuration and the permeability configuration resulting from the subsurface stratigraphy. Figure 6.23 shows three potential field diagrams which demonstrate the effect of topography and geology on regional flow systems. The equipotential net (dashed lines) has been produced from the numerical solution and the flow lines have been sketched in to indicate the direction of flow. The water table configuration is shown by the solid line at the top of the contoured section.

Figure 6.23a shows the flow through a homogeneous isotropic medium and illustrates that the existence of a hummocky water table configuration results in numerous subbasins within the major groundwater system. Figure 6.23b shows the effect on the potential field of a lenticular body of high permeability, given the same water table configuration as in Fig. 6.23a. In this case the flow lines are not shown although they may be readily envisaged by the reader. Finally, Fig. 6.23c shows the regional flow pattern in an area of sloping stratigraphy. In this case just two flow lines have been drawn in to illustrate that the difference of a few metres in the point of recharge will make the difference between the water entering a minor subbasin or the major regional system of groundwater flow.

6.5.8 *Groundwater flow in coastal aquifers*

The discharge of groundwater directly to the sea is very small in comparison with the outflow of rivers—amounting to only about 5 per cent of the total continental fresh water outflows (Zekster and Dzhamalov, 1981). It is, however, of much greater importance than this low value suggests, due to the high density of human population in coastal areas and

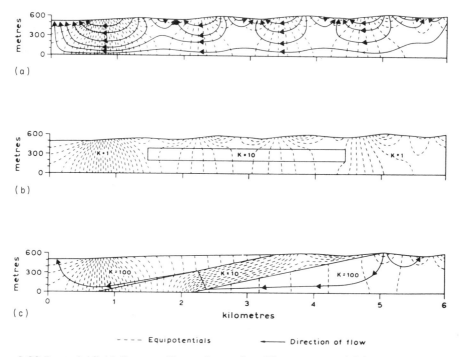

Figure 6.23 *Potential field diagrams illustrating regional flow patterns: (a) homogeneous, isotropic medium with a hummocky water table, (b) as above, but with a lens of high-permeability material and (c) an area of sloping stratigraphy. (From original diagrams by R. A. Freeze and P. A. Witherspoon,* Water Resources Research, **3**, *p. 625, Fig. 1, p. 629, Fig. 4, p. 630, Fig. 5. © 1967 American Geophysical Union.)*

the susceptibility of coastal aquifers to degradation by salt water intrusion (UNESCO, 1987).

In natural conditions there is a hydraulic gradient towards the coast and the resulting seaward groundwater flow effectively limits the subsurface landward encroachment of saline water. Thus in normal unconfined groundwater conditions, with a water table sloping towards sea level at the coast, the groundwater body takes the form of a lens of fresh water 'floating' on more saline water beneath. The position of the interface between the fresh and salt water was investigated independently by Badon Ghyben (1889) and Herzberg (1901). Assuming hydrostatic conditions and a negligible mixing zone the Ghyben–Herzberg relationship may be written as

$$h_s = \frac{\rho_f}{\rho_s - \rho_f} h_f = \alpha h_f \qquad (6.6)$$

where h_s is the depth of the fresh water below sea level, ρ_f is the density of fresh water, ρ_s is the density of sea water and h_f is the height of the water table above sea level. This relation indicates that groundwater is encountered, not at sea level but at a depth below sea level equivalent to about α times the height of the water table above sea level (Fig. 6.24a).

This represents the condition of approximately hydrostatic equilibrium between the

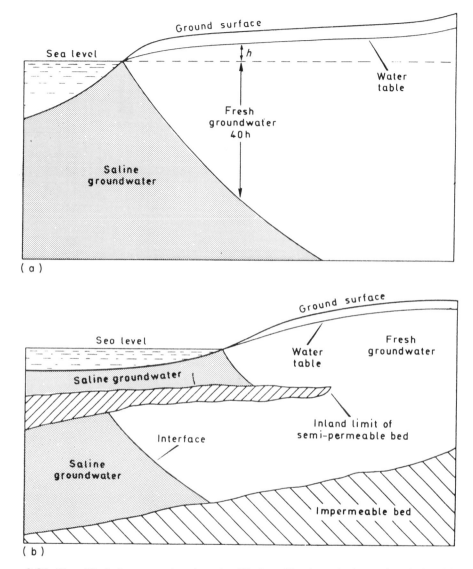

Figure 6.24 *Simplified diagrams showing the Ghyben–Herzberg hydrostatic relationship in (a) homogeneous coastal aquifer and (b) layered coastal aquifer. (The latter is based on an original diagram by M. A. Collins and L. W. Gelhar,* Water Resources Research, **7**, p. 972, Fig. 1. © 1971 *American Geophysical Union.)*

lighter fresh groundwater (density 1.00 g/cm³) and the heavier saline groundwater (density 1.02–1.03 g/cm³, depending on temperature and salinity), and for $\rho_s = 1.025$ the ratio $\alpha = 40$. In stratigraphically layered aquifers a number of investigators have indicated the presence of multiple saline wedges (Schmorak and Mercado, 1969; Collins and Gelhar, 1971). Figure 6.24b illustrates a simple situation in the presence of a limited semi-permeable layer and assuming steady flow and a sharp interface between fresh and saline

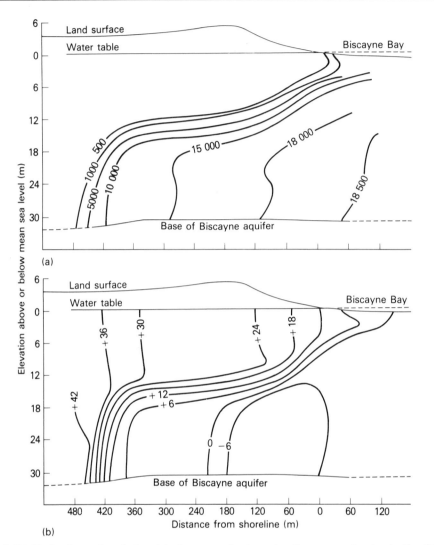

Figure 6.25 *Hydrodynamic relationship between fresh and saline groundwater in the Biscayne aquifer, Florida: (a) isolines salt concentration (ppm Cl⁻) illustrating the zone of diffusion and (b) equipotential lines (cm) in the zone of diffusion (from original diagrams by Kohout and Klein, 1967, and by Stringfield and Legrand, 1969).*

groundwater. Collins and Gelhar (1971) pointed out that whereas the saline wedge in the upper unconfined aquifer must intrude inland from the shoreline, the wedge in the lower aquifer may intrude even further inland or, as shown in the diagram, may exist seaward of the shoreline where a sufficiently large freshwater head is available at the landward end of the semi-permeable layer.

The assumption of a sharp interface between saline and fresh groundwaters is, in most cases, unrealistic. Tidal fluctuations as well as variations in recharge and discharge continually disturb the balance between the fresh water and the sea water and cause the interface to fluctuate. As a result of these fluctuations and the diffusion of salt water, the sharp interface is destroyed and a transitional zone of diffusion of brackish water is created (Stringfied and Legrand, 1971). This is well illustrated in Fig. 6.25a, which shows the zone of diffusion in the Biscayne aquifer near Miami, Florida.

Hubbert (1940) demonstrated that a state of dynamic rather than static equilibrium must exist at the interface or there would be no way for fresh water to discharge into the sea and that the depth to the interface is more than that indicated by the simple Ghyben–Herzberg relationship. For low hydraulic gradients the differences are small but for steep gradients the difference may be substantial. Cooper (1959) and Kohout (1960) described the flow pattern associated with this discrepancy as involving a continual flow of salt water from the sea floor into the zone of diffusion. Only in this way can the continuous discharge of salty water from the zone of diffusion into the sea be explained, the concentration gradient across the zone of diffusion being too weak to account for the transport of salts by means of dispersion or diffusion. Thus the salts are transported into the zone of diffusion largely by a flow of salt water with a consequent loss of head in the salt water environment (Cooper *et al.*, 1964). This is clearly illustrated by the equipotentials in the Biscayne aquifer (Fig. 6.25b) which shows that, in fact, negative heads in terms of sea water densities exist in the sea water environment. In this situation the interface between saline and fresh water will clearly occupy a position seaward of that estimated from the Ghyben–Herzberg relationship.

It will be apparent that any sustained reduction in fresh groundwater flow and the associated lowering of water table and potentiometric levels resulting from the excessive abstraction of groundwater from coastal aquifers will readily lead to the incursion of saline groundwater and to the long-term contamination of the aquifer. This problem has been investigated in many areas including the Netherlands, Spain, Israel, France, the USA, Italy and Britain (UNESCO, 1987). Numerical solutions for the movement and position of the interface between saline and fresh water include those of Pinder and Cooper (1970), Bear and Kapuler (1981) and Andrews (1981).

The existence of saline groundwater in some coastal aquifers may represent the residue of an earlier invasion of sea water, which occurred when the land was relatively lower and which now takes the form of a wedge of highly saline water beneath a superficial seaward-flowing body of fresh groundwater. The presence of saline water in the chalk along parts of the Humberside, Norfolk and Suffolk coasts of England may probably be explained in this way (Buchan, 1963), as may some of the saline groundwater along the Atlantic coast of the United States (Siple, 1965).

References

Andrews, R. W. (1981) Salt water intrusion in the Costa de Hermosillo, Mexico, *Ground Water*, **19**: 635–47.

Badon Ghyben, W. (1889) Nota in verband met de voorgenomen putboring nabij Amsterdam (Notes on the probable results of the proposed well drilling near Amsterdam), *Tijdschrift van het Koninklijt Inst. van Ingenieurs*, The Hague, pp. 8–22. Quoted by Todd (1980).

Bear, J. and I. Kapuler (1981) Numerical solution for the movement of an interface in a layered coastal aquifer, *J. Hydrol.*, **50**: 273–98.

Bergado, D. T., A. S. Balasubramanian and W. Apaipong (1986) Fluid monitoring of subsidence effects at AIT campus, Bankok, Thailand, *IAHS Publ.*, **151**: 391–404.

Bergstrom, R. E. and R. E. Aten (1965) Natural recharge and localization of fresh groundwater in Kuwait, *J. Hydrol.*, **2**: 213–31.

Brown, R. H., A. A. Konoplyantsev, J. Ineson and V. S. Kovalensky (eds) (1972) Groundwater studies, in *Studies and reports in hydrology*, No. 7, UNESCO, Paris.

Buchan, S. (1963) Geology in relation to groundwater, *Journal of the Institution of Water Engineers*, **17**: 153–64.

Buchan, S. (1965) Hydrogeology and its part in the hydrological cycle. Informal discussion of the Hydrological Group, *Proc. ICE*, **31**: 428–31.

Carbognin, L. and P. Gatto (1986) An overview of the subsidence of Venice, *IAHS Publ.*, **151**: 321–8.

Childs, E. C. (1969) *An introduction to the physical basis of soil water phenomena*, John Wiley and Sons, London, 493 pp.

Childs, E. C. and N. Collis-George (1950) The permeability of porous materials, *Proc. Roy. Soc., A*, **201**: 392–405.

Clarke, W. E. (1967) Computing the barometric efficiency of a well, *Amer. Soc. Civ. Engrs.*, **93**, HY4, 93–98.

Collins, M. A. and L. W. Gelhar (1971) Seawater intrusion in layered aquifers, *WRR*, **7**: 971–9.

Cooper, H. H. (1959) A hypothesis concerning the dynamic balance of fresh water and salt water in a coastal aquifer, *JGR*, **64**: 461–7.

Cooper, H. H., F. A. Kohout, H. R. Henry and R. E. Glover (1964) Sea water in coastal aquifers, relation of salt water to fresh groundwater, *USGS Wat. Sup. Pap.*, 1613C, 84 pp.

Darcy, H. (1856) *Les Fontaines publiques de la ville de Dijon*, V. Dalmont, Paris.

Davis, S. N. and R. J. M. De Wiest (1966) *Hydrogeology*, John Wiley and Sons, New York, 463 pp.

Domenico, P. A. (1972) *Concepts and models in groundwater hydrology*, McGraw-Hill, New York, 405 pp.

Downing, R. A. and G. P. Jones (1985) Hydrogeology—some essential facets. Hydrogeology in the service of man, *18th Congress of the International Association of Hydrogeologists*, pp. 1–16.

Downing, R. A., D. B. Smith, F. J. Pearson, R. A. Monkhouse and R. F. Otlet (1977) The age of groundwater in the Lincolnshire Limestone, England and its relevance to the flow mechanism, *J. Hydrol.*, **33**: 201–16.

Dupuit, J. (1863) *Etudes théoriques et pratiques sur le mouvement des eaux*, 2nd edn, Dunod, Paris.

Eckis, R. (1934) *Bulletin California Div. Water Res.*, **45**: 279.

Figueroa Vega, G. E. (1977) Subsidence of the City of Mexico, a historical review, *IAHS Publ.*, **121**: 35–8.

Forchheimer, P. (1914) *Hydraulik*, Teubner, Leipzig and Berlin.

Freeze, R. A. (1969) The mechanism of natural groundwater recharge and discharge I. One-dimensional, vertical, unsteady, unsaturated flow above a recharging or discharging groundwater flow system, *WRR*, **5**: 153–71.

Freeze, R. A. and J. Banner (1970) The mechanism of natural groundwater recharge and discharge 2. Laboratory column experiments and field measurements, *WRR*, **6**: 138–55.

Freeze, R. A. and J. A. Cherry (1979) *Groundwater*, Prentice-Hall Inc., Englewood Cliffs, N.J., 604 pp.

Freeze, R. A. and P. A. Witherspoon (1966) Theoretical analysis of regional groundwater flow, 1. Analytical and numerical solutions to the mathematical model, *WRR*, **2**: 641–56.

Freeze, R. A. and P. A. Witherspoon (1967) Theoretical analysis of regional groundwater flow, 2. Effect of water-table configuration and subsurface permeability variation, *WRR*, **3**: 623–34.

Freeze, R. A. and P. A. Witherspoon (1968) Theoretical analysis of regional groundwater flow, 3. Quantitative interpretation, *WRR*, **4**: 581–90.

Godwin, H. (1931) Studies in the ecology of Wicken Fen, 1. The groundwater level of the Fen, *Journal of Ecology*, **19**: 449–73.

Guangxiao, D. and Z. Yiaoqi (1986) Land subsidence in China, *IAHS Publ.*, **151**: 405–14.

Habermehl, M. A. (1985) Groundwater in Australia. Hydrogeology in the service of man, *18th Congress of the International Association of Hydrogeologists*, pp. 31–52.

Heath, R. C. (1983) Basic groundwater hydrology, *USGS Wat. Sup. Pap.*, 2220, 84 pp.

Herzberg, B. (1901) Die Wasserversorgung einiger Nordseebader (The water supply to some North Sea resorts), *J. für Gasbeleuchtung und Wasserversorgung*, **44**: 815–19. Quoted in Todd (1980).

Hewlett, J. D. and A. R. Hibbert (1967) Factors affecting the response of small watersheds to precipitation in humid areas, in *Forest hydrology*, W. E. Sopper and H. W. Lull (eds), Pergamon, Oxford, pp. 275–90.

Hillel, D. (1982) *Introduction to soil physics*, Academic Press, New York, 364 pp.

Horton, J. H. and R. H. Hawkins (1965) Flow path of rain from the soil surface to the water table, *Soil Sci.*, **100**: 377–83.

Hubbert, M. K. (1940) The theory of groundwater motion, *Journal of Geology*, **48**: 785–944.

Hubbert, M. K. (1956) Darcy's law and the field equations of the flow of underground fluids, *Transactions American Institute of Mining Metal Engineers*, **207**: 222–39.

Hurst, C. W. and W. B. Wilkinson (1986) Rising groundwater levels in cities, in *Groundwater in engineering geology*, J. C. Cripps, F. G. Bell and M. G. Culshaw (eds), Engineering Geology Special Publication 3, Geology Society, pp. 75–80.

Ineson, J. (1956) Darcy's law and the evaluation of 'permeability', *IASH Symposia Darcy*, **2**: 165–72.

Ineson, J. (1962) Fluctuations of groundwater levels due to atmospheric pressure changes from nuclear explosions, *Nature*, **195**: 1082.

Ineson, J. (1963) Applications and limitations of pumping tests: (b) Hydrogeological significance, *JIWE*, **17**: 200–15.

IH (1988) *Hydrometric Register and Statistics 1981–85*, Hydrological data UK series, Institute of Hydrology, Wallingford, 178 pp.

Jacob, C. E. (1940) On the flow of water in an elastic artesian aquifer, *Trans. AGU*, **21**: 574–86.

Jacob, C. E. (1944) Correlation of groundwater levels and precipitation on Long Island, New York, *Trans. AGU*, **25**: 928–39.

Kharchenko, S. I. and T. Maddock Jr (1982) Investigation of the water regime of river basins affected by irrigation, in *Technical documents in hydrology*, UNESCO, Paris, 63 pp.

Kohout, F. A. (1960) Cyclic flow of salt water in the Biscayne aquifer of south-eastern Florida, *JGR*, **65**: 2133–41.

Kohout, F. A. and H. Klein (1967) Effect of pulse recharge on the zone of diffusion in the Biscayne aquifer, *IASH Symp. Haifa, Publ.*, **72**: 252–70.

Legrand, H. E. and V. T. Stringfield (1973) Karst hydrology—a review, *J. Hydrol.*, **20**: 97–120.

Lloyd, J. W. (1986) A review of aridity and groundwater, *Hydrology Processes*, **1**: 63–78.

Lohman, S. W. (1961) Compression of elastic artesian aquifers, *USGS Prof. Pap.*, **424-B**: 47–8.

Lohman, S. W. (1972) (Chairman) Definitions of selected groundwater terms—revisions and conceptual refinements. Report of the Committee on redefinition of groundwater terms, *USGS Wat. Sup. Pap.*, 1988, 21 pp.

McWhorter, D. B. and D. K. Sunada (1977) *Groundwater hydrology and hydraulics*, Water Resources Publications, Fort Collins, Colorado, 290 pp.

Margat, J. (1985) Groundwater conservation and protection in developed countries. Hydrogeology in the service of man, *18th Congress of the International Association of Hydrogeologists*, pp. 270–93.

Marsh, T. J. and P. A. Davies (1984) The decline and partial recovery of groundwater levels below London, *Proc. ICE*, **74**: 263–76.

Maxey, G. B. (1968) Hydrogeology of desert basins, *Groundwater*, **6**: 10–22.

Maxey, G. B. (1969) Subsurface water—groundwater, in *The Progress of Hydrology*, University of Illinois, Urbana, Ill., **2**: 787–815.

Maxey, G. B. and R. N. Farvolden (1965) Hydrogeologic factors in problems of contamination in arid lands, *Groundwater*, **3**: 29–32.

Meinzer, O. E. (1917) Geology and water resources of Big Smokey, Clayton, and Alkali Spring Valleys, Nevada, *USGS Wat. Sup. Pap.*, 423.

Meinzer, O. E. (1923) Occurrence of groundwater in the United States, *USGS Wat. Sup. Pap.*, 489, pp. 1–321.

Meinzer, O. E. (1928) Compressibility and elasticity of artesian aquifers, *Economic Geology*, **23**: 263–91.

Meinzer, O. E. (1932) Outline of methods for estimating groundwater supplies, *USGS Wat. Sup. Pap.*, 638, pp. 99–144.

Meyboom, P. (1963) Patterns of groundwater flow in the prairie profile, in *Groundwater, Proc. Hydrology Symposium, No. 3*, National Research Council of Canada, Ottawa, pp. 5–20.

Meyboom, P. (1965) Three observations on streamflow depletion by phreatophytes, *J. Hydrol.*, **2**: 248–61.

Meyboom, P. (1966) Unsteady groundwater flow near a willow ring in hummocky moraine, *J. Hydrol.*, **4**: 38–62.

Meyboom, P. (1967a) Groundwater studies in the Assiniboine River drainage basin, *Geological Survey Canadian Bulletin, 139*, 64 pp.

Meyboom, P. (1967b) Mass transfer studies to determine the groundwater regime of permanent lakes in hummocky moraine of western Canada, *J. Hydrol.*, **5**: 117–42.

Miller, R. J. and P. F. Low (1963) Threshold gradient for water flow in clay systems, *Proc. SSSA*, **27**: 605–9.

Nace, R. L. (1969) Human use of groundwater, in *Water, earth and man*, R. J. Chorley (ed.), Methuen, London, pp. 285–94.

Parker, G. G. and V. T. Stringfield (1950) Effects of earthquakes, trains, tides, winds and atmospheric pressure changes on water in the geologic formations of Southern Florida, *Economic Geology*, **45**: 441–60.

Pinder, G. F. and H. H. Cooper (1970) A numerical technique for calculating the transient position of the saltwater front, *WRR*, **6**: 875–82.

Poiseuille, J. L. M. (1846) Recherches expérimentales sur le mouvement des liquides dans les tubes de très petit diamètre, *Roy. Acad. Sci. Inst. France Math. Phys. Sci. Mem.*, **9**: 433–543.

Poland, J. F. (1961) The coefficient of storage in a region of major subsidence caused by compaction of an aquifer system, *USGS Prof. Pap.*, **424-B**: 52–4.

Poland, J. F. (1984) Guidebook to studies of land subsidence due to groundwater withdrawal, *Studies and Reports in Hydrology*, No. 40, UNESCO, Paris, 305 pp.

Poland, J. F. and G. H. Davis (1969) Land subsidence due to withdrawal of fluids, *Reviews in Engineering Geology*, **2**: 187–269.

Price, M. (1985) *Introducing groundwater*, George Allen and Unwin, London, 195 pp.

Price, M. (1987) Fluid flow in the chalk of England, in *Fluid flow in sedimentary basins and aquifers*, J. C. Goff and B. P. J. Williams (eds), Society of London Special Publication, 34, pp. 141–56.

Price, M. and D. W. Reed (1989) The influence of mains leakage and urban drainage on the groundwater levels beneath conurbations in the UK, *Proc. ICE*, **86**: 31–9.

Price, M., M. J. Bird and S. S. D. Foster (1976) Chalk pore-size measurements and their significance, *Water Services*, October, pp. 596–600.

Roberts, I. (1883) On the attractive influence of the sun and moon causing tides . . . in the underground water in porous strata, *Report of British Association*, p. 405.

Robinson, T. W. (1939) Earth-tides shown by fluctuations of water-levels in wells in New Mexico and Iowa, *Trans. AGU*, **20**: 656–66.

Scheidegger, A. E. (1960) *The physics of flow through porous media* (revised edition), University of Toronto Press, 313 pp. (first edition, 1957).

Scheidegger, A. E. (1974) *The physics of flow through porous media*, 3rd edn, University of Toronto Press, 353 pp.

Schmorak, S. and A. Mercado (1969) Upconing of fresh water–sea water interface below pumping wells, field study, *WRR*, **5**: 1290–311.

Scott, J. S. and F. W. Render (1964) Effect of an Alaskan earthquake on water levels in wells at Winnipeg and Ottawa, Canada, *J. Hydrol.*, **2**: 262–8.

Siple, G. E. (1965) Salt-water encroachment in coastal South Carolina, *Proceedings of Conference on Hydrologic Activities in the South Carolina Region*, Clemson University, pp. 18–33.

Slichter, C. S. (1902) The motions of underground waters, *USGS Wat. Sup. Pap.*, 67.

Stephenson, D. A. (1971) Groundwater flow system analysis in lake environments, with management and planning implications, *Water Research Bulletin*, **7**: 1038–47.

Stringfield, V. T. and H. E. Legrand (1969) Relation of sea water to fresh water in carbonate rocks in coastal areas, with special reference to Florida, U.S.A. and Cephalonia (Kephallinia), Greece, *J. Hydrol.*, **9**: 387–404.

Stringfield, V. T. and H. E. Legrand (1971) Effects of karst features on circulation of water in carbonate rocks in coastal areas, *J. Hydrol.*, **14**: 139–57.

Swartzendruber, D. (1962) Non Darcy behaviour in liquid saturated porous media, *JGR*, **67**: 5205–13.

Theis, C. V. (1935) Relation between the lowering of the piezometric surface and the rate and duration of discharge of a well using groundwater storage, *Trans. AGU*, **16**: 519–24.

Todd, D. K. (1980) *Groundwater hydrology*, John Wiley and Sons, New York, 535 pp.

Toth, J. (1962) A theory of groundwater motion in small drainage basins in central Alberta, Canada, *JGR*, **67**: 4375–87.

Toth, J. (1963) A theoretical analysis of groundwater flow in small drainage basins, *JGR*, **68**: 4795–812.

Troxell, H. C. (1936) The diurnal fluctuation in the groundwater and flow of the Santa Ana River and its meaning, *Trans. AGU*, **17**: 496–504.

UNESCO (1984) Groundwater in hard rocks, in *Studies and Reports in Hydrology 33*, UNESCO, Paris, 228 pp.

UNESCO (1987) Groundwater problems in coastal areas, *Studies and Reports in Hydrology 45*, UNESCO, Paris, 596 pp.

Walton, W. C. (1970) *Groundwater resource evaluation*, McGraw-Hill, New York, 664 pp.

Wellings, S. R. (1984) Recharge of the upper chalk aquifer at a site in Hampshire, England: 1. Water balance and unsaturated flow, *J. Hydrol.*, **69**: 259–73.

Winter, T. C. (1976) Numerical simulation analysis of the interaction of lakes and groundwater, *USGS Prof. Pap.*, 1001, 45 pp.

Wright, E. P., A. C. Benfield, W. M. Edmunds and R. Kitching (1982) Hydrogeology of the Kufra and Sirte basins, eastern Libya, *QJEG*, **15**: 83–103.

Zekster, I. S. and R. G. Dzhamalov (1981) Groundwater discharge into the world oceans, *Nature and Resources*, **17**: 20–2.

7. Runoff

7.1 Introduction

Runoff or streamflow comprises the gravity movement of water in channels which may vary in size from the one containing the smallest ill-defined trickle to the ones containing the largest rivers such as the Amazon, the Congo, and the Mississippi. As well as *streamflow*, runoff may be variously referred to as *stream* or *river discharge*, or *catchment yield*, and is normally expressed as a volume per unit of time. The *cumec*, i.e. one cubic metre per second, and *cumecs per square kilometre* are commonly used units. Runoff may also be expressed as a depth equivalent over a catchment, i.e. millimetres per day or month or year. This is a particularly useful unit for comparing precipitation and runoff rates and totals since precipitation is almost invariably expressed in this way. Alternative runoff expressions still found in the literature include millions of gallons per day (m.g.d.) and, particularly in American irrigation literature, acre-feet, i.e. the volume of water that would cover one acre to a depth of one foot.

At a general level the relationship between streamflow and precipitation can be expressed in terms of the continuous circulation of water through the hydrological cycle. More specifically, we can recognize that in natural conditions each river and stream receives water only from its own drainage basin or catchment area. Each catchment can, therefore, be regarded as a system receiving inputs of precipitation and transforming these into outputs of evaporation and streamflow. Allowing for changes of storage within the system, input must be equalled by output. In all but the driest areas output from the catchment system is continuous but the inputs of precipitation are discrete and often widely separated in time. As a result (see Fig. 7.1), the annual hydrograph typically comprises short periods of suddenly increased discharge associated with rainfall or snowmelt and intervening, much longer, periods when streamflow represents the outflow from water stored on and below the surface of the catchment and when the hydrograph therefore takes the exponential form of the typical exhaustion curve.

7.2 Quickflow and slowflow

The immediacy of streamflow response seems to indicate that some part of the initial precipitation takes a rapid route to the stream channels (i.e. *quickflow*); the continuity of

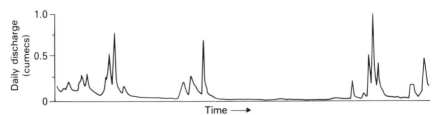

Figure 7.1 *Annual hydrograph for the Catchwater Drain, North Humberside, 1967.*

flow through often prolonged dry periods seems equally clearly to indicate that some part of the initial precipitation takes a much slower route, which in the interests of consistency should be called *slowflow* but which is usually referred to as *baseflow*. These two fundamental components of flow are apparent in rivers of all sizes. However, in large river systems lag effects, both within and outside channels, and the multiplicity of flow contributions to the main channel from numerous tributary streams make it more difficult to interpret hydrograph response to precipitation. Accordingly much of the initial discussion in this chapter attempts to explain the response to precipitation of *headwater streams* draining catchment systems that are comparatively small and simple.

In such situations the response of catchments to precipitation is often very rapid but is rarely the same, i.e. the proportion of precipitation that appears quickly as streamflow under the storm hydrograph differs from storm to storm. Figure 7.2 shows graphs from a pioneering paper by Ramser (1927), which emphasize the immediacy and the variability of

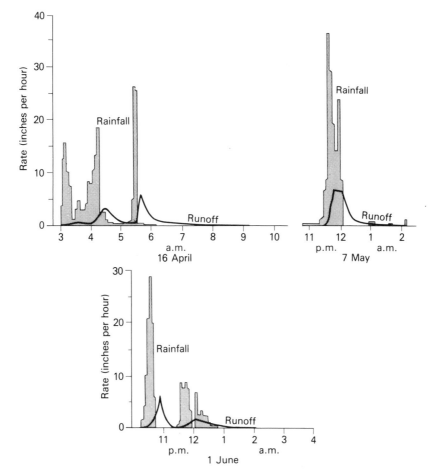

Figure 7.2 *Graphs of rainfall and runoff for three storm events in Tennessee, USA (from original diagrams by Ramser, 1927).*

streamflow response to precipitation, and which, because rainfall and streamflow are correctly plotted on the same scale, also emphasize the low percentage of rainfall that appears as quickflow under the storm hydrograph. Subsequently, Woodruff and Hewlett (1970) mapped this percentage value for the eastern USA, showing that there are distinct spatial as well as temporal variations, the former exhibiting an alignment with the structural grain and having a weighted mean value of about 10 per cent, i.e. on average, throughout the eastern USA about 10 per cent of total precipitation appears quickly under the storm hydrograph. It is estimated that globally some 34 per cent of the total precipitation falling on the land areas reaches the oceans as runoff so that if Woodruff and Hewlett's figures are applied generally it can be seen that runoff comprises quickflow, representing 10 per cent of precipitation, and slowflow, representing 24 per cent of precipitation.

7.3 Sources and components of runoff

Traditionally the temporally and spatially variable response of streamflow to precipitation has been explained in terms of divergent and contrasting flowpaths of precipitation towards the stream channels. Figure 7.3 shows that precipitation may arrive in the stream channel by one of a number of flowpaths: (a) direct precipitation onto the water surface; (b) overland flow; (c) shallow subsurface flow (throughflow); and (d) deep subsurface flow (groundwater flow). Snowfall and snowmelt will, in due course, follow one of these four flowpaths.

These terms are used widely and relatively unambiguously in the literature. Persistent misuse of other terms such as surface runoff and direct runoff has resulted in unnecessary

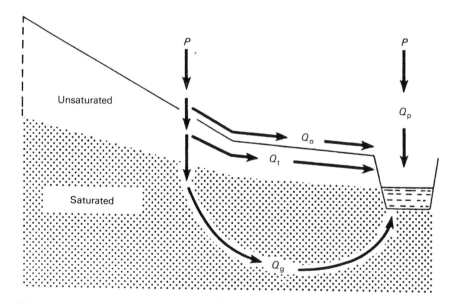

Figure 7.3 *Flowpaths of the sources of streamflow: Q_p is direct precipitation onto the water surface, Q_o is overland flow, Q_t is throughflow and Q_g is groundwater flow.*

confusion and ambiguity, and accordingly Fig. 7.4 attempts to provide a consistent and logical terminology. From this it will be seen that surface runoff is that part of total runoff that reaches the drainage basin outlet via overland flow and the stream channels, although it may in some circumstances also include throughflow that has discharged at the ground surface at some distance from the stream channel. Subsurface runoff is the sum of throughflow and groundwater flow and is normally equal to the total flow of water arriving at the stream as saturated flow into the stream bed itself and as percolation from seepage faces on the stream bank (Freeze, 1972). Quickflow, or direct runoff, is the sum of channel precipitation, surface runoff and rapid throughflow, and will represent the major runoff contribution during storm periods and most floods. It will be observed that quickflow and surface runoff as defined above are not synonymous.

Baseflow or base runoff may be defined as the sustained or fair-weather runoff (Chow, 1964), and is the sum of groundwater runoff and delayed throughflow, although some hydrologists prefer to include the total throughflow as illustrated by the broken line in Fig.

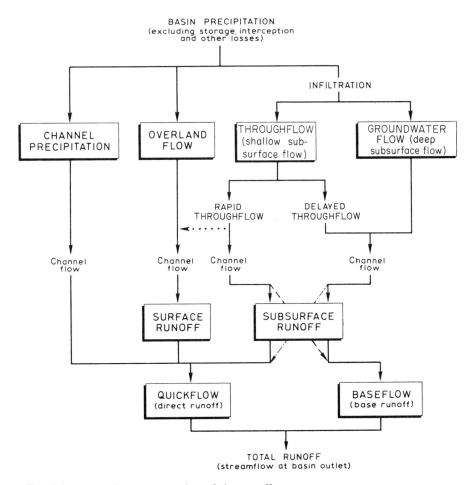

Figure 7.4 *Diagrammatic representation of the runoff process.*

7.4. Again baseflow and groundwater flow, as defined above, are not synonymous; indeed Hewlett (1961b) and Hewlett and Hibbert (1963) demonstrated that baseflow from steep mountain drainage basins may consist almost entirely of unsaturated lateral flow from the soil profile.

7.3.1 *Channel precipitation*

Not all of these sources of runoff are of equal importance. For example, the contribution of direct precipitation on to the water surface is normally small simply because the perennial channel system occupies only a small percentage of catchment area, cf. about 1 per cent in the eastern USA (Hewlett and Nutter, 1969). However, this will increase significantly during a prolonged storm and Rawitz *et al.* (1970) estimated that on occasions channel precipitation accounted for as much as 61 per cent of total runoff in a small Pennsylvania catchment. Where catchments contain a large area of lakes or swamps channel precipitation may be persistently important.

7.3.2 *Overland flow*

Overland flow comprises the water that, failing to infiltrate the surface, travels over the ground surface towards a stream channel either as quasi-laminar sheet flow or, more usually, as flow anastomosing in small trickles and minor rivulets. The main cause of overland flow is the inability of water to infiltrate the surface, and in view of the high value of infiltration characteristic of most vegetation-covered surfaces it is not surprising that overland flow is a rarely observed phenomenon (except on laboratory models!). Conditions in which it assumes considerable importance include the saturation of the ground surface, the hydrophobic nature of some very dry and sodic soils, the deleterious effects of many agricultural practices on infiltration capacity and freezing of the ground surface.

7.3.3 *Throughflow*

Water that infiltrates the soil surface and then moves laterally through the upper soil horizons towards the stream channels, either as unsaturated flow or, more usually, as shallow perched saturated flow above the main groundwater level, is known as throughflow. Alternative terms found in the literature include interflow, subsurface stormflow (Hursh, 1936), storm seepage (Barnes, 1939) and secondary baseflow (Barnes, 1938). The general condition favouring the generation of throughflow is one in which lateral conductivity in the surface horizons of the soil is substantially greater than the overall vertical hydraulic conductivity through the soil profile. Then during prolonged or heavy rainfall water will enter the upper part of the profile more rapidly than it can pass vertically through the lower part, thus forming a perched saturated layer from which water will 'escape' laterally, i.e. in the direction of greater hydraulic conductivity.

Except where conditions have been artificially disturbed, e.g. surface compaction, the situation described above is the one most commonly found. Even in a deep relatively homogeneous soil profile hydraulic conductivity will tend to be greater in the surface layers than deeper down in the profile, thereby encouraging the generation of through-flow. However, still more favourable conditions exist where a thin permeable soil overlies impermeable bedrock, with a markedly stratified soil profile, or where an iron-pan occurs a short distance below the surface. There may be several levels of throughflow below the surface corresponding to textural changes between horizons and to the junction between weathered mantle and bedrock. In addition, there is evidence that water may travel downslope through old root holes and animal burrows and other subsurface pipes and that in some circumstances a high degree of soil biological activity may play a role as important in runoff generation (Bonell *et al.*, 1984) as in drainage basin erosion (Jungerius, 1985). In view of the variety of possible throughflow routes it is to be expected that some will result in a more rapid movement of water to the stream channels than will others, so that it is sometimes helpful to distinguish between rapid and delayed throughflow (see Fig. 7.4). It is almost certain, however, that apart from pipe flow the very rapid arrival of throughflow at the stream channels, which has been observed by some investigators, must result from 'piston displacement' (Section 7.4.2). Further complica-tions result from the fact that some throughflow does not discharge directly into the stream channel but comes to the surface at some point between the stream and the catchment divide, and may then continue to move over the soil surface to the stream channel. Freeze (1972) suggested that this should be considered as part of the total subsurface runoff; others have considered it as a contribution to overland flow and surface runoff, an assumption indicated by the dotted line in Fig. 7.4.

The role of throughflow in total runoff will be discussed in more detail in subsequent sections of this chapter, although it is interesting to note at this stage that experimental evidence has long indicated that it may account for up to 85 per cent of total runoff (Hertzler, 1939).

7.3.4 *Groundwater flow*

Away from the relatively steeply sloping terrain of the headwaters, where subsurface runoff is dominated by throughflow, most of the rainfall that infiltrates the catchment surface will percolate through the soil layer to the underlying groundwater and will eventually reach the main stream channels as groundwater flow through the zone of saturation. Since water can move only very slowly through the ground, the outflow of groundwater into the stream channels may lag behind the occurrence of precipitation by several days, weeks or even years. Groundwater flow also tends to be very regular, representing as it does the outflow from the slowly changing reservoir of moisture in the soil and rock layers. It must not be inferred from this that groundwater may not show a rapid response to precipitation. Indeed, the 'piston displacement' mechanism frequently results in a rapid response of groundwater flow to precipitation during individual storm periods, and especially on a seasonal basis, and this is represented by the chain-dot line in Fig. 7.4. Since this can operate only in moist soil and subsoil conditions, however, the replenishment of large moisture deficits created particularly during summer conditions may result in a considerable lag of groundwater outflow after precipitation during and immediately following prolonged dry periods. In general, groundwater flow represents the main long-term component of total runoff and is particularly important during dry spells when surface runoff is absent.

7.4 Event-based variations

Some of the flow paths described above are quicker than others, e.g. direct precipitation and overland flow are likely to deliver water to the stream channels more rapidly than is deep groundwater flow. Understandably, therefore, early attempts to explain the variation of streamflow with time, especially through a precipitation event, concentrated almost exclusively on the overland flowpath.

7.4.1 *The Horton hypothesis*

This was the view expressed clearly by R. E. Horton (1933) who proposed quite simply that the soil surface partitions falling rain so that one part goes rapidly as overland flow to the stream channels and the other part goes initially into the soil and thence either through gradual groundwater flow to the stream channel or through evaporation to the atmosphere. The partitioning device is the infiltration capacity of the soil surface which Horton (1933) defined as '. . . the maximum rate at which rain can be absorbed by a given soil when in a given condition'.

Reference to Fig. 7.5 shows that during that part (t) of a storm when rain falls at a rate (i) that is greater than the rate (f) at which it can be absorbed by the ground surface there will occur an excess of precipitation (P_e) which will flow over the ground surface as overland flow (Q_o). No overland flow will occur if the rainfall intensity is lower than the infiltration capacity; instead the infiltration that takes place will first top up the soil-moisture reservoir until the so-called moisture capacity is attained, after which further infiltration through the ground surface will percolate to the groundwater reservoir thereby increasing the groundwater flow (Q_g) to the stream channel.

Horton (1933) suggested that infiltration capacity (f) would pass through a fairly definite cycle for each storm period. Starting with a maximum value at the onset of rain, f would decrease, rapidly at first, as a result of the compaction of the soil surface by falling raindrops, colloidal swelling of the soil, which would close sun-cracks and other interstices, and the clogging of interstices by the inwashing of fine particles. After the initial rapid decline infiltration capacity would become stable or decline only very slowly

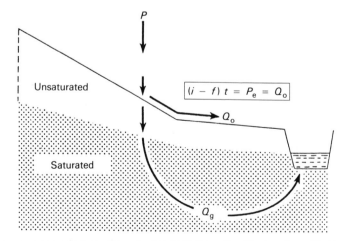

Figure 7.5 *The response of streamflow to precipitation: the Horton hypothesis.*

for the remainder of the storm and would begin to recover immediately after the end of the storm. He maintained that this cycle of infiltration capacity resulted from the operation of processes '. . . confined to a thin layer at the soil surface'.

Subsequent work, although confirming the general shape of the infiltration capacity curve, has demonstrated that much greater importance should be attached to factors acting *within* the soil profile, especially the lengthening flowpath of percolating rainfall, the initial moisture gradient and variations of hydraulic conductivity with depth.

Nevertheless, for some soils the Horton concept remains broadly applicable. For example, in small 'water harvesting' plots in the Negev Desert, Israel, where sodic soils crust easily during rainfall, often only the first few millimetres of rainfall are able to infiltrate through cracks, etc., before the crust begins to shed water with about the same efficiency as an asphalt road surface (van der Molen, 1983).

Whether the infiltration capacity cycle results from the dominance of surface or subsurface controls, its implications for the short-term variation of runoff are quite clear. Rain of high intensity may generate precipitation excess and therefore overland flow throughout a storm; rain of moderate intensity will not generate overland flow until the initially high infiltration capacity has declined; and rain of low intensity may fail to generate overland flow at all. Furthermore, since infiltration capacity is likely to show a continued decrease through a sequence of closely spaced storms, we would expect a given amount of rainfall falling late in the storm sequence to generate more overland flow and therefore a greater streamflow response than the same amount of precipitation falling early in the storm sequence.

Horton's explanation of river response to precipitation was based on three assumptions, the first of which was that infiltration capacity can be measured easily and that, knowing infiltration capacity and rainfall intensity, it would then be a straightforward matter to calculate precipitation excess and thus overland flow. In fact it is very difficult to extrapolate satisfactorily infiltrometer or small basin measurements of infiltration to larger areas. Second, it was assumed that the soil surface can act as a plane of hydrological separation whereas there is normally at the soil surface a transition between the soil mass and the overlying atmosphere such that porosity and hydraulic conductivity increase in a direction away from the soil mass. Accordingly, only in cases of severe compaction or crusting will the soil surface approximate to the Hortonian concept of a plane of separation. Third, it was assumed that a sheet of water could accumulate on and flow over this hypothetical plane surface. As Chorley (1978) observed, Horton devoted considerable attention to the hydraulics of this postulated thin film of overland flow and also, significantly, to explaining why it so frequently remained unobserved in the field.

It follows, then, that the Horton hypothesis is most likely to be applicable in conditions of sparse vegetation cover, and especially in semi-arid and arid climates and where sodic soils exhibit rapid crusting. Indeed, aerial photographs of desert areas frequently show surface flow patterns and associated geomorphological features of erosion and deposition which result from water flowing over the surface (van der Molen, 1983).

7.4.2 *The Hewlett hypothesis*

An alternative hypothesis emerged in the 1960s as a result of doubts raised by US Forest Service hydrologists perplexed by their failure to observe Hortonian overland flow even during intense rainstorms that resulted in a rapid river response—doubts that had been expressed clearly in the literature over a period of almost 60 years from 1902 (cf. Fernow, 1902; Zon, 1927; Hursh, 1944) and that were subsequently substantiated by clear evidence that rainfall intensity has no appreciable effect on storm flows (Hewlett et al., 1977, 1984). In this hypothesis, which was first articulated cogently by J. D. Hewlett and A. R. Hibbert (Hewlett, 1961a; Hewlett and Hibbert, 1967), it is argued that over much of a catchment area, even during intense and prolonged precipitation, all precipitation infiltrates the soil surface (Fig. 7.6). Then as a result of the combined processes of infiltration and throughflow through the soil profile, first the shallow water table areas immediately adjacent to the stream channels and subsequently the lower valley slopes become saturated as the water table rises to the ground surface. In these surface-saturated areas infiltration capacity is zero so that all precipitation falling on them, at whatever intensity, is excess precipitation (P_e) or overland flow, which we can term *saturated overland flow* ($Q_o(s)$), in contrast to *Hortonian overland flow* (Q_o) as previously defined.

It is only the saturated area of the catchment that can act as a source of quickflow; all other areas of the catchment absorb the rain that falls and either store it or transfer it

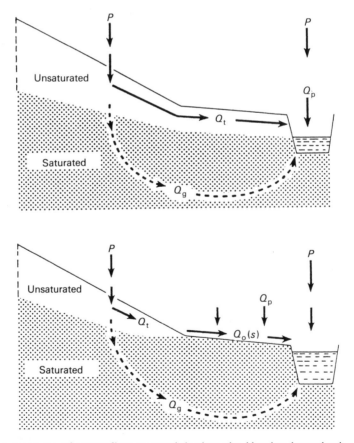

Figure 7.6 *The response of streamflow to precipitation: the Hewlett hypothesis.*

beneath the ground surface. Note also that the source area for quickflow is of variable size, growing as rainfall proceeds. According to Hewlett and Hibbert (1967), 'It would not be impossible to treat overland flow as an exceptionally rapid extension of the channel system into areas where the soil cannot transmit water as subsurface flow. This seems more appropriate than treating all direct runoff as overland flow.' Day (1978) cautioned, however, on the basis of field observations in a small rural catchment near Armidale, NSW, that notions of a simple expanding and contracting system need '. . . review and refinement'.

It is disappointing that a combination of semantic confusion over the interchangeable terms 'channel expansion' and 'saturated overland flow', personal bias in literature citation and just plain ignorance on the part of many hydrologists has obscured the fact that the essential concept of variable source areas was first expounded in a few brief paragraphs by Hewlett (1961a) and was subsequently modified only marginally, if at all, by hydrologists such as R. P. Betson and T. Dunne, to whom credit for these ideas is often wrongly ascribed (cf. Anderson and Burt, 1978; Freeze, 1980).

Hewlett's hypothesis was qualitatively supported by separate investigations of subsurface water movement through valley slope profiles. Field experiments in Iowa, USA, by Kirkham (1947), for example, had indicated that during intense precipitation water infiltrated on the slopes and then moved horizontally through the middle slope material and vertically upwards near the base of slopes as a result of the pattern of pre-water pressure which developed during the storm. This basic pattern was confirmed by Toth (1962) and generalized in his mid-line model. A similar pattern was simulated by Klute *et al*. (1965) and was further substantiated by additional field evidence at a catchment scale, such as that presented by Freeze (1967). In other words, it seems clear that the increase of pressure potential with depth in the lower slopes represents a condition that facilitates rapid saturation of the surface layers when even modest quantities of water are added to the soil profile by infiltration or shallow throughflow. Interestingly, this possibility had been anticipated two decades earlier by Vaidhianathan and Singh (1942) who explained the surprisingly large response to small inputs of rainfall of the free water level within the 'capillary fringe', where there is a marked gradient of pressure potential towards the surface, in terms of the flattening of the concave menisci at the soil surface and the consequent increase of pressure potential. Much later the physical basis of the capillary fringe effect was described in detail by Gillham (1984).

In Hewlett's original concept of variable source areas it was implied that these would be contiguous with the stream channels. Later work has sought to establish that areas of saturation overland flow may occur widely within a catchment area, often in locations far removed from the stream channels and that, furthermore, if such disjunct areas have effective hydrological connections with the valley bottoms or lower slopes they, too, may contribute quickflow to the stream channel. Leaving aside those (exceptional) areas where because of severe compaction, sparse vegetation cover or thin, degraded soils (all frequently the result of human interference), genuine Hortonian overland flow may indeed be produced as described by Betson (1964), the search for source areas of quickflow became linked rapidly with evidence of flow convergence. Kirkby and Chorley (1967) suggested three probable types of location where convergence of flow might lead to surface saturation and to saturation overland flow, in addition to contiguous channel-side areas. Figure 7.7 shows that these are: (a) slope concavities in plan where convergence

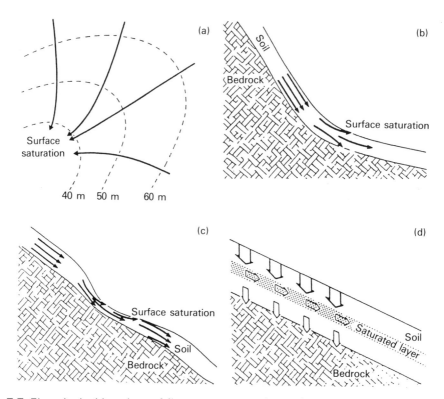

Figure 7.7 *The principal locations of flow convergence in catchment areas.*

leads to subsurface flow rates that may exceed the transmission capacity of the porous medium and lead, therefore, to the emergence of flow at the soil surface in the central areas of the concavities; (b) slope concavities in section where, assuming uniform hydraulic conductivity throughout the section, subsurface flow rates will be directly proportional to the hydraulic gradient so that water will enter a concavity from upslope areas more rapidly than it can leave downslope; and (c) areas of thinner soil whose water holding and transmitting capacity is low.

Although the third type of location may be as important as the other two, assessments of slope concavity are more easily made, either in the field or from maps and aerial photographs, than assessments of soil depth. Inevitably, therefore, more attention has been devoted to the effects of slope concavity, both in field investigations and in modelling procedures (cf. Dunne and Black, 1970; Anderson and Burt, 1978; Beven, 1978; Pierson, 1980; Burt and Butcher, 1985). Zaslavsky and Sinai (1981) presented field evidence from a number of European locations where concentration of moisture in concave areas has been observed, including some of their own results from a site near Beer-Sheba, Israel; O'Loughlin (1981) showed by analysis that the size of saturated zones on undulating hillslopes depends strongly on topographic convergence or divergence. The dynamic nature of quickflow-contributing areas and their relationship to the geomorphological structure of catchments was stressed by Beven and Wood (1983) and their relationship to vegetation by Gurnell (1981) and Gurnell *et al.* (1985). Clearly a major difficulty is to identify and to quantify the contributing areas. Van de Griend and Engman (1985) emphasized the importance to date of detailed field surveys, but noted also the potential role of remote sensing techniques which now permit the measuring and monitoring of soil moisture over areas rather than at a single point.

A fourth type of flow convergence, illustrated in Fig. 7.7d, occurs as water percolates vertically through a soil profile. Partly because of the reduced hydraulic gradient as the flowpath of the percolating water lengthens and partly because most soils, whether layered or not, exhibit a reduction of hydraulic conductivity with depth, rates of inflow decrease with depth, leading to the development of a layer, or layers, of temporary saturation. This has been a well-documented phenomenon for many years (cf. Kidder and Lytle, 1949). Normally the downslope hydraulic gradient will result in the removal of this accumulation of water as throughflow before the build-up of saturation reaches the soil surface. In flat areas, however, or in sloping areas having very high rainfall amounts and intensities, saturation overland flow will be produced. This has been found especially where an impeding layer occurs at shallow depths in the soil profile, as in the pseudo-gley soils of central Europe, so that water accumulates on a Bt-horizon in the soil profile during wet periods, or with a high-intensity rainfall on the upper slopes of a tropical rainforest catchment investigated by Bonell and Gilmour (1978).

Such examples of disjunct variable source areas resulting from a variety of convergence phenomena extend the Hewlett concept in the manner proposed by Jones (1979), provided that it is possible to demonstrate satisfactory hydrological linkages resulting in the rapid transmission of water between disjunct areas and the stream channel. Various mechanisms for such linkages have been proposed. In their model of variable source areas Engman and Rogowski (1974) favoured an overland flow connection. Assuming an initial channel-side source area and a disjunct source area farther upslope which results from flow convergence in an area of thinner soils, it was suggested that during precipitation a surface water connection is established permitting the upslope source area to contribute quickflow directly to the stream channel. Those who have walked the moors and fells of upland Britain in typical weather conditions may recognize here an illustration of observed reality. These are often areas where the combination of heavy rainfall and shallow soils of variable depth exemplify the conditions described by Engman and Rogowski.

As in most recently glaciated areas, soils in Britain are notoriously shallow. Elsewhere, where soils are thicker or of more uniform depth, or where vegetation cover is denser, it will be necessary to seek other linkage mechanisms. Thus more than 40 years ago in the southern Appalachians, USA, where soils average 2 m in depth (Hewlett, 1961b) and where no surface flow was observed but at a time when it seemed inconceivable that the acknowledged slow movement of water through the weathered mantle could deliver an appreciable amount of water to the stream channel during a storm period, Hursh (1944) proposed his pipeflow theory in which turbulent flow through large, quasi-cylindrical conveyances, such as animal burrows or decayed root channels, could lead subsurface stormflow rapidly through the slope material. This process was confirmed for a forested catchment in Luxembourg by Bonell *et al.* (1984). However, biotic voids have been regarded by some hydrologists as 'pseudo-pipes' in contrast to the more widespread and hydrologically important pipes formed by hydraulic and hydrological processes (Jones,

1981). This latter type has been found in a wide range of locations (cf. Jones, 1971, 1981; Jones and Crane, 1984; Tanaka, 1982; Tsukamoto *et al.*, 1982; Walling and Burt, 1983) and may increase the quickflow contributing area to as much as two or even five times that identified from surface contours (Jones, 1986, 1987).

The high velocity of conduited, macropore subsurface flow (see also Section 5.5.2 under 'Macropores') means that the water arriving in the stream channel by this route will almost certainly be 'new' water, i.e. water added by the current storm, rather than 'old' water, i.e. water that was stored in the catchment prior to the current storm event. However, field experiments in a variety of environments using natural isotope tracers appear increasingly to support the view of Hewlett that 'old' water dominates the storm runoff hydrograph, even in areas where the existence of macropores is well established. This was convincingly demonstrated by Sklash *et al.* (1986) and Pearce *et al.* (1986) for catchments in New Zealand in which earlier analyses by Mosley (1979) had, in contrast, suggested that 'new' water dominated the storm runoff.

It therefore seems that the existence of macropores is not in itself evidence of a significant role in the runoff process. Instead, the reexamination of older ideas (cf. Hursh, 1936; Barnes, 1938) about the possible role of diffuse, rather than conduited, subsurface flow, including both shallow throughflow and deeper groundwater flow, now appears to offer a more plausible basis for explaining aspects of streamflow response to precipitation. Initially resistance to the notion of throughflow occurring in substantial quantities reflected, first, a firmly rooted belief that water movement along such a flow path would be far too slow to deliver quickflow to the stream channel and that only overland flow could provide a reasonable explanation of the storm hydrograph and, second, an implicit faith in the verticality of the infiltration–percolation route from the soil surface to the water table. Not until an improved understanding had developed of the anisotropic nature of the soil profile was it realized that the vertical flow path may in fact be a least-likely option and that instead '... water responds to changing hydraulic gradients and flows more or less parallel to the slope surface, depending on local moisture contents, soil conductivities and the steepness of gradients' (Hewlett and Troendle, 1975).

The 'thatched roof' analogy of Zaslavsky and Sinai (1981) is a helpful explanatory aid (Fig. 7.8). No hydrologist, having measured the infiltration characteristics of bundles of straw, would recommend their use as a roofing material. And yet, even in the heaviest rain, the building remains dry, no water runs over the thatch as 'overland' flow, there is no 'groundwater' and no evidence of zones of 'temporary saturation', i.e. all the rainfall is evacuated along the narrow layer of the thatch itself. The thatched roof works because the alignment of the straw imparts a preferential permeability along the stems and because the roof slopes; it would not work if the straw bundles were placed vertically or if the roof were flat. In the case of the soil which covers the ground surface we know that, whether or not an impeding layer exists beneath the surface, there is normally a preferential hydraulic conductivity through the more open-textured upper layers parallel to the surface so that where the soil occurs on a uniform slope it would act hydrologically like a thatched roof, i.e. there may be no need to postulate either overland flow or groundwater recharge.

Hewlett's experiments with sloping soil models (Hewlett, 1961a, 1961b; Hewlett and Hibbert, 1963, 1967) indeed indicated water flow paths resembling the behaviour of a thatched roof, and led Hewlett (1961a) to prepare the diagram shown in Fig. 7.9. Note that Hewlett shows no overland flow and no deep groundwater recharge; this implies that

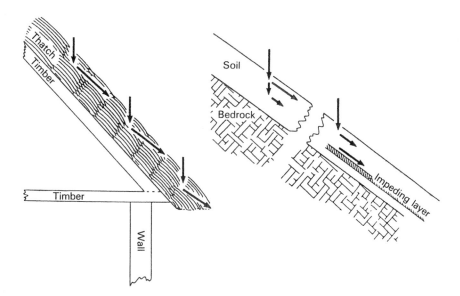

Figure 7.8 *The thatched roof analogy of water movement through the soil profile.*

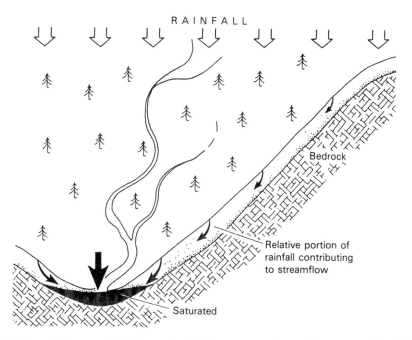

Figure 7.9 *The relative contributions of rainfall to streamflow (from an original diagram by Hewlett, 1961a).*

all rainfall infiltrates and is then transmitted through the soil profile, contributing preferentially to storm flow with upslope rainfall recharging the soil-moisture store in preparation for succeeding days and weeks of baseflow whereas downslope rainfall and channel precipitation provide most of the storm flow. Although it may be appropriate now to partially reinterpret this pioneering scheme, there seems little doubt that, in due course, this diagram will be seen to represent one of the single greatest conceptual advances in the history of hydrology.

A major outstanding problem, though, was the acknowledged slow rate of water movement through the soil and weathered mantle. How could this flow through the soil 'thatch' arrive at the stream channel as stormflow if its maximum rate of movement was only 5–6 m per day? Hewlett and his coworkers adopted various strategies, none of them entirely satisfactory, in order to overcome this major obstacle. For example, subsurface flow routes were kept as short as possible, as in Fig. 7.9, which shows most of the throughflow contribution coming from the areas closest to the channel. In addition, as we have already seen, the channel was considered to expand during precipitation, thereby, enabling it '. . . to tap the subsurface flow systems which, for whatever reason, have over-ridden their capacity to transmit water beneath the surface' (Hewlett and Nutter, 1970).

Again, throughflow was considered to include any water reaching the channel as overland flow provided that at some point in its flowpath, for however short a distance, it had infiltrated and later reemerged at the surface, since as a result of a subsurface route of 'only a few feet' the water's arrival at the stream channel '. . . will often be sufficiently delayed to form a second peak on the hydrograph' (Hewlett and Nutter, 1970).

This particular argument was, of course, a double-edged sword since it could be regarded as damning admission of the inefficacy of throughflow. In order to counter this Hewlett and his colleagues advanced the notion of 'translatory flow', whereby each new increment of rainfall displaces all preceding increments, causing the oldest to exit from the bottom end of the system.

Using tagged water in a laboratory rig, Horton and Hawkins (1965) confirmed earlier theoretical suggestions that this would occur in a vertical soil column subjected to successive inputs of simulated rainfall. Ignoring the effects of dispersion, which blurred slightly the following outline, each new rainfall displaced the tagged water and resulted in the outflow of untagged water from the bottom of the soil column until eventually the 'final' rainfall resulted in the outflow of the tagged water itself.

Regarding the soil profile as an inclined column receiving inputs of rainfall, Hewlett and Hibbert (1967) adduced the displacement mechanism to suggest why each input of rainfall could be accompanied by a virtually instantaneous outflow of subsurface water at the slope foot. The weakness of this explanation is that an input will result in an equivalent output only if the available moisture storage capacity within the system is already filled or nearly full. In drier conditions rainfall inputs and/or displacements will be used to 'top up' the soil moisture store rather than to maintain the chain of displacements. This means that the mechanism will be effective most frequently after a period of rain and/or on the lower (i.e. moister) slopes, thereby confirming the result if not the reasoning of Hewlett's original diagram.

Hewlett and Hibbert sought further confirmation of the dominant role of throughflow from the sloping soil models referred to earlier. A sloping soil block was thoroughly wetted, covered to prevent evaporation and then allowed to drain, during which time the outflow was measured continuously. The outflow pipe established a free water table, which was used as the zero datum for all measurements, and below it a saturated wedge. The soil moisture and outflow data were interpreted as showing that unsaturated drainage from the soil mantle was alone sufficient to account for the entire recession limb of the storm hydrograph in steep forested headwater catchments and that the saturated wedge was not of itself a source but rather a conduit '. . . through which slowly draining soil moisture passes to enter the stream' (Hewlett and Hibbert, 1963). Two decades later Boughton and Freebairn (1985) presented recession data from five small agricultural catchments in south-east Queensland which showed that rates and sources of through-flow evident in the recorded hydrographs were possible in the plough depth of the soil.

7.4.3 *The role of groundwater*

Subsequently the role of the saturated slope–foot wedge in the total outflow of water from the slope profile has been questioned by some workers (cf. Anderson and Burt, 1977). Although Hewlett and Hibbert's original data and conclusions were totally correct, it is clear that the relationship between saturated and unsaturated flow components will be partly a function of slope angle. Hewlett (1982) provided a succinct summary as follows (our italicization):

The steeper the soil body is inclined, the greater will be the contribution of unsaturated flow to sustained outflow. In flat basins, groundwater storage represents a large percentage of total storage; in steep basins, the soil moisture store is much the larger one. *In the long run, the water comes from where the water was.*

Understandably, therefore, the potentially important role of groundwater in explaining the storm hydrograph has been a matter of interest and investigation for many years. Indeed, De Zeeuw (1966) introduced the concept of 'variable source areas' without being aware of Hewlett's work. He argued that in the Netherlands the response of drain and ditch flow to precipitation depends on the number of drains and ditches that are deep enough to cut the water table and that will, therefore, receive the more rapid field (i.e. local) discharge compared with the slower seepage (i.e. regional) groundwater flow. As groundwater levels rise, so more drains and ditches receive the quicker local flow. Subsequently, De Vries (1976, 1977) demonstrated the role of groundwater flow in the evolution of stream networks in the Netherlands and Ernst (1978) provided a quantitative model of such a system. O'Brien (1977) showed that groundwater accounted for 93 per cent of the total annual discharge from two small wetland catchments in Massachusetts, USA, and Sklash and Farvolden (1979) argued from tracer and piezometric measurements in two small Canadian catchments for the acceptance of the '. . . active, responsive and significant role of groundwater in storm runoff'.

In highly permeable catchments and of course in the lowland areas of larger drainage basins, groundwater has long been accepted as the major component of streamflow, although in such areas there has been an implied association between the magnitude of the groundwater contribution and the diminution of the response of rivers to precipitation. The concern here is rather with the situation where groundwater may make a major contribution to the storm hydrograph in a wide range of hydrogeological and relief conditions and where the response of rivers to precipitation is both rapid and pronounced. Hursh and Brater (1941) had indeed advocated such a role for groundwater near the stream channels, although some three decades were to elapse before other investigators (e.g. Pinder and Jones, 1969; Dincer *et al.*, 1970; Martinec *et al.*, 1974) used field and particularly tracer measurements to support the view that groundwater is a major and active component of storm runoff. The question of how groundwater could appear sufficiently rapidly in the stream channels was plausibly resolved by Sklash and Farvolden (1979) in terms of a large and rapid increase in groundwater potential near the stream channels, reflecting the formation of a groundwater ridge. The resulting steepened hydraulic gradient and increased groundwater discharge area together would then be capable of producing large groundwater contributions to the stream channel. A similar groundwater ridge has been observed by Ragan (1968) and Hewlett (1969) who referred to '... an ephemeral rise in the groundwater table' near the stream channel which '... helps produce the storm hydrograph'. Zaltsberg (1987) analysed 15 years of data for

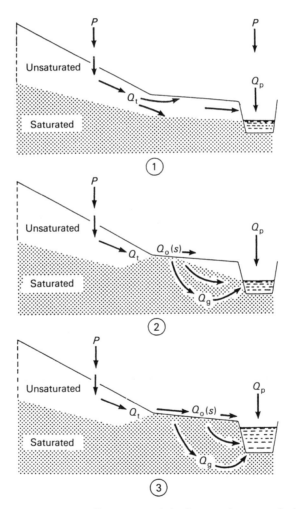

Figure 7.10 *The response of streamflow to precipitation: an integrated view.* Q_t *includes both matrix flow and pipeflow.*

the Wilson Creek basin in Manitoba and showed that, on average, groundwater contributed about 30 per cent to summer storm runoffs.

The mechanism of near-stream groundwater ridge formation is probably that proposed by Vaidhianathan and Singh (1942) since, as was shown earlier (Section 7.4.2), it is in the lower-slope areas that the gradient of moisture potential will facilitate rapid increases of potential as a result of comparatively modest surface inputs of water, an effect subsequently confirmed by Abdul and Gillham (1984). However, a further important factor is the possibility of preferential groundwater recharge in zones of convergence, as proposed by Zaslavsky and Sinai (1981), who argued that convergence may lead not only to the emergence of flow lines at the ground surface but also to a corresponding downward deflection leading to concentrated groundwater recharge. Since the lower valley sides are often concave in profile and will therefore be convergence zones, we have now arrived at the situation summarized in Fig. 7.10. Water does indeed infiltrate the slope surface and move as *throughflow* (including pipeflow) in the slope mantle. Convergence and infiltration in the lower-slope areas will lead to surface saturation *and* groundwater recharge which will create both an *overland flow* and a *groundwater* contribution to the storm hydrograph, with the groundwater ridge merging eventually in some locations into a wider riparian area of surface saturation. Figure 7.10 may be considered then as a visual summary of the preceding discussion and as a conceptual framework, which goes some way towards integrating the apparently conflicting hypotheses of more than half a century.

7.4.4 *Exceptions to the Hewlett hypothesis?*

The strengths of the Hewlett hypothesis are that it accommodates a broad diversity of field observations of runoff, that it incorporates realistically the important dynamic aspects of the runoff process, e.g. the inherently non-linear effect of a variable source area, and that it appears to accommodate the entire range of runoff formations from Horton overland flow, including the extreme case of the car-park hydrograph at one end of the range to the deep porous basin, with stable channel length, in which total runoff is derived almost entirely from subsurface flow components at the other end of the range. Horton overland flow should, in other words, be considered properly as an example of Hewlett saturated overland flow when infiltration rates are so much lower than rainfall intensities that vertical flow convergence results in rapid surface ponding. However, some hydrologists prefer to treat it as a 'different' process in those environmental conditions where infiltration rates are exceptionally low or rainfall intensities exceptionally high.

Thus, as Morin and Jarosch (1977) observed, for most types of soil in the Mediterranean area and other semi-arid areas where infiltration rate drops rapidly during rainstorms, runoff consists of overland flow and contains virtually no subsurface or throughflow components. In these areas of predominantly sodic soils the rapid drop in infiltration rate is due mainly to the formation of a crust whose hydraulic conductivity is several orders of magnitude lower than the subsurface conductivity (McIntyre, 1958; Morin *et al.*, 1981). The crust forms as a result of both a physical disintegration of the soil aggregates and their compaction by raindrop impact and also a chemical dispersion of the clay particles and the formation of a clogged layer (Agassi *et al.*, 1985). However, field experiments in Spain (Scoging and Thornes, 1979) showed that quickflow resulted from profile-controlled saturated overland flow and not from Hortonian overland flow.

In contrast, it should be noted that, in catchment-scale hydrological investigations extending over several decades in some steeply sloping drainage basins in the semi-arid south-west of the USA, widespread overland flow has never been observed. On the other hand, in one of the most sophisticated and ingenious field experiments so far established in a tropical semi-arid area of gently sloping terrain, Bonell and Williams (Bonell and Williams, 1986; Williams and Bonell, 1987; Bonell *et al.*, 1987b) not only confirmed the existence of localized overland flow but demonstrated a field measurement technique, using adjacent offset troughs, which appears to afford an opportunity to study both its temporal and spatial variability.

Tropical rainforest environments are commonly associated with high rainfall intensities and are believed by some hydrologists to offer conditions conducive to the formation of Horton overland flow. However, in a long-term intensive catchment experiment in north-east Queensland, where daily falls in excess of 250 mm are common, Bonell *et al.* (1983a, 1983b, 1987a, 1987b) found that widespread overland flow produced in undisturbed forest areas was *saturation* overland flow resulting from saturation in the top soil layers caused when rainfall rates exceeded the saturated hydraulic conductivity of the profile below 0.2 m. As a result, no change in runoff hydrology occurred following logging of the forest, although there was a massive increase in the production of suspended solids.

Cassells *et al.* (1985) considered that the major significant difference between runoff processes in temperate and tropical forest catchments is that, in the latter, wet areas are widespread throughout the catchments during storm events rather than being concentrated in riparian areas. Because of the high wet-season soil moisture contents such areas can redevelop almost instantaneously with the onset of intense storms. As a result of surface outflows from these widespread areas, quickflow inevitably accounts for a large proportion of total streamflow. Interestingly, though, field experiments in a small tropical catchment in Brazil by Nortcliffe and Thornes (1984) showed that quickflow there is almost entirely the result of saturated overland flow from floodplain areas immediately adjacent to the stream channel. The authors felt that these findings did not support the view that '. . . a dichotomy exists between alternative models, but rather that a variety of flow-generating mechanisms will occur in different environments and in the same environment at different times'.

These findings in semi-arid and rainforest conditions seem to confirm the intuitive conclusion that, in general, slope, slope material and slope vegetation are in such an equilibrium that all precipitation is able to infiltrate. Only where one or more of these factors has been drastically modified, usually by man or during the course of 'catastrophic' meteorological events, is widespread overland flow generated. Were this not so the entire land surface would be scarred by gullies. The concept of variable (or partial) source areas is an attempt to reconcile the absence of widespread overland flow with the spatial variability of channel flow (cf. Huff *et al.*, 1982) and the rapid response of most streams to precipitation by postulating that over-the-surface movement of water is restricted to limited areas of a drainage basin.

7.4.5 *Hydrograph separation*

In many hydrological analyses it is necessary to separate the volume of discharge under the hydrograph into a quickflow and a slowflow component. Many techniques of 'hydro-

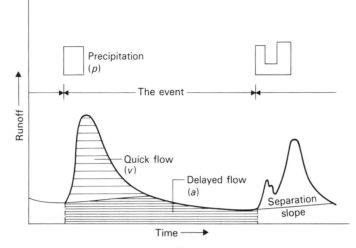

Figure 7.11 *Hydrograph separation into quickflow and delayed flow using a constant separation slope of 0.000 546 cumecs/(km². h), i.e. 0.0472 mm/day. (From an original diagram by J. D. Hewlett and A. R. Hibbert, in* Forest hydrology. © *1967, Pergamon Press Ltd.)*

graph analysis' were predicated on the assumption that it was possible to separate genetically the components of flow on the basis that overland flow would arrive most rapidly at the stream channel, throughflow next, and groundwater flow at the slowest rate. The analysis of runoff variations in this chapter, however, has demonstrated that quickflow may have taken a surface and/or a subsurface route to the stream channel and that slowflow or baseflow may consist of throughflow from the soil profile and/or deeper groundwater flow.

On the basis of time of arrival, therefore, it is feasible to make an arbitrary separation only between quickflow and slowflow. A widely used approach is that proposed by Hewlett and Hibbert (1967) in which 'quick flow' is separated from 'delayed flow' by a line of constant slope (0.05 c.f.s. per square mile per hour or 0.000 546 cumecs per square kilometre per hour or 0.0472 mm/day) projected from the beginning of a stream rise to the point where it intersects the falling limb of the hydrogen (Fig. 7.11). This value was chosen because it was greater than the normal diurnal fluctuation of flow, gave a relatively short time base to the largest single-peaked hydrographs in the study area and permitted large storms separated by a period of about three days to be calculated as separate events. The same value has, in fact, been used by subsequent workers in entirely different areas, cf. Walling (1971) in south-west England. Hornbeck (1973) used a value of 1.25 mm/day in New Hampshire and Woodruff and Hewlett (1970) slightly modified the technique for use in the wider area of the eastern USA.

Other arbitrary separation techniques are illustrated in Fig. 7.12. For example, quickflow and baseflow may be separated by drawing a straight line from the sharp break of slope X where discharge begins to increase to some arbitrarily chosen point (Z) on the recession limb of the hydrograph. Point Z may be located at the point of greatest curvature near the lower end of the recession limb (line 1) (Hursh and Brater, 1941) or at a given time

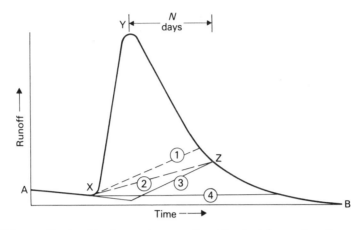

Figure 7.12 *Other methods of hydrograph separation. See text for explanation.*

interval after the occurrence of peak flow (line 2). The time interval (N) may be determined from hydrograph inspection or from a simple empirical equation:

$$N = A^{0.2} \tag{7.1}$$

where N is in days and A is the drainage area in square miles (Linsley *et al.*, 1958). Alternatively, the pre-storm baseflow recession curve (AX) may be projected forwards in time to a point beneath the peak of the hydrograph and then connected by another straight line to the arbitrarily chosen point Z (line 3). Finally, the simplest approximation is that of a horizontal line drawn from point X to its intersection with the recession limb (line 4).

In conclusion, although the genetic separation of the storm hydrograph has been shown to be unrealistic at the present time, it may yet prove feasible, with further research developments, to identify the phase relationships of the various components of the hydrograph, either by measurements of water temperature (Smith, 1968) or by the analysis of water chemistry. Early work on the chemical analysis of runoff was done by Russian workers (Voronkov, 1963) and a useful method was proposed by Pinder and Jones (1969) in which a chemical mass balance was used to calculate the groundwater flow component of total streamflow. Kunkle (1968), Toler (1965) and Nakamura (1971) used conductivity to estimate the groundwater contribution to baseflow.

More recently, Anderson and Burt (1982), using specific conductance as an index of solute concentration, employed a chemical mixing model of the Pinder and Jones type to predict throughflow and overland flow in a catchment where both flow components were also measured. The conclusion was that the model was too simple to predict accurately a complex runoff response. Similarly, Calles (1985) used conductivity to determine the groundwater contribution to a small stream in Sweden. Duysings *et al.* (1983), however, found that the use only of conductivity measurements gave limited or even misleading information about hydrochemical processes. For a forested lowland catchment in the Netherlands their statistical analyses showed significant relationships and different behaviour of three groups of solutes: K, NH_4, NO_3/Mg, Na, Cl/Ca, Si, SO_4, each group having a different response to rainfall. On this basis they were able to distinguish three types of runoff event, i.e. winter and spring storms generating direct runoff *and* baseflow, summer and autumn storms generating only direct runoff and snow melt runoff.

Increasingly during the past two decades, natural isotopic tracers such as deuterium (D) and oxygen-18(^{18}O) have been used to solve a number of hydrological problems including the identification of old and new water contributions to streamflow on the basis that new water (rain or snowmelt) and old water (soil moisture and groundwater) have distinct isotopic signatures (cf. Dincer *et al.*, 1970; Sklash and Farvolden, 1979; Rodhe, 1981; Sklash *et al.*, 1986). Between storm events the stream comprises baseflow which carries the isotopic signature of old water. During runoff events, however, the isotopic character of the stream is diluted by the addition of new water. Deuterium and oxygen-18 were used very successfully by Sklash *et al.* (1986) to identify the source of new water contributions to quickflow in a hydrologically responsive, forested headwater catchment in New Zealand and to confirm the unsuitability for storm runoff studies of chemical tracers, such as chloride, which show a 'flushing' effect (see also Section 8.5.4).

7.5 Daily flow variations

For many purposes variations of runoff with time are studied using runoff values for uniform calendar time intervals (days, weeks, months, years) rather than for runoff events of non-uniform duration. In the case of major continental rivers, where the passage of flood peaks through the system is measured in months, weekly flow values are often appropriate. For the comparatively small catchments of the British Isles, however, which respond rapidly to precipitation/melt events, hydrographs of daily flow values provide a more useful visual comparison of runoff variations. The four examples shown in Fig. 7.13 illustrate a range of flow conditions from the flashy behaviour of the Lune, with its dominant quickflow component, to the subdued behaviour of the Avon, with its very large slowflow or baseflow component. Flow conditions in the Dee and the Thames are clearly intermediate in character.

This long-term relationship between quickflow and baseflow provides a basis for classifying streams as ephemeral, intermittent or perennial. *Ephemeral* streams consist solely of quickflow and therefore exist only during and immediately after a precipitation/ melt event. There are usually no permanent or well-defined channels and the water table is always below the bed of the stream. Ephemeral streams are therefore typical of arid and semi-arid areas and in these conditions are characterized by large transmission losses, as flood waves, generated by storm rainfall, reduce downstream as they are absorbed by the dry stream beds (Lane *et al.*, 1980). In the 150 km² Walnut Gulch catchment in Arizona only about 15 per cent of the water entering the channels as runoff actually leaves the catchment as streamflow (Renard, 1979). *Intermittent* streams, which flow during the wet season and dry up during the season of drought, consist mainly of quickflow but baseflow makes some contribution during the wet season, when the water table rises above the bed of the stream. A particular case occurs in high-latitude areas when flow ceases as groundwater freezes during the winter. *Perennial* streams flow throughout the year because, even during the most prolonged dry spell, the water table is always above the bed of the stream, so that groundwater flow can make a continuous contribution to total runoff. Rarely is it possible to classify the entire length of a stream under only one of

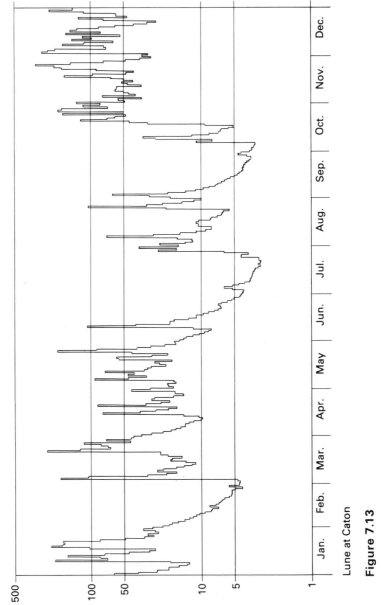

Lune at Caton

Figure 7.13

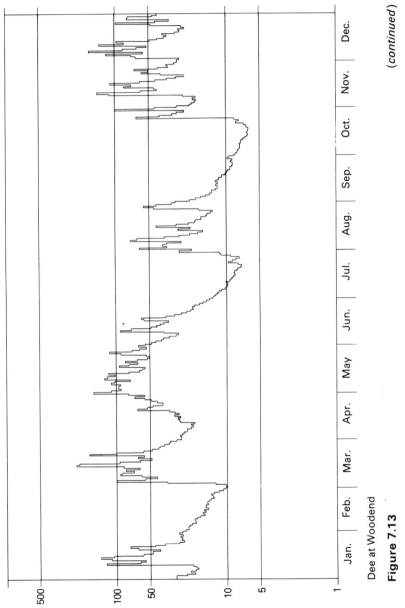

Dee at Woodend

Figure 7.13

(continued)

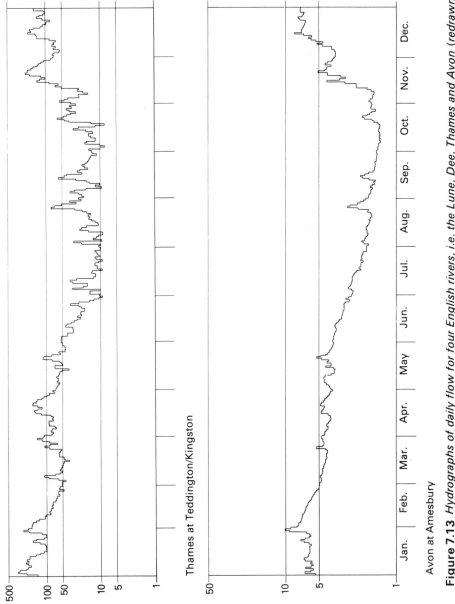

Thames at Teddington/Kingston

Avon at Amesbury

Figure 7.13 *Hydrographs of daily flow for four English rivers, i.e. the Lune, Dee, Thames and Avon (redrawn from one of the standard retrieval options offered by the Surface Water Archive, Institute of Hydrology, Wallingford, OX10 8BB).*

these three headings. A chalk bourne, for example, is normally intermittent in its upper reaches but perennial farther downstream; many other streams are ephemeral in their upper reaches but intermittent downstream.

The contrasting relationships between quickflow and baseflow contributions to total runoff, which is so evident in the daily flow hydrographs in Fig. 7.13, reflect the integrated operation of a wide range of topographical, pedological, vegetational and geological factors that condition the runoff processes described earlier in this chapter. The extremes of flow associated with a flashy stream and the more muted variation of a baseflow-dominated stream may be quantified and compared more conveniently if the daily flow values are arranged according to their frequency of occurrence and plotted as a flow duration curve, i.e. a curve showing the percentage of time that specified flows are equalled or exceeded during a given year or period of years. Flow-duration curves for selected British rivers are published routinely in the Surface Water Archive's new-style Yearbooks (cf. IH/BGS, 1985) and are available for all archived UK rivers as one of the standard available retrieval options. For ease of comparison a flow-duration curve may be plotted in dimensionless form, as in Fig. 7.14, by dividing the daily discharge values by the

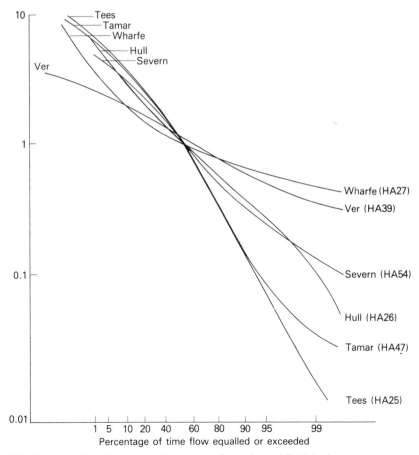

Figure 7.14 *Dimensionless flow-duration curves for selected British rivers.*

average daily discharge value for the period under study. The technique thereby combines in one curve the entire range of stream flows, although not arranged chronologically, and the shape and slope of the curve will reflect the complex combination of hydrological and catchment factors that determine the range and variability of stream discharge.

Flow-duration curves that slope steeply throughout, cf. Tees and Tamar in Fig. 7.14, denote highly variable flows with a large quickflow component, whereas gently sloping curves, cf. Ver, indicate a large baseflow component. In addition, the slope of the lower end of the flow-duration curve characterizes the perennial storage in the catchment (Searcy, 1959) so that a flat lower end indicates a large amount of storage.

Since the slope of a flow-duration curve is a useful measure of streamflow variability various slope indexes have been derived. The US Geological Survey has long related the flow available 50 per cent of the time to the flow available 90 per cent of the time (Searcy, 1959). In Britain, Hall (1968) found that for a wide range of rivers the mean daily flow was exceeded 30 per cent of the time and the modal flow was exceeded 70 per cent of the time, and suggested that the ratio of mean and modal flows could be related to geological and other catchment characteristics. For 58 widely distributed British rivers the median 30:70 per cent ratio was found to be 2.6 (Ward, 1981) with 32 rivers having 30:70 per cent ratios of 2–2.99, 12 falling into each of the adjacent classes, i.e. 1–1.99 and 3–3.99, and the remaining two having ratios greater than 4.0. Flow-duration curves representative of the largest group (2–2.99) are those for the Hull, Wharfe and Severn in Fig. 7.14, which can, therefore, be considered as 'typical' British rivers. Low values are exemplified by the Ver (1.79) and high values by the Tamar (4.32) and the Tees (4.42). In contrast the median 30:70 per cent ratio for the coastal rivers of northern New South Wales, Australia, is 4.4 with a range from 2.8 to 8.4, and for rivers in the Murray–Darling basin the median value is 6.7 with a range from 4.7 to 13.0. For the same Australian rivers the median 50:90 per cent ratios are 5.1 for the coastal rivers and 18.2 for those of the Murray–Darling basin (Ward, 1984).

7.6 Seasonal variations

As Figs 7.1 and 7.13 indicate, streams and rivers may exhibit not only event-based variations of flow but also seasonal variations which are largely a reflection of climate and, in particular, of the balance between rainfall and evaporation. The pattern of seasonal variations which tends to be repeated year after year is often known as the *regime* of the river or stream. Thus equatorial rivers tend to have a fairly regular regime, tropical rivers show a marked contrast between runoff in the rainy and dry seasons, while in other climatic areas complications may arise from the fact that precipitation falls as snow and does not, therefore, contribute directly to runoff until melting occurs. In this way river regimes may be considered in relation to the climatic zones from which they principally derive (Beckinsale, 1969; Guilcher, 1965; Lvovitch, 1958). Beckinsale (1969) presented a useful climatic classification of river regimes based directly upon the Koppen climatic classification. This is illustrated and very briefly summarized in Fig. 7.15, in which it will be seen that Beckinsale made some allowance for the difficult cases of rivers that originate in mountain snow and ice environments over a wide range of climatic types.

Inevitably, however, the larger and therefore the more important rivers cross one or perhaps several significant climatic boundaries and are difficult to analyse in this way. This

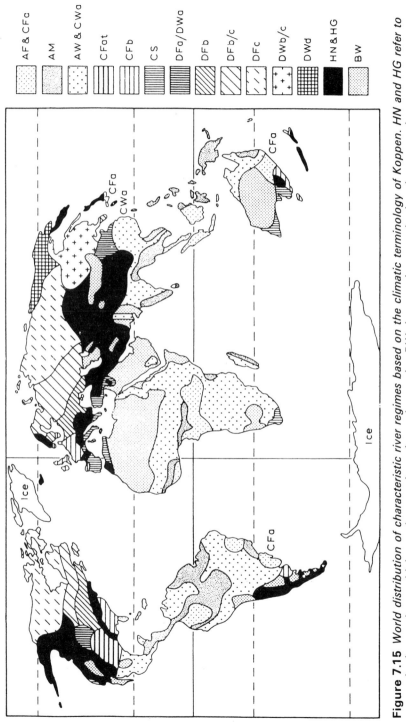

Figure 7.15 *World distribution of characteristic river regimes based on the climatic terminology of Koppen. HN and HG refer to mountain river regimes which are shown where the scale permits; BW denotes desert and other dry areas where streams cannot originate (from an original diagram by Beckinsale, 1969).*

problem was largely overcome by Pardé, who was one of the earliest geographers to attempt a synthetic classification of river regimes based upon the nature, rather than upon the causes, of the seasonal variations in flow. Pardé, in fact, distinguished three main types of river regime: simple, complex I and complex II, in his book *Fleuves et Rivières*, first published in 1933 (Pardé, 1955).

7.6.1 *Simple regimes*

Simple regimes are those variations of river flow throughout the year in which a simple distinction may be made between one period of high water levels and high runoff and one period of low water levels and low runoff. Such regimes may result from one of several contrasting factors (see Fig. 7.16): thus, many European mountain rivers have a high water level period in July and August when the glaciers feeding them melt most rapidly and a very low or zero flow during the winter months when temperatures are low and icemelt is negligible. Again, in many of the oceanic areas of Europe, rainfall is fairly evenly distributed throughout the year but the peak of evaporation during the summer months results in low runoff during this season, in contrast to high runoff values during the winter months when evaporation is small. In tropical areas, on the other hand, evaporation tends to be high throughout the year, so that the rainfall distribution is the main determinant of the river regimes, with high runoff occurring as a result of the summer rains. Finally, simple regimes may result from the melting of a snow cover, either in mountainous areas during the early summer or over the great plains areas, such as those of Eurasia or North America, in late spring.

7.6.2 *Complex I regimes*

Complex I regimes are characterized by at least four, and sometimes as many as six, hydrological phases, although normally there are two low runoff and two high runoff periods. In the case of European streams, the first high runoff period, resulting perhaps from snowmelt, may occur in spring and then be followed by a period of low runoff. Later in the year, a second period of high water levels and runoff may occur in the summer as a result of, say, convectional rainfall over a 'continental' area, or in the autumn as a result of Mediterranean storms, or in the winter as a result of an excess of rainfall over evaporation in an oceanic area. This sequence results, then, in two periods of peak runoff which are separated by two periods of lower discharge, giving four distinct hydrological phases through the year.

7.6.3 *Complex II regimes*

Complex II regimes form the third and probably the most important group in Pardé's classification and are found on most of the world's large rivers. Since these normally flow through several distinct relief and climatic regions, and may receive the waters of large tributaries which themselves flow over varied terrain, rivers comprising this group normally have simple or complex I regimes in their headwater reaches but downstream are gradually influenced by a variety of factors such as snow or glacier melt, rainfall and

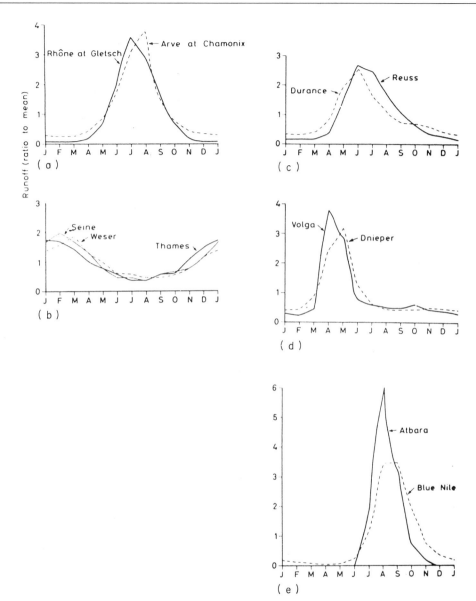

Figure 7.16 *Examples of simple regimes: (a) glacier melt; (b) oceanic rainfall–evaporation; (c) mountain snowmelt; (d) plains snowmelt; (e) tropical rainfall (drawn from data in Pardé, 1955).*

evaporation regimes which may emphasize the trends found in the headwater regime or which, because they work in opposite ways, may cancel each other out.

Three examples of the main types of complex II regime are shown in Fig. 7.17. The Rhine typifies those rivers that are definitely glacier or snowmelt streams in their headwater reaches, but which become increasingly influenced downstream by a single type of rainfall regime. Thus, at Kehl, the Rhine has a simple meltwater regime with a

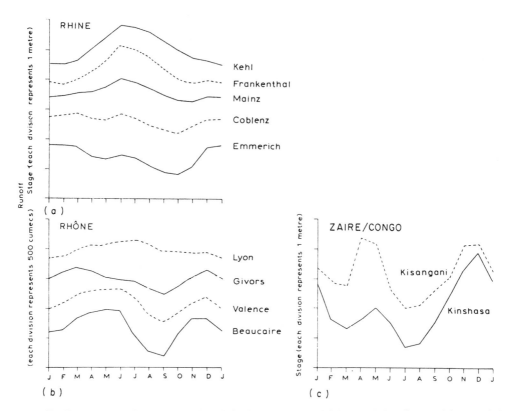

Figure 7.17 *Regimes of (a) the Rhine, (b) the Rhône and (c) the Zaire/Congo (from original diagrams by Pardé, 1955).*

summer maximum; after the confluence of the Neckar at Frankenthal, the slightest signs of a winter rainfall peak become apparent, and this feature is further strengthened after the confluence of the Main at Mainz. After the entry of yet another large tributary, the Moselle, the regime is again altered considerably, and at Coblenz, the runoff peak attributable to the excess of winter rainfall over evaporation almost equals the summer meltwater peak. Finally, at Emmerich, on the Dutch border, the winter oceanic rainfall regime has become the dominant factor and the contribution of meltwater now gives rise only to a secondary peak in the summer months.

The Rhône is an interesting example of those rivers that are influenced initially by meltwater and then, farther downstream, by at least two types of rainfall regime. At Lyon, a glacier melt summer maximum is apparent, but at Givors, below the confluence of the Saône, an oceanic rainfall–evaporation regime becomes dominant. Farther downstream, however, the influx of water from the Isere, above Valence, reestablishes the dominance of the meltwater peak, although there is still a secondary winter maximum. Finally, in the lower reaches, as at Beaucaire, a Mediterranean rainfall regime is superimposed upon all this, resulting in a rapid increase of runoff in the autumn.

Thirdly, the Zaire/Congo typifies those rivers whose regimes are influenced only by rainfall—but rainfall of two or more distinct climatic zones. Throughout its length, the Zaire/Congo has a regime of double high waters corresponding to the double rainfall maxima of equatorial regions; but complications arise, partly because the Zaire/Congo basin covers such a vast area and partly because the greater part of it lies south of the equator, so that the runoff peak resulting from the inflow of the southern hemisphere tributaries is larger than that resulting from the northern hemisphere tributaries. At Kisangani, just north of the equator, the regime is most typical of the equatorial type, but by the time Kinshasa is reached, the vast inflow of waters from the southern hemisphere tributaries has resulted in a very prominent December maximum.

7.6.4 *River regimes in Britain*

Apart from papers by Pardé (1939) and Ward (1968) and a brief discussion in Rodda *et al.* (1976), little has been written about British regimes. Analysis of 37 regimes, calculated from a 10-year run of data, showed that 29 fell clearly within Pardé's 'simple' classification of the oceanic rainfall–evaporation type with summer minima and winter maxima (Ward, 1968). This is to be expected since even the largest rivers, such as the Severn and Thames, are short by world or even by European standards so that there is little possibility of the regimes being complicated by a succession of contrasting influences. The remaining eight regimes showed certain characteristics of Pardé's 'complex I' category, having either a definite period of high water in both summer and winter, or an extended period of winter high water together with a tendency towards a period of high water in summer, or, finally, two periods of autumn and winter high water. These complex regimes occur on rivers in the north and west and can probably be accounted for by a combination of geological and climatological factors, including spring snowmelt and reduced summer evaporation.

There is also clear evidence that the time of occurrence of runoff extremes becomes successively later towards the south and east of Great Britain. This is illustrated in Fig. 7.18 which shows that mean monthly maximum runoff in north-west Scotland and the western extremities of England and Wales occurs in December but is as late as March in parts of East Anglia and Hertfordshire. Similarly, the mean minimum monthly runoff occurs in June over Scotland, northern England and northern Wales, and even in May in the case of some Cheshire rivers, but is as late as September or even October in parts of eastern and central southern England. These runoff trends partly reflect the pattern of evaporation but particularly the greater water-holding capacity and the consequently later release of accumulated water from the sedimentary rocks of the English plain. An equally clear geological influence has been noted at a subregional scale. In the south-east, for example, there are obvious contrasts between 'chalk' and 'clay/urban' catchments in terms of the time of occurrence of minima flows (Marsh, 1988).

Monthly mean flows for the period of record for UK rivers are available in the form of computer graphs or data tabulations as one of the many retrieval options offered by the IH Surface Water Archive in its series of annual yearbooks *Hydrological Data UK* (cf. IH/BGS, 1985) and in the valuable five-year catalogue of river flow gauging stations, groundwater level recording sites and statistical summaries of data, the first of which, *Hydrometric Register and Statistics 1981–85*, appeared in 1988 (IH/BGS, 1988).

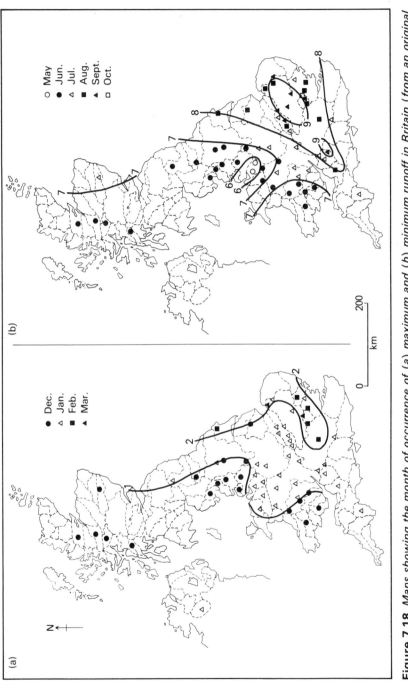

Figure 7.18 Maps showing the month of occurrence of (a) maximum and (b) minimum runoff in Britain (from an original diagram by Ward, 1981).

7.7 Long-term variations of flow and flow variability

By definition river regimes are an expression of seasonal conditions averaged over many years. Since similar seasonal patterns will tend to occur in both wet and dry years, regime graphs tend to suggest a stability of long-term runoff which may be grossly misleading. Annual runoff values exhibit a variability that reflects closely the variability of precipitation and that, as with precipitation, is approximately inversely related to the annual totals (shown for runoff in Fig. 7.19). There is therefore a marked contrast in the variability of runoff between humid and dry areas.

Comparisons between British rivers and those of the semi-arid Karoo of South Africa and the Murray–Darling basin in south-eastern Australia illustrate significant differences. For example, the coefficient of variation of annual flow for the Thames (1883–1986) was 0.29 and for the Severn (1922–1985) was 0.20, but for six Karoo rivers having records of between 30 and 55 years the average coefficient of variation was 0.89 (Gorgens and Hughes, 1982). However, for the Barwon at Walgett (1886–1974) and the Darling at Menindee (1881–1959) the coefficients of variation of annual flow were 1.30 and 1.46 respectively. With such a high degree of variability several centuries of flow data would be necessary in order to derive satisfactory estimates of mean flow conditions. Indeed, in such circumstances, the concept of 'average' flow conditions becomes relatively meaningless, as is further illustrated in Fig. 7.20. The bar graphs emphasize the great variability of flow from year to year, especially in the case of the Barwon and Darling, and also illustrate the alternation of sequences of wet and dry years and their contrasting relative importance in the two areas. For example, the Barwon and Darling are characterized by sustained periods of low flow interspersed with shorter periods of high flow, whereas the Thames and Severn are characterized by long periods of medium to high flow and shorter intervening periods of low flow. This bunching or grouping of wet and dry conditions has been referred to as 'persistence' or the 'Hurst phenomenon', and is an important complicating factor in the stochastic variation of precipitation driving the runoff process. Even a cursory examination of Fig. 7.20c, for example, discloses a sequence of approximately 30 'dry' years to about 1910, followed by approximately 30 'wet' years to the early 1940s, and finally a period of 'intermediate' years since then. Furthermore, the Hurst coefficient, which indicates the tendency towards persistence, itself appears to vary with time, and for three consecutive 30-year periods of Thames flow data was 0.51, 0.10 and 0.54 (Francis, 1973).

For most areas of the world, flow records are much shorter than those for the Thames or Darling which have been discussed here. The opportunities for identifying trends in the variation of runoff with time are therefore extremely limited. However, for some areas there have been comparatively successful reconstructions of runoff from the very much longer records of precipitation. An interesting example was provided by Marsh and Littlewood (1978) who estimated runoff from England and Wales for the period 1728–1976.

Eleven-year running means of these values are shown in Fig. 7.21 together with those for mean annual rainfall over England and Wales as presented by Rodda *et al.* (1978). The graphs show very clearly the alternation of wet and dry periods already evident in Fig. 7.20. Superimposed on this pattern are the 50-year means and long-term trend lines. Although the 249-year mean runoff value is 435 mm, the five successive 50-year means (49 years for the final period) are 413, 417, 453, 447 and 443 mm, indicating either a slight upward trend through the period of record or, as the trend lines suggest, possibly a

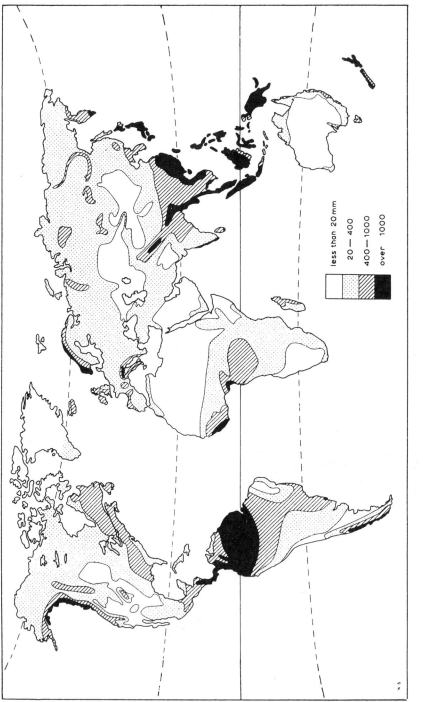

Figure 7.19 *Simplified world map of mean annual runoff. (From an original map by M. I. Lvovitch, Transactions of American Geophysical Union, **54**, p. 34, Fig. 1. © 1973 American Geophysical Union.)*

less than 20 mm

20 — 400

400 — 1000

over 1000

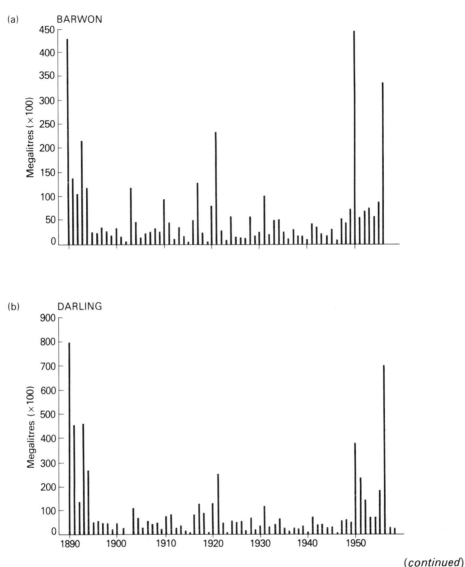

(continued)

Figure 7.20 *Annual flow for (a) the Barwon and (b) the Darling rivers in Australia and (overleaf) (c) the Thames and (d) the Severn in Britain.*

period of about 150 years in which there was a clear upward trend followed by a period of about 100 years, since the late 1870s, in which there has been a scarcely discernible downward trend. Since the mid-1970s runoff variability has been significantly greater than average—markedly so by comparison with the previous 15 years or so. During this period runoff has been greater than average, especially in Scotland, and within-year variability has been characterized by lower low flows and higher high flows, if not by an increased frequency of major floods (Marsh, 1988).

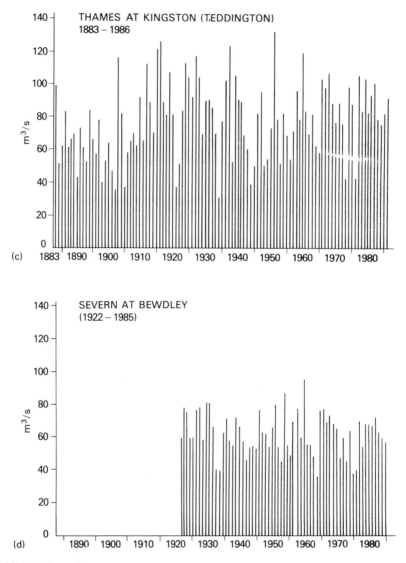

Figure 7.20 (*continued*)

7.8 Extremes of runoff

To some extent persistence accentuates the contrast between the extremes of flow illustrated in Fig. 7.20. Flood conditions resulting from a given precipitation event tend to be more severe if that precipitation event occurs at the end of a long sequence of such events. Similarly, low flow or drought conditions will intensify as the preceding dry period is prolonged. Thus in many parts of England and Wales the extreme low flows recorded in August 1976 (see Fig. 7.30) came at the end of the driest 17-month period on record. However, the comparisons between extreme high and low flows should not be pursued

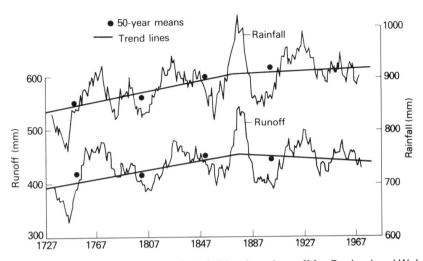

Figure 7.21 *Eleven-year running means of rainfall (top) and runoff for England and Wales, 1728–1976, with 50-year means and estimated trend lines (based on data presented by Rodda et al., 1978, and Marsh and Littlewood, 1978).*

too far since although extreme flow events *are* totally dependent on antecedent conditions, extreme flood events are much more directly dependent on the severity of the causal precipitation/melt event. Severe floods can, and of course frequently do, occur in deserts and other persistently low rainfall areas.

7.8.1 *Flood flows*

Flood definitions
Apart from the rare effects of landslides or dam failures, river floods are caused almost entirely by excessively heavy and/or excessively prolonged rainfall or, in areas of snow or ice accumulation, by periods of prolonged and/or intense melt. In each case the operative processes result in a large input of quickflow to the stream channel. The severity of the flooding will, therefore, reflect the severity of quickflow-forming processes which, as Fig. 7.22 indicates, comprise not only the amount and intensity of precipitation and melt but also the effects of a number of intensifying conditions, most of which operate to speed up the movement of water within a catchment area, i.e. to reduce the time of concentration when this is defined as the time required for water falling or melting on the most remote part of the catchment to contribute to streamflow at its outlet. The time of concentration is thus the shortest time in which the whole of the catchment contributes to runoff either directly, as water moves over and through the surface layers, or indirectly through the piston displacement mechanism, whereby water infiltrating on the higher parts of the catchment will cause a related amount of water to be exfiltrated lower down the slope. The main flood intensifying conditions are illustrated in Fig. 7.22. Some are comparatively stable and conservative; others are relatively more variable.

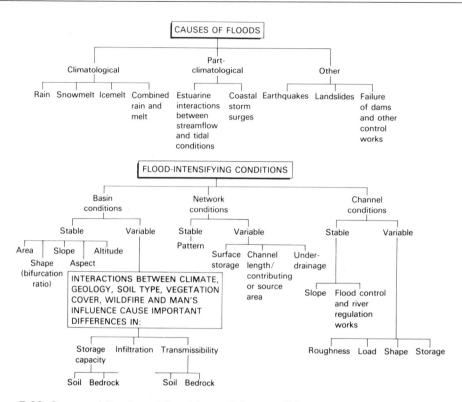

Figure 7.22 *Causes of floods and flood-intensifying conditions.*

Catchment conditions

Of the stable catchment characteristics *area* affects not only the time of concentration but also the total volume of runoff generated by a catchment-wide precipitation event. *Shape* is also important, particularly in association with the disposition of the channel network. For example, where a high bifurcation ratio R_b is associated with a long, narrow catchment, flood peaks may be low and attenuated, especially with a storm moving upstream, whereas a comparatively circular catchment with a low birfucation ratio will generate higher and sharper flood peaks (Fig. 7.23a). Water movement, both over and through the slope materials, will increase with *slope*, while *aspect* and *altitude* affect both the amount and type of precipitation as well as the extent to which its effectiveness is modified by evaporation. The integration of stable catchment characteristics effectively defines drainage basin geomorphology which has been shown by numerous workers to relate significantly to flood response.

Variable catchment characteristics are numerous and their effects on flood hydrology are often complex. However, three important secondary characteristics result from the complex interactions between, for example, climate, geology, soils, vegetation and the effects of man in burning, clearing, cultivation and urbanization. First, the capacity for water *storage* of the soil and deeper subsurface layers may affect the timing and magnitude of flood response to precipitation, with low storage capacity increasing

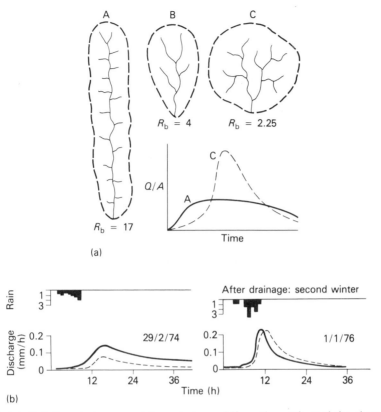

Figure 7.23 (*a*) *Relations between catchment shape, bifurcation ratio and the shape of the flood hydrograph (from an original diagram by Strahler, 1964); (b) hydrographs from a small clay loam catchment before and after field drainage (from an original diagram by Robinson* et al., *1985).*

quickflow production. Second, low *infiltration* rates encourage a high proportion of overland flow. Third, even in a catchment where most of the precipitation infiltrates the soil surface, flood response will be modified greatly by subsurface *transmissibility*. Considerable success has been achieved in Britain, through the preparation of a map of Winter Rain Acceptance Potential (WRAP), in relating the important hydrological properties of the soil to its potential for generating flood flows (Farquharson *et al.*, 1978).

Channel conditions
Channel and channel network characteristics are essentially dynamic and variable, often changing markedly within a few hours. *Drainage pattern* is probably the most conservative network characteristic, its effect closely related to that of bifurcation ratio and catchment shape. Generally speaking, dendritic patterns, which result in the coalescence in the lower catchment of flood flows from a number of major tributaries, are associated with sharp, high-magnitude flood peaks at the catchment outlet. Patterns that permit the evacuation from the catchment of flood flows from the downstream tributaries before those from the upstream tributaries have arrived often result in a more muted flood response.

The relatively more variable network characteristics may be very important. Initially, large volumes of unconnected *surface* (*depression*) *storage* may act as a storm reservoir which can contribute quickflow to the stream channels only when the necessary interconnections have been made by continued precipitation. Indeed, as earlier discussions have shown, it is the total area of interconnected saturated surface within the catchment that basically determines the volume of quickflow produced by a precipitation/melt event. During the early stages of precipitation, especially after a prolonged dry period, these areas may be restricted to the water surface of the channel network, but as precipitation continues so the *source areas* of quickflow expand until in extreme flood conditions the entire catchment may be contributing quickflow to the stream channel.

The effects on flood flows of *artificial drainage*, such as the furrowing often associated with afforestation or the underdrainage of arable farmland, appear to be variable and are not yet fully understood. In some clay soils the magnitude of flood peaks is increased, at least temporarily (see Fig. 7.23b) by rapid water movement through fissures induced by the mechanical or desiccating effects of artificial drainage (Reid and Parkinson, 1984; Robinson *et al.*, 1985). In the uplands of mid-Wales, however, Newson and Robinson (1983) found that drainage resulted in lowered soil water levels and reduced peak flows.

With reference to relatively stable channel characteristics, it is clear that the passage of a flood peak will be faster in unregulated steep channels and slower in well-regulated, relatively flat ones. The downstream velocity and magnitude of a flood peak will also be determined by variable channel characteristics, particularly roughness, which depends largely on constituent bed and bank materials and on vegetation growth, as well as channel shape and storage properties, which may vary rapidly with changing flow and load conditions. The load being carried in solution or suspension or as bedload within the channel is also important for the augmenting effect it has on total water depth and discharge, and may range from virtually zero in some clearwater streams to 100 per cent or more in a mudflow.

Spatial patterns of flood discharge in Britain

Two aspects of the spatial distribution of flood discharges in Britain are illustrated in Fig. 7.24. The first is an index of flood potential in the form of an estimate of mean annual flood (Fig. 7.24a) and the second is an indication of flood experience during the period of recorded flow data in the form of the highest instantaneous gauged discharge (Fig. 7.24b).

The best estimate of mean annual flood (BESMAF) was derived by the NERC flood studies team (NERC, 1975) by extending the recorded flood experience by correlation with nearby flood records. When such extension was not possible the arithmetic mean of annual maximum floods was used or, when the period of record was too short, some other technique was employed, e.g. the use of a peak-over-threshold series. For ungauged catchments equations were developed relating the mean annual flood to several catchment characteristics. The six-variable equation is

$$\bar{Q} = C.AREA^{0.94}STMFRQ^{0.27}S1085^{0.16}SOIL^{1.23}RSMD^{1.03}(1 + LAKE)^{-0.85} \quad (7.2)$$

where C is a coefficient which varies regionally, AREA is catchment area (in km²), STMFRQ is the number of stream junctions (per km²), S1085 is the slope from 10 to 85 per cent of the main stream length (in m/km), SOIL is an index based on the winter rain

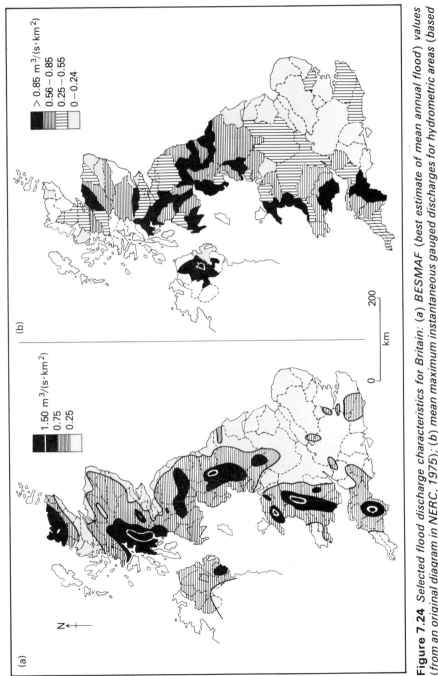

Figure 7.24 *Selected flood discharge characteristics for Britain: (a) BESMAF (best estimate of mean annual flood) values (from an original diagram in NERC, 1975); (b) mean maximum instantaneous gauged discharges for hydrometric areas (based on data in DOE, 1978).*

acceptance rate of the catchment, RSMD is the net daily rainfall having a return period of five years (in mm) and LAKE is the proportion of the catchment draining through lakes. All the catchment characteristics incorporated in the equation can either be read directly from topographic maps or obtained from standard tables. Subsequent work has shown that the six-variable equation is not equally suited to all areas of Britain, and Hanna and Wilcock (1984), for example, suggested that a modified form of the four-variable equation of the *Flood Studies Report* should be used in Northern Ireland.

The pattern of isopleths drawn through the BESMAF values is a fairly predictable one. Apart from small isolated areas of higher ground or relatively impermeable surfaces, lowland Britain has BESMAF values substantially below 0.25 cumecs/km². High values, above 0.75 cumecs/km², are restricted to Dartmoor, the Welsh mountains, areas of the Pennines and Cumbria, the Southern Uplands and parts of western and northern Scotland.

Despite the difference in cartographic presentation, a comparison of BESMAF with the distribution of highest instantaneous gauged discharges, averaged for each hydrometric area (Fig. 7.24b), reveals a striking similarity. In both cases it should be noted that the comparatively low values of flood flow per unit area over south-eastern England may be misleading in the sense that they are, to some extent, compensated by the large area of the catchments concerned. The highest instantaneous gauged discharge for the Thames at Teddington, for example, is approximately 790 cumecs, which exceeds that for the Tees at Broken Scar of 679 cumecs or for the Clyde at Blairston of 577 cumecs. The usefulness of maximum gauged discharges as an index of flood potential obviously depends upon the length of record, since long records are more likely than short ones to incorporate rare, high-magnitude events. Extension of the comparatively brief gauging records of most British rivers will, therefore, gradually increase the values currently mapped in Fig. 7.24b until they approach useful 'design flood' dimensions in terms of their magnitude–frequency relationships.

It was found that the relationship between mean annual flood ($\bar{Q}$) and the flood of a given return period (Q_t) could be generalized for each of the defined regions shown in Fig. 7.25a and the factor by which the mean annual flood value must be multiplied to obtain Q_t for selected return periods is tabulated in Fig. 7.25b. Thus Fig. 7.25a and b may be used in conjunction to obtain approximate values of floods having those specified return periods. Furthermore, analyses using the large European database coordinated as part of the FREND (Flow Regimes from Experimental and Network Data) project (IH, 1987) have confirmed that in areas having the same flood-producing mechanism it is possible to 'pool' or average flood frequency curves so as to produce inter- as well as intranational comparisons, cf. Fig. 7.25c.

7.8.2 *Low flows*

At the other extreme the problems posed by low flows, although different, are equally varied and severe. Low flows not only reduce the amount of water available for supply but also lead to water quality degradation, as the diluting and reaerating capability of the stream is reduced, and thence to the aesthetic degradation of the affected channel reach. The ASCE Task Committee on Low Flow (ASCE-TASK, 1980) drew attention to these problems and noted the inadequacy of methods for estimating low-flow characteristics

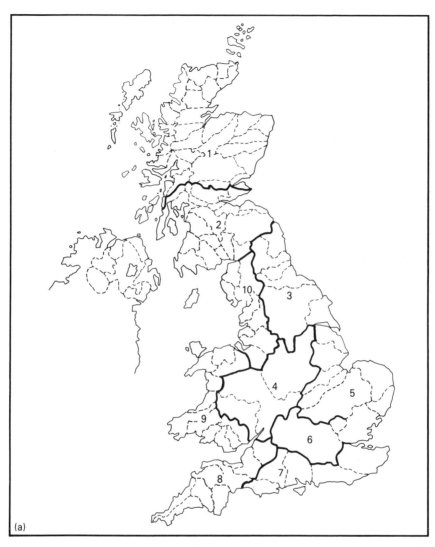

(a)

(continued)

Figure 7.25 (a) *Regions defined by the NERC flood studies team (NERC, 1975); (b) mean annual flood multipliers for deriving Q_t for selected return periods (from data in NERC, 1975); (c) average flood frequency curves for selected European areas (from an original diagram in IH, 1987).*

and the need for further studies to indicate appropriate ways of using low-flow information.

Low-flow definitions
At least in part the inadequacy of methods for estimating low flows derives from the failure of hydrologists to standardize low-flow definitions. There is, for example, no low-flow equivalent of the PMF. The probable *minimum* flow that can be experienced is obviously zero but in most small catchments and many larger ones (cf. the Darling River draining an

Region	Return period						
	2	5	10	25	50	100	200
1	0.90	1.20	1.45	1.81	2.12	2.48	2.89
2	0.91	1.11	1.42	1.81	2.17	2.63	3.18
3	0.94	1.25	1.45	1.70	1.90	2.08	2.27
4	0.89	1.23	1.49	1.87	2.20	2.57	2.98
5	0.89	1.29	1.65	2.25	2.83	3.56	4.46
6/7	0.88	1.28	1.62	2.14	2.62	3.19	3.86
8	0.88	1.23	1.49	1.84	2.12	2.42	2.74
9	0.93	1.21	1.42	1.71	1.94	2.18	2.45
10	0.93	1.19	1.38	1.64	1.85	2.08	2.32
Great Britain	0.89	1.22	1.48	1.88	2.22	2.61	3.06
Ireland	0.95	1.20	1.37	1.60	1.77	1.96	2.14

(b)

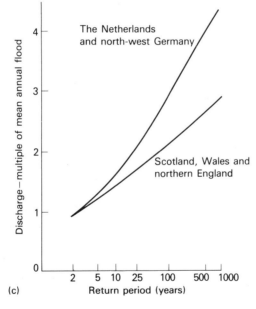

(c)

Figure 7.25 (*continued*)

area of 570 000 km²) zero flows occur frequently, and it is then the *frequency* of zero flows that is the more useful index of low-flow conditions. In virtually all rivers, of course, the problems associated with low flows are manifest long before a flow of zero is attained, and are intensified as the duration of a given low-flow discharge is prolonged. Hydrologists are therefore concerned primarily with defining selected critical low-flow discharges and with identifying the duration and frequency of spells of low flow.

As Pirt and Douglas (1982) observed, there are several flow conditions that can be defined as 'low'. One is the lowest flow ever experienced, a condition that was probably achieved over much of southern and central England in the late summer of 1976. A more usual measure, however, is the 95 per cent exceedance flow (Q95), i.e. the flow that plots at the 95 per cent position on a flow-duration curve and is therefore exceeded 95 per cent of the time or on average on all but 18 days of the year.

Because, unlike extreme high flows, extreme low flows tend to be prolonged rather than instantaneous, the daily flow interval is less appropriate than a period flow interval of, say, 7 days or 10 days. Thus in the Institute of Hydrology's *Low flow studies report* (IH, 1980) equations were derived for estimating Q95(10), i.e. the 95 percentile 10-day flow.

Alternative flow *frequency* characteristics may be used rather than flow-duration characteristics to define low-flow conditions. The *Low flow studies report* (IH, 1980) used the mean annual minimum 10-day flow, MAM(10), although an older and apparently more useful measure (Pirt and Douglas, 1982; Pirt and Simpson, 1982) is the condition defined by Hindley (1973) as Dry Weather Flow (DWF), i.e. the mean annual minimum 7-day flow. This is approximately the driest week in the average summer and is exceeded from 89 to 93 per cent of the time depending on the type of catchment (Pirt and Douglas, 1982). DWF obviously has a return period of 2.33 years (Q7, 2.33) but in the USA it is the 10-year return period 7-day flow (Q7, 10) which is the most widely used index of low flow (ASCE-TASK, 1980). Some analyses have been based on very extreme flows, cf. the 7-day 20-year flow (Q7, 20), but few gauging stations are designed to measure extreme low flows and so recorded data may be subject to large measurement errors (Pirt, 1983). Loganathan *et al.* (1985) assessed various frequency analysis methods as applied to low-flow data for streams in Virginia, USA.

Residual flow diagrams
In comparison with flood flows, low flows may be substantially affected by human activity, especially the discharge of effluents into stream channels and the abstraction of water from stream channels for supply purposes. In both cases the added or extracted water may represent a very large percentage of the natural flow and must therefore be taken into account if analyses of low-flow data are to shed light on natural low-flow processes. Lloyd (1968) suggested the residual flow diagram as a simple graphical method of accounting for artificial interference with low flows, and the technique was again argued convincingly by Pirt (Pirt and Douglas, 1982; Pirt, 1983).

A residual flow diagram (RFD) separates streamflow at any point into its natural and artificial components. In Fig. 7.26, for example, distance downstream is shown on the vertical axis and flow on the horizontal axis. Natural runoff is shown to the left of the vertical axis and the artificial flow component to the right, so that total flow is represented by the distance between the natural and artificial lines, e.g. A'–A". If the artificial line crosses the central axis this indicates that total flow at that point is less than the natural flow would have been in the absence of human interference. Thus in the hypothetical example shown in Fig. 7.26 the river consists of natural flow only in reaches 1 and 2, with a natural tributary entering in reach 2. The first artificial component is shown in reach 3, representing perhaps industrial effluent or compensation water from a reservoir. In reach 4 abstraction of water for supply depletes the total flow and reduces natural flow below its 'untouched' level. A major tributary enters in reach 5 and since it contains both natural and artificial flow components both sides of the RFD are affected. Another tributary enters in reach 6, but in this case abstraction has taken place from the tributary as indicated by the

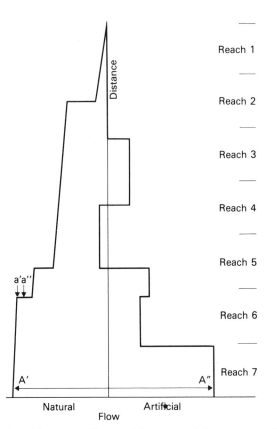

Figure 7.26 *A simplified residual flow diagram (from an original diagram by Pirt, 1983).*

reduced artificial component, and finally in reach 7 there is a large increase in artificial flow in the main stream.

A residual flow diagram can be constructed for any selected flow condition, e.g. Q7, 2.33 or Q7, 10, and then used as the basis for investigating low-flow hydrology, particularly in terms of catchment controls. For example, the portion to the left of the central axis represents only the natural runoff of the main stream and its tributaries. It is, therefore, possible to divide the net addition to natural runoff between any two points on the main stream by the contributing catchment area and so calculate the natural runoff yield per unit area or natural runoff coefficient. This was done for contributing subcatchments within the Severn–Trent basin by Pirt and Simpson (1982) and the coefficients mapped as in Fig. 7.27. Using relevant geological and topographic information it was then possible to interpolate runoff coefficients for ungauged catchments and to provide a first estimate of Q7, 2.33 for any ungauged point solely from a knowledge of catchment area.

Factors affecting low flows
In broad terms low flows are determined by the balance between precipitation and evaporation and are therefore particularly susceptible to persistence when this results in

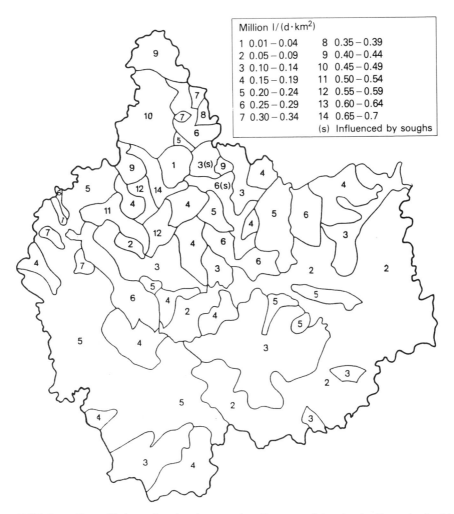

Million l/(d·km²)

1 0.01 – 0.04	8 0.35 – 0.39
2 0.05 – 0.09	9 0.40 – 0.44
3 0.10 – 0.14	10 0.45 – 0.49
4 0.15 – 0.19	11 0.50 – 0.54
5 0.20 – 0.24	12 0.55 – 0.59
6 0.25 – 0.29	13 0.60 – 0.64
7 0.30 – 0.34	14 0.65 – 0.7
	(s) Influenced by soughs

Figure 7.27 *Runoff coefficients for the dry weather flow condition in the Trent basin (from an original diagram by Pirt, 1983).*

the bunching of a sequence of dry years. Within a drainage basin experiencing essentially uniform climatological conditions, however, such as that of the Trent shown in Fig. 7.27, other more local or catchment controls play a significant role in determining the detailed pattern of low-flow variation. Inevitably, since low flows comprise baseflow, subsurface factors such as soil and geology are likely to play an important part. For example, Fig. 7.28a, based on FREND project data (IH, 1987), shows the dimensionless annual minimum series for two catchments. The strong control of geology is clearly identified, with groundwater discharge in the chalk catchment sustaining low flows even in extreme droughts, whereas low flows from the clay catchment are at very low rates throughout the range of flow magnitudes. A similar contrast is shown by the flow-duration curves in Fig. 7.28b, which are also based on FREND data. Pirt and Douglas (1982) and Pirt and

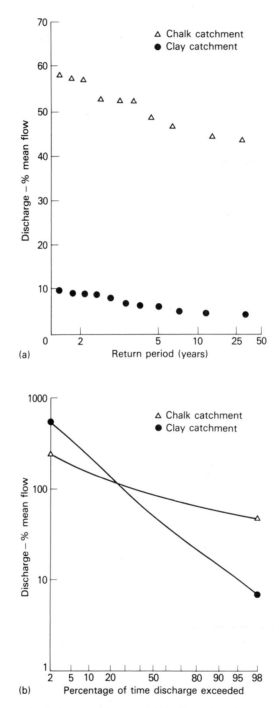

Figure 7.28 (a) *Annual minimum series and* (b) *flow-duration curves for clay and chalk catchments* (*from original diagrams in IH, 1987*); (c) *set of normalized flow-duration curves for Carboniferous rocks* (*from an original diagram by Pirt, 1983*).

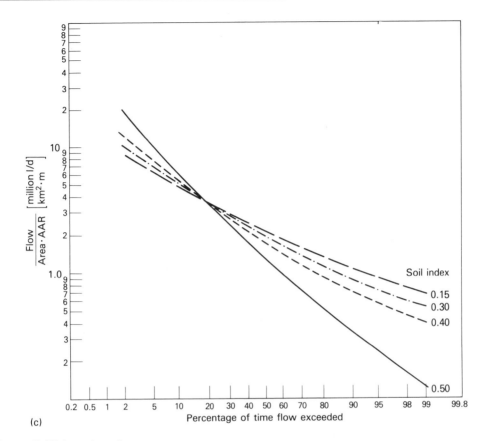

Figure 7.28 (*continued*)

Simpson (1982) developed methods for extrapolating the RFD/runoff coefficient analysis, discussed in the preceding subsection, in order to quantify geological influences on low flows.

It was acknowledged that although RFD and runoff coefficient maps relate to only one specified flow conditions, e.g. Q7, 2.33, many users would prefer to compare the data to another statistic, e.g. Q7, 10 or Q7, 20. This was done in two ways. In the first approach a family of naturalized flow-duration curves were normalized by dividing by area (km²) and mean annual effective rainfall (m), and the variations between them were found to depend almost entirely on catchment geology and soil. The various curves were then averaged to provide a set of master curves relating to specific geological and soil types using the NERC *flood studies report* soil index (NERC, 1975), cf. Fig. 7.28c for Carboniferous rocks. These can then be used to derive flow-duration curves for ungauged catchments for which area, effective rainfall, geology and soil type are known, by selecting the appropriate master curve, multiplying a number of points on the curve by effective precipitation and catchment area, and finally adjusting for any artificial effects.

In the second approach multiple regression analysis of eight catchment variables produced an equation to predict the difference (Q_{diff}) between the discharge per unit area for Q7, 2.33 and that for Q7, 20, and thus the slope of the normalized flow-frequency curve. Flow-frequency curves for ungauged catchments can then be developed for the map of runoff coefficients (Fig. 7.27) by multiplying the relevant runoff coefficients by catchment area. This gives the 2.33 year flow which is then plotted on Gumbel probability paper. Q_{diff} is calculated using the equation referred to above and is subtracted from the 2.33 flow value to give the value of Q7, 20 which is plotted at the 0.95 probability position on the Gumbel paper. A straight line drawn through the two points then permits either the interpolation of 7-day flows having intermediate return periods or extrapolation to higher return periods. Examples of such plots for the Henmore Brook at Ashbourne and the River Sence at Blaby are shown in Fig. 7.29, together with comparative plots derived from measured data. The results indicate that good estimates of flow-frequency curves can be obtained in this way for ungauged sites provided that maps of DWF runoff coefficients are available (Pirt and Simpson, 1982).

An alternative approach to the study of the low-flow characteristics of British rivers is the development of the baseflow index (BFI) by the Institute of Hydrology (cf. Beran and Gustard, 1977). BFI is an index of hydrograph behaviour which is related to catchment

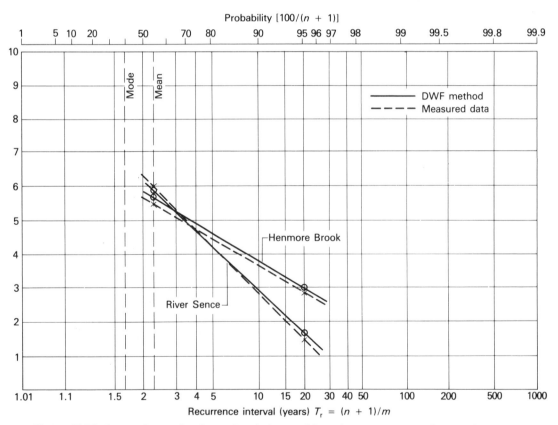

Figure 7.29 *Comparisons of estimated and observed flow-frequency curves for gauging stations at Ashbourne and Blaby (based on data in Pirt, 1983).*

geology and recent work in Scotland has resulted in the production of a stream network map of BFI at a scale of 1 : 62 500 (IH, 1984) which can be used to estimate low-flow statistics at ungauged sites.

Spatial patterns of low flow in Britain

Low-flow studies received a tremendous stimulus in Britain as a result of the 1975–6 drought which generated a substantial literature of its own, culminating in the publication of a comprehensive *Atlas of drought in Britain, 1975–6* (Doornkamp *et al.*, 1980). In terms of streamflow, the effects of the drought were quite dramatic, resulting over much of

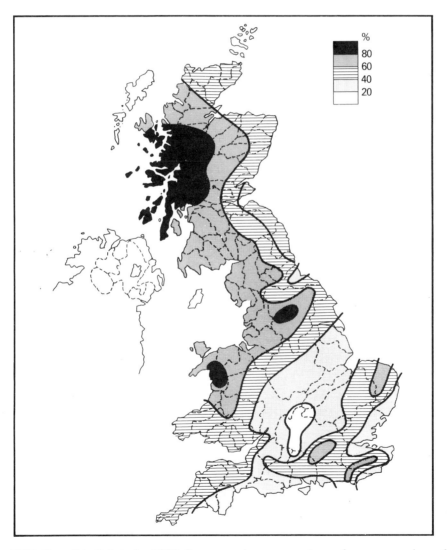

Figure 7.30 *Runoff in Britain for 1975–6 expressed as a percentage of mean annual runoff (from an original diagram in Ward, 1981).*

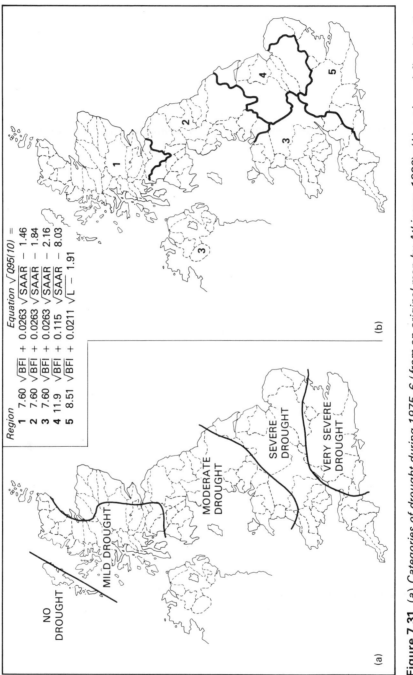

Figure 7.31 (a) Categories of drought during 1975–6 (from an original map by Atkinson, 1980); (b) regions defined by the NERC low-flow study (based on data in IH, 1980).

southern England in runoff values less than 40 per cent of the long-term mean, although even within this area there were 'islands' of comparatively higher runoff associated with precipitation and catchment conditions (see Fig. 7.30).

Atkinson's (1980) spatial categorization of the 1975–6 drought is shown in Fig. 7.31a and it is interesting to compare this with the pattern of regions demarcated by the IH low-flow study (IH, 1980) estimates of Q95(10) in Fig. 7.31b, which are based on BFI and either annual average precipitation (SAAR) or main stream length (L).

The most recent drought in the UK was in 1984. The development and severity of this drought were examined by Harrison (1985) and by Marsh and Lees (1985) with particular reference to regional variations in its intensity, and valuable statistical and cartographical comparisons were made between conditions in 1984 and those in 1975–6.

7.9 Runoff from snow-covered areas

At a superficial level runoff from permanently or seasonally snow-covered areas differs from rainfall-driven runoff in that the period of low flow often coincides with the period of maximum precipitation and the periods of high flow with the frequently much drier period of maximum melt. In detail, streamflow from a snow-covered catchment will reflect not only the complex interaction of factors discussed earlier but also the effects of additional factors, of which the most important are (a) the water equivalent of the snowpack, i.e. how much water would be released if the snowpack melted completely; (b) the rate of melting; and (c) the physical characteristics of the snowpack, e.g. ice-banding, differential compaction, ripeness, etc.

Clearly the spatial variation of each of these factors (cf. Ferguson, 1985a), together with the spatial variability both of melt-season rain falling upon the snowpack and also of subsnowpack terrain conditions, combine to create an enormously complex runoff system which looks deceptively simple viewed from an aircraft or a space satellite. Broad-ranging reviews of snow hydrology (cf. Colbeck et al., 1979; Barry, 1981, 1983; Rango, 1985) emphasized that despite great advances in instrumentation and remote sensing, scientific progress in this field is limited by fundamental weaknesses in the quantity and quality of our data.

The primary areas of outstanding research interest are concerned with snowpack surveys, energy exchanges and snowmelt in the snowpack, and the relationship between snowpack properties and water movement.

7.9.1 *Snowpack surveys*

Of primary concern in snow hydrology is the estimation of the water equivalent of a snowpack and the spatial variation of water equivalent within a catchment area. For this purpose point measurement, e.g. using pressure sensing devices such as snow pillows or metal pressure tanks, is still the most reliable and most widely used source of data. The problem of adequate sampling, however, is acute since snow covers tend to exhibit much more spatial variability than rainfall distribution because of the differential effects of drifting and partial melt (see also Section 2.7.2).

Understandably, therefore, much hope is being placed in the development of remote sensing techniques whereby an airborne or spaceborne sensor could map synoptically the areal extent, thickness and water equivalent of the snowpack over large areas on a periodic, repeating basis (cf. van de Griend and Engman, 1985).

Water equivalent surveys in areas of low relief have been carried out to an accuracy of about 1 cm water equivalent using low-flying aircraft carrying gamma-ray detectors which monitor natural radiation from the soil (Larson, 1975), but much further work remains to be done, as was emphasized by Barry (1983) when he noted that on several major snowpack mapping problems using remote sensing techniques progress was being made '. . . including the discrimination of snow/ice from cloud'!

7.9.2 *Energy exchange and snowmelt*

Snowmelt is the result of many different processes which result in a net transfer of heat to the snowpack (see also Section 2.7.3). The principal energy-balance fluxes involved were discussed by Wilson (1941) and subsequent investigators have confirmed the relative importance of these energy sources for snowmelt. According to Colbeck *et al.* (1979) the most important are solar radiation, long-wave radiation (especially at night), sensible heat transfer from the air to the snow by convection and conduction, and latent heat transfer by evaporation and condensation (and occasionally rain) at the snow surface. Conductive heat flow from the ground to the base of the snowpack is of comparatively minor importance, but together with the penetration of short-wave radiation into the top few centimetres of the snowpack and the much deeper penetration of rainfall, it exemplifies the partially distributed nature of snowmelt energy sources. Nevertheless, as Colbeck *et al.* (1979) emphasized, meltwater is predominantly generated at the snowpack surface, and this has been confirmed by sequential density profiles of melting snowpacks.

In forested areas or other situations where the vegetation cover protrudes through the snowpack the energy exchange is very much more complex than in the open bare snow situation, and many of the relevant components, e.g. radiative and sensible heat transfer and evaporation, are almost impossible to measure accurately. Even in situations away from the complicating effects of vegetation, where energy flux measurements can be made it is only possible to make those measurements 'at-a-point'. In order to develop satisfactory flood forecasts or soil water forecasts these point measurements must be extrapolated to areas of several hundreds of square kilometres. Although some progress has been made in respect of albedo and net radiation (cf. Storr, 1972), much work still remains to be done on the extrapolation from point measurements of long-wave radiation and of sensible and latent heat fluxes.

In terms of the generation of runoff from snow-covered areas, the situation is complicated further by the fact that snow covers which may look alike will produce widely different streamflow contributions from a given energy input because of the amount by which the snowpack temperature falls below the melting point (0°C). Baseline conditions for snowmelt calculations assume: (a) the latent heat of ice is 3.35×10^5 J/kg (80 cal/gm); (b) the snow is pure ice; and (c) the snow temperature is 0°C.

Often, however, especially in the winter months, snowpack temperatures are well below 0°C so that initially heat is required to raise the temperature to melting point. Until that temperature is reached no meltwater runoff will occur. On the other hand, during the melt season the snowpack may not only be isothermal at 0°C but may also contain some liquid

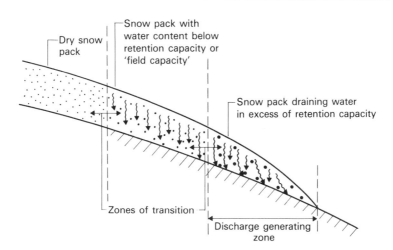

Figure 7.32 *The distribution of snowpack conditions in a mountain area (from an original diagram by van de Griend, 1981).*

water in the interstices of the ice matrix up to a water content equivalent to the retention capacity (cf. 'field capacity') of the snowpack, i.e. the snowpack is 'ripe'. Liquid water in excess of the snowpack's retention capacity will drain through the pack as gravity flow.

Figure 7.32 suggests that in mountainous areas these different snowpack conditions will be distributed in relation to elevation. In snowpack areas in plains, however, the distribution of ripe and unripe snow will be much less predictable. In either situation it will be clear that when a snowpack containing liquid water is melted, the liquid water is also released, resulting in a total outflow from the snowpack which exceeds that derived simply from energy-balance considerations.

It has been suggested that in extreme circumstances snowpack 'collapse' may occur where an ultraripe snowpack, comprising a blend of ice matrix, snow and the maximum possible amount of liquid water, is subjected to rapid thaw conditions, e.g. sudden temperature rise plus heavy warm rainfall. Then the matrix of the snowpack may rupture sending a slush flow or slush avalanche, i.e. a cascade of snow, ice and water downslope to the channel system. An early paper by Wolf (1952) discussed this component of the Glen Cannich floods in Scotland. Subsequently slushflows and slush avalanches have been widely documented (cf. Washburn, 1980; Hestnes, 1985; Onesti, 1985).

7.9.3 *Snowpack properties and water movement*

The rate at which meltwater, generated near the surface of a snowpack, can move through the pack to the underlying ground surface and thence to the channel system will be greatly affected by the structure and stratification of the snowpack. Most snowpacks develop a layering of less permeable ice strata within the generally more permeable snow matrix. According to Colbeck *et al.* (1979), this layering arises from the sequential nature of snow deposition and/or the preferential retention and refreezing of water in fine-grained layers such as wind crusts. The layers divert the percolating meltwater so that complicated flow paths develop which delay and diffuse the outflow of meltwater.

During the initial stages of infiltration, if the retention or 'field' capacity of the snowpack is not exceeded, no meltwater runoff will occur at all. Subsequently the downward movement of the wetting front into a naturally stratified snowpack is accompanied by delays and ponding at the ice layers and the development of flow fingers which eventually penetrate to the base of the snowpack (Wankiewicz, 1978; Colbeck, 1975, 1979; Marsh and Woo, 1984, 1985). These flow fingers appear to remain as zones of higher permeability even after snowpack ripening. Thus, from the early stages of melting and despite a large portion of the snowpack remaining 'unripe', rapid flow paths for meltwater may be opened up.

Similarly, beneath the snowpack, the initial routing of meltwater early in the melt season is through a saturated layer overlying the soil surface. Then as meltwater channels develop a more rapid drainage of the snowpack occurs because the flow path length for meltwater has been greatly reduced. Colbeck *et al.* (1979) observed that as a result of flow finger development through the pack and melt channel development beneath the pack, a snow-covered catchment '... makes a gradual transition from snow-controlled to terrain-controlled water movement' and expressed surprise that '. . . virtually no information exists about the temporal and spatial distribution of these features of the snow cover'.

7.9.4 *Modelling runoff from snow-covered areas*

Because of the complex dynamics of processes operating within and below the snowpack during the melt period, it is unsurprising that there have been few successful attempts to model runoff from snow-covered areas. Indeed, Price and Hendrie (1983) emphasized that, even in a hydrological system with apparently uniform soil, vegetation and snowpack characteristics, a great variability of process was observed during snowmelt, including a variation in the runoff coefficient from 0 to 60 per cent.

The major elements in most models of snowmelt runoff will be snow accumulation, interception (in situations where vegetation protrudes through the snowpack), heat-balance components, water retention and movement within the snowpack, and the associated subsnowpack and channel processes. In addition, an ideal model should accommodate the spatial variability of all these elements. These elements were identified, in an excellent review of snowmelt runoff modelling (WMO, 1986), as the components of a series of interlinked submodels, viz. a meteorological, a snowmelt and a transformation model. These deal respectively with the meteorological inputs to the snowpack, the amount of meltwater delivered to the ground under the snowpack and the output of streamflow resulting from this predicted meltwater value. Ferguson and Morris (1987) suggested the need for a fourth submodel, a depletion submodel, to predict the areal extent of the snowpack which plays such an important part in determining the volume of meltwater produced.

Although most models of runoff from snow-covered areas incorporate a fairly sophisticated snowmelt submodel, their success in incorporating the other three submodels varies very widely. Three contemporary examples may serve to illustrate this point. Archer (1983) applied a five-parameter model for forecasting snowmelt flood runoff to a number of small upland catchments in north-east England. Akan (1984) presented a physics-based mathematical model of springtime runoff from snow-covered hillslopes which accounted for the processes of liquid water flow, heat conduction and vapour diffusion within the snowpack and coupled runoff from the unsaturated zone of the snowpack with a saturated basal flow. Finally, Marsh and Woo (1984) developed a model that combined the wetting front and flow finger processes described earlier, the growth of ice layers after ponding at pre-melt horizons and at the base of the snowpack, infiltration into frozen soils and heat conduction within the snow and underlying soil.

Models of runoff from snow-covered areas are still in a comparatively early stage, especially those involving detailed consideration of the energy balance. Considerable improvement will undoubtedly be sought not only in the modelling of at-a-point conditions but, more importantly still, in integrating at-a-point conditions into a catchment-wide pattern of spatial variations. However, at present, except for very small catchments, Ferguson and Morris (1987) questioned whether or not '... the best available meteorological and transformation models are still too inaccurate to make it worthwhile using a physics-based snowmelt model'.

7.9.5 *Runoff from glacierized areas*

In high mountain areas the presence of glaciers as well as of seasonal or permanent snow cover adds a further dimension to runoff variations. These areas are widely distributed throughout the world from the high Arctic to the tropics and in both maritime and continental situations (Young, 1985), and were considered by Roots and Glen (1982) to play '... an extremely important and distinctive role in the hydrological processes of the planet, and in the regional hydrology of all continents'. Acknowledgement of the hydrological importance of glacierized areas has been reflected in a succession of international symposia on, for example, the hydrology of glaciers (IAHS, 1973), snow and ice (IAHS, 1975), hydrological aspects of alpine and high mountain areas (IAHS, 1982) and techniques for the prediction of runoff from glacierized areas (Young, 1985), and more recently in an edited volume of papers on glaciofluvial sediment transfer (Gurnell and Clark, 1987).

As in snow-covered areas, the hydrology of glacierized basins is to a large extent thermally controlled. The interaction between variations in energy supply and variations in precipitation amounts and type (rain or snow) leads to variations either in the production of meltwater or in the storage of ice and snow within the drainage basin. Changes in the balance between meltwater production and storage from year to year mean that annual total runoff may be greater than or less than annual precipitation, i.e. glacierized basins are characterized by changes in snow and ice storage between years.

As well as snow and ice storage, however, *liquid* water may be stored within glaciers and in marginal lakes alongside them. Furthermore, the changing morphology of the glacier ice matrix within which water flows and is stored and the dynamic interaction between the glacier ice and the liquid water stored within it, complicate further the variations of runoff from glacierized basins. As Young (1985) observed, a characteristic of runoff from these areas is the outburst floods created by the sudden release of large quantities of water stored within, under or alongside glaciers. Understandably, such events add further to the difficulties of modelling runoff from glacierized basins.

Characteristics of glacier runoff
Glacier runoff is characterized by two main components, i.e. a *periodic*, thermally driven meltwater regime, which produces distinctive diurnal and seasonal variations of flow, and

an *aperiodic* component, which results from the occurrence of either extreme meteorolog-ical events or sudden releases of water from the glacial drainage system. In addition, *long-term* runoff variations may reflect aspects of climatic variation and climatic change.

Figure 7.33 shows the expected *diurnal cycle* of meltwater runoff. It also illustrates, however, that during the first part of the melt season there is an increasing 'baseflow' component of runoff upon which this diurnal cycle is superimposed. Rothlisberger and Lang (1987) suggested that baseflow from glacierized basins comprised groundwater runoff, runoff from water-filled cavities within the ice, runoff from the melt-fed firn water aquifer in the accumulation area of the glacier and regular drainage from lakes. They further suggested that the diurnal 'quickflow' component whose variation is super-imposed upon that of baseflow, consists of the rapidly draining component of that

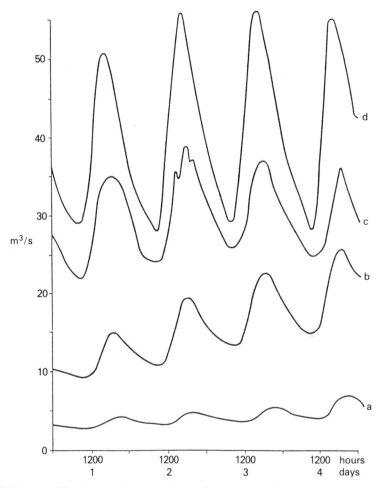

Figure 7.33 *The runoff hydrograph of a glacierized basin, the Matter-Vispa, during four periods in 1959:* (a) *17–20 May,* (b) *14–17 June,* (c) *23–26 June and* (d) *19–22 July* (*from an original diagram by Elliston, 1973*).

day's meltwater, i.e. meltwater draining supra- and subglacially from the lowest parts of the basin and meltwater from the snow-free part of the glacier which drains via short conveyances to the main subglacial conduits.

The *seasonal variation* of runoff is exemplified in Fig. 7.16a. The increase of discharge during the summer reflects in general the seasonal increase of available melt energy, although the detailed timing of variations in flow reflects also the progressive development of the glacier drainage system during the ablation season and the build-up of baseflow from storage as described above.

Aperiodic variations of runoff from glacierized basins are exemplified by unusually high flood discharges resulting from (a) periods of very rapid melt over a week or more, which permit high rates of baseflow as well as of quickflow, (b) the occurrence of extreme high-intensity rainfall, especially late in the afternoon when meltwater runoff is at a maximum, or (c) sudden releases of water ('yokulhlaup' is the Icelandic term) which has either been held in storage within the glacier, or as surface lakes on or adjacent to the ice, or has been dammed back by ice in tributary valleys. In some cases, where such damming results from the seasonal advance of a glacier, the flood discharge may show a marked periodicity, as in the regular annual drainage of the Gornersee which develops at the junction of the Gorner and the Grenz glaciers in Switzerland (Bezinge, 1987).

Finally, *long-term variations* of glacier runoff, which result mainly from long-term changes of climate, are an amalgam of conflicting influences. Periods of persistently warmer summer weather result in increased ablation and high runoff values, whereas a series of cool summers favours increased storage and low runoff. However, continued ice removal causes shrinkage of the lower sections of glaciers and since these are the zones of highest melt rate, their progressive disappearance results in a corresponding loss of potential for meltwater yield (Rothlisberger and Lang, 1987). Thus in the European Alps a period of high average runoff in the 1940s, the warmest decade since records began, is superimposed upon a clear declining trend in summer runoff which has accompanied the reduction in glacier area since the last maximum of the Little Ice Age.

Modelling glacier runoff

Fountain and Tangborn (1985) summarized a number of glacier runoff models using hourly or daily data input which has been considered by the IAHS Working Group on Prediction of Runoff from Glacierized Areas. The simplest of these assume a linear relationship between air temperature and melt rate, with meltwater contributing immediately to runoff. The most sophisticated models incorporate detailed energy balances and route meltwater internally through the glacier. All have two essential components, one representing meltwater production and the other the drainage of water from the glacier, and it is the latter component that tends to result in most of the recognized model deficiencies. As Fountain and Tangborn (1985) observed, not only is the internal hydrology of glaciers still poorly understood, despite much theoretical study, but also theories about internal drainage are difficult, if not impossible, to verify experimentally.

At a much larger scale Ferguson (1985b) presented a conceptual model of runoff from glacierized mountains in which basin meltwater yield in a particular year is related to off-glacier snowmelt, glacier snowmelt and icemelt. The model was developed to accommodate year-to-year variations, the effects of snow cover and the effect of summer weather, and was applied successfully to contrasting tributaries of the upper Indus.

7.10 Man's influence on runoff

There are very few areas of the world in which runoff is not affected to some extent by the influence of man. In remote uplands dams have been constructed for water supply and hydroelectric power generation. In the tropics vast areas of rainforest have been removed. Elsewhere, former grasslands have been ploughed up, moorlands have been forested, semi-desert areas have been irrigated, swamps have been drained, and everywhere there has been a great increase in urbanization, and the resulting spread of artificial, impermeable surfaces. In all these ways the response of catchment areas to rainfall and consequently the pattern and distribution of runoff has been changed.

7.10.1 *Hydraulic structures*

The flow of many of the world's large rivers is controlled or modified by dams and reservoirs which have been constructed for power, water supply, irrigation or flood-control purposes. Beaumont (1978) presented graphs showing the dramatic increase in dam building world-wide between 1945 and about 1970. During this period more than 8000 major dams were built and it seems inevitable that, as a result, the hydrology of many river systems was changed significantly. Beaumont quoted data from Lvovitch (1973), reproduced here as Table 7.1, which shows the amount of 'stable' runoff, i.e. largely groundwater-derived baseflow, that is regulated by lakes and reservoirs. It can be seen that man's impact is particularly marked in Africa and North America, where about 20 per cent of stable runoff is regulated by reservoirs, and in Europe and Asia, where the figures are 15 and 14 per cent respectively.

The effect of these structures has been similar to that of natural lakes to the extent that flood peaks are normally 'absorbed' by the artificial lakes and subsequently released gradually. An additional effect, in some instances, has been a marked diminution of flow, particularly where multipurpose schemes are in operation in which the impounded water is used not only for water supply and power but also for irrigation, and related changes in channel form and sediment characteristics (Petts and Lewin, 1979).

Structures located away from the stream channels may also influence runoff. For example, the use of snow fences to control the spatial accumulation of snowfall and thereby also the timing and yields of water from subsequent melting were described by Berndt (1964) and Martinelli (1964).

Table 7.1 'Stable' runoff (km³) of the continents (based on a table in Lvovitch, 1973).

| | Of under-ground origin | Regulated by | | Total | Regulated by reservoirs as % of total | Total stable runoff as % of total runoff |
		Lakes	Reservoirs			
Europe	1065	60	200	1325	15.1	43
Asia	3410	35	560	4005	14.0	30
Africa	1465	40	400	1905	21.0	45
North America	1740	150	490	2380	20.6	40
South America	3740	—	160	3900	4.1	38
Australasia	465	—	30	495	6.1	25

7.10.2 *River diversion*

Streamflow may also be affected by artificial modifications to stream channels, especially in areas that are prone to flooding. Such modifications commonly include straightening and enlargement and the construction of relief and bypass channels in order to reduce both discharge and water levels at critical points.

Diversions of flow from one river to another are now an accepted and commonly implemented strategy in water resources development. In Britain the Water Resources Board strategic plan for England and Wales (Water Resources Board, 1973) incorporated inter-river transfers, e.g. from the North Tyne to the Wear and Tees, already in operation as part of the massive Kielder reservoir scheme, and from the Severn to the Thames. Elsewhere in the world such transfers have taken place over much greater distances. An outstanding example is provided by the California State Water Project in which the California Aqueduct, 714 km (444 miles) in length and with a water-carrying capacity of 165.5 cumecs (4.2 m acre-feet per year), can carry more water a longer distance than any conduit so far constructed.

Of even greater significance are the reversals of flow which have been implemented or are planned in areas where the topography is suitable. On a comparatively small scale, a northward-flowing stream in southern Ontario has been dammed at Longlac and its flow reversed through a comparatively small breach excavated in the shallow drainage divide (see Fig. 7.34). This results in a significant increase in discharge in the southward-flowing Aguasabon River, which is used to float logs down to a major pulping plant at Terrace Bay on Lake Superior. On an entirely different scale are the Russian proposals to divert the flow of some of the major northward-flowing rivers such as the Ob-Irtysh, Yenisey, Pechora and Northern Dvina (Micklin, 1981). In the case of the Ob/Irtysh, this would be achieved by constructing a large dam just below the confluence of the two rivers and leading the water 2600 km (1600 miles) southward to the Amu Darya where it would be utilized in regional irrigation projects in the Aral–Caspian basin (see Fig. 7.35). Such is the scale of these proposals that the full implications for the changed salinity of the Arctic ocean are still uncertain. International concern on ecological and climatological grounds was described by Micklin (1981) and appears to have led to a complete reappraisal of the proposals by the Soviet government.

7.10.3 *Agricultural techniques*

A further aspect of human influence on runoff may result from the application of specific agricultural techniques and practices, particularly where these cause a sudden change in catchment characteristics, e.g. vegetation cover. Some of these changes have been brought about deliberately, as *conservation measures*, and in this context the important work carried out over a long period of time by the Tennessee Valley Authority affords an obvious example. It has been estimated that more than 4000 km² of marginal cropland in the Great Plains of the USA have been converted to permanent grass, as a result of Federal conservation programmes, with a corresponding significant reduction in quickflow. Three years after conversion, runoff was representative of that from native meadow (Dragoun, 1969).

Other changes have brought about 'accidental' and sometimes initially surprising hydrological results. In the Soviet Union, for example, the impact of autumn ploughing,

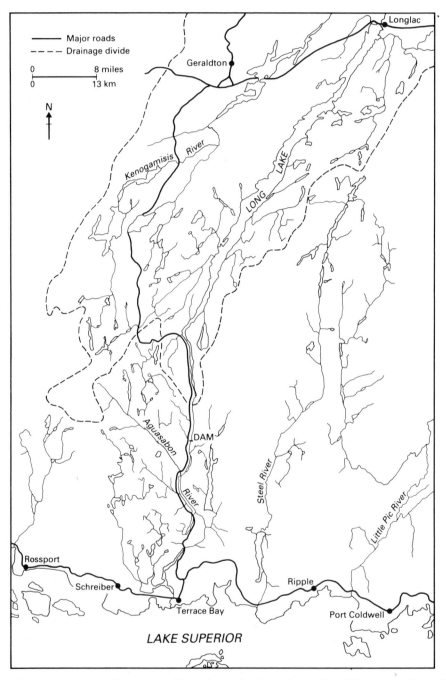

Figure 7.34 *Map showing flow reversal into the basin of the Aguasabon River in southern Ontario.*

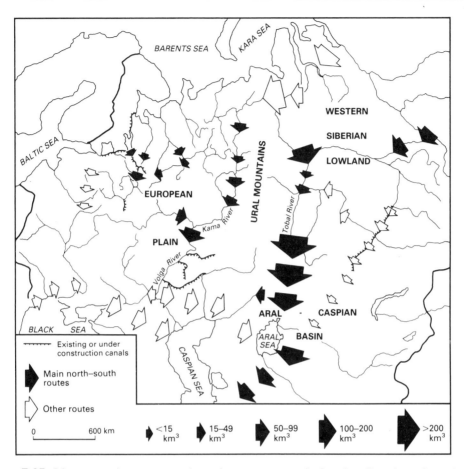

Figure 7.35 *Diagrammatic representation of some proposals for the diversion of northward-flowing rivers in the USSR. (From an original diagram by P. P. Micklin, EOS, **62**, pp. 489–93.* © *1981 American Geophysical Union.)*

which has been widespread since the 1920s, is summarized in Fig. 7.36. According to Lvovitch (1980), autumn ploughing has reduced surface runoff by 1.5–2 times in the forest zone, by 2–4 times in the forest–steppe zone and by 4–8 times in the steppe zone. In each zone this represents an absolute decrease of 30–40 mm, which was accompanied by a significant increase in groundwater recharge. In southern Ontario ploughed soils in good management yielded much less runoff during the winter months than grass-covered areas (Ayers, 1965). It was suggested that this was because the rough texture of the ploughland results in extensive narrow ridges of dark soil being exposed to solar energy. Heat is then absorbed and transmitted readily by these patches to melt the snow, and in addition a large potential for depression storage occurs on ploughed land. Finally, Lvovitch (1980) quoted some interesting analyses of the impact of agriculture on runoff in Germany where Keller (1970) showed that an increase in crop yields in West Germany

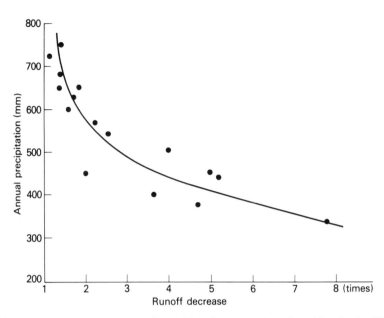

Figure 7.36 *Reduction of surface runoff resulting from autumn ploughing in the USSR (from an original diagram by Lvovitch, 1980. Blackwell Scientific Publications Ltd.).*

between 1931 and 1960 was accompanied by increased evaporation and reduced runoff despite an increase in annual precipitation.

Some of the most dramatic land-use changes are those associated with *afforestation* and *deforestation*, the hydrological effects of which continue to engender controversy. Numerous experiments have demonstrated that the volume and timing of runoff may be modified substantially by forest cutting and removal procedures, e.g. clear cutting, block or strip cutting, and selective thinning. Hibbert (1967) provided an early review of forest experiment data which indicated that most first-year streamflow increases were 300 mm or less and that generally speaking the effect of vegetation treatment declined with time as revegetation occurred. Afforestation of previously non-forested areas seemingly resulted in an average decrease of streamflow of about 220 mm. A subsequent review of universal results by Bosch and Hewlett (1982) was further updated by Trimble *et al.* (1987) (see Fig. 7.37).

A major objective, particularly in important water supply catchments, is to manage the forests so as to permit optimum yields of both timber and water. Research has emphasized the clear relationships between forest cutting and increased runoff, and the possibility of developing sound forest management practices that are also sound hydrological practices. Other possibilities include the replacement of one type of forest cover by another or the replacement of forest by grassland or other agricultural crops. In general it has been shown that runoff is reduced when deciduous trees are replaced by conifers and increased when forest is replaced by lower-growing vegetation such as grass or crops.

In many areas, and certainly in North America, upland drainage basins contain some of the most important reserves of exploitable timber and are also the major water supply

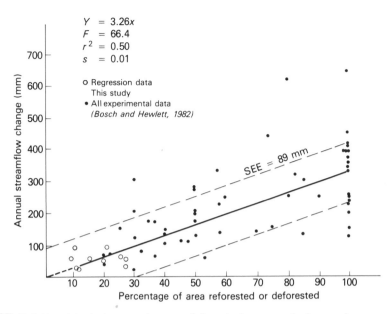

Figure 7.37 *Relationship between change of forested area and change in annual streamflow* (*From an original diagram by Bosch and Hewlett, 1982*).

catchments. Optimization procedures are therefore essential if the development of one resource is not to affect adversely the development of the other. Ideally, management practices would aim to achieve the highest timber yields, using economically viable harvesting methods, consistent with the highest water yields, distributed as uniformly as possible in time. In view of the marked seasonality of snow accumulation and melt this involves the manipulation of vegetation cover and snow accumulation in such a way that melting is spread over an extended period of time. Considerations of water and energy balance in high-altitude forested areas confirm the long-held conclusion, noted by Costin *et al.* (1961), that a forest honeycombed with numerous small openings or cleared strips is the most efficient in accumulating and conserving snow and, by implication, extending the period of snowmelt.

A major problem is that although it is comparatively easy to identify the main ways in which forest and other vegetation covers are hydrologically different (e.g. in terms of soil stabilization and infiltration characteristics, improved soil water storage occurring beneath vegetation covers having high interception and evaporation losses, and spatial and temporal effects on snow accumulation and melt), most of the experimental work that has been done has been of the 'black-box' variety which does not permit these various differences to be individually quantified. Thus vegetation changes within a catchment area have been related to the modified output of runoff from the catchment, and only a few attempts (cf. Hewlett and Helvey, 1970) have been made to detail the hydrological effects or to relate the effects of land use manipulation to modern ideas on the runoff process. Often the hydrological effects of afforestation in upland Britain are complicated by the effects of ploughing and draining prior to planting (Robinson, 1986) and by the fact that in large catchments afforestation is fragmentary and partial, rather than total, and has been achieved over many years (Acreman, 1985).

In low-lying areas, where the water table is close to the ground surface, crops can often be introduced in place of rough marshy grazings, or the yields of existing crops can be improved as a result of *artificial drainage*. Both open ditches and buried tile drains are used extensively in, for example, the low flat areas bordering much of the southern part of the North Sea, in the eastern counties of England, and in the Netherlands. Here, the aims of agricultural management are to maintain the water table at a height that permits the optimum growth of crops but, although groundwater control is the objective, the secondary effects of artificial drainage upon runoff are often quite considerable.

Closely spaced open drains are often believed to behave in a similar way to a high-density natural drainage network, i.e. lateral flow distances are comparatively short so that water should reach the channels quickly. In addition, since the water table is lowered evaporation is reduced and dry period baseflow should be increased. Tile drains, buried below the water table, behave in a similar way, and by encouraging lateral movement of water at some small distance below the ground surface, especially in wet conditions when groundwater levels are high, their effects on the runoff hydrograph may resemble those of throughflow. Lawrence (1981) in fact differentiated between 'top flow', i.e. the rapid horizontal flow, which takes place in the top 30–45 cm (see Fig. 7.38), and the more classical, slower groundwater flow to the tile drain, which occurs with lower water table levels.

Recently Water Authorities in England have begun to express concern that extensive ditching or 'gripping' in moorland areas has significantly affected the runoff response of rivers draining large areas of the Pennines, increasing quickflow and flood potential. Detailed investigation of hydrological data on moorland gripping (cf. Robinson, 1985) and on lowland tile and mole drainage in the UK has improved our understanding of the flow processes that operate and has helped to explain some of the apparently anomalous effects on streamflow. There is evidence that in some upland situations the artificial drainage of mineral soils results in a *reduction* of peak rates of outflow to the stream network (Newson and Robinson, 1983). This reduction results because the lowered water table increases available moisture storage at times of rainfall and encourages the deeper movement of water through less permeable horizons. In contrast, gripping in peat areas affects water table levels only immediately adjacent to the drains. The corresponding-ingly limited effect on soil moisture storage may not then be sufficient to counter the faster flows through the artificial network which results in *higher* peak discharges.

Furthermore, it has been shown that in clay areas the effects of artificial drainage on streamflow may vary seasonally (cf. Robinson and Beven, 1983; Robinson et al., 1987). In dry conditions, especially in the autumn, macropores opened up by shrinkage permit rapid flows and higher discharges from drained than from undrained areas, whereas in wet conditions in winter and early spring the hydraulic efficiency of macropores is reduced by swelling of the clay, and because of the more rapid development of surface saturation in undrained areas peak discharges now tend to be higher from undrained than from drained land.

Finally, Parkinson and Reid (1986) showed that the response of tile drains in heavy clay land is directly related to local ground slope so that drainage efficiency and peak drain discharge are inversely related to slope.

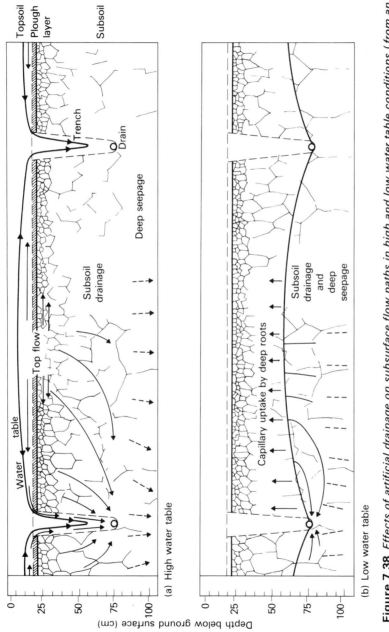

Figure 7.38 *Effects of artificial drainage on subsurface flow paths in high and low water table conditions (from an original diagram by Lawrence, 1981).*

7.10.4 *Urbanization*

The effects on runoff of the spread of settlement and ancillary features such as roads, pavements and airfields has been discussed in detail (Bauer, 1969; Savini and Kammerer, 1961; Hall, 1985) and needs little emphasis. Over large areas infiltration capacity is reduced; falling precipitation is caught by rooftops and roads and is passed through drainage systems which have been designed to dispose of it into nearby streams as rapidly as possible. The result is that, immediately below large urban areas, there tends to be a marked and rapid build-up of quickflow which will be accentuated where slopes are steep. Increases in the magnitude of peak flows are thus a result partly of an increase in the volume of quickflow and partly of the more rapid movement of runoff which is possible in an urbanized area. Anderson (1968) found that these two factors combined to increase flood peaks in northern Virginia by between two and eight times while in Texas Espey *et al.* (1966) found that urban development resulted in peak discharges which were from 100 to 300 per cent greater than those from undeveloped areas. The problems of modelling the relationship between the increase in impervious area associated with urbanization and the changed flood magnitude and frequency characteristics of urbanized streams was discussed, with reference to the work of the Institute of Hydrology, by Packman (1979).

Apart from peak flows, urbanization also affects water quality and a wide range of other hydrological variables. Some of these effects are summarized in the matrix diagram in Fig. 7.39. Not the least important is the reduced soil moisture and groundwater recharge which is an obvious corollary of the increased quickflow component of runoff. Lvovitch (1980) presented comparative data for the centre of Moscow and the Moscow basin downstream of Zvenigrod which demonstrated that surface runoff in the former is 3–5 times higher than that in the Moscow River basin and groundwater recharge is reduced by 50 per cent.

Although urban areas account for only a small percentage of the total land area they accommodate a large and growing percentage of the population. In the USA, for example, it was estimated that in 1971 urban areas comprised less than 2 per cent of the land area but housed more than 75 per cent of the population—a figure that could grow to 85 per cent by 2000 AD. As a result urban hydrology has matured as an identifiable topic within hydrology, with its own specialized literature, concerned largely with the application and adaptation of basic hydrological principles in urban areas. That maturity was manifested by the spate of textbooks which emerged in the late 1970s and early 1980s (cf. Lazaro, 1979; Kibler, 1982; Sheaffer *et al.*, 1982; Whipple *et al.*, 1982; Hall, 1985). Even so, as Delleur (1983) observed, much remains still to be done, particularly regarding the response of urban streams to variations of runoff and pollutant loads, the development of appropriate models and the resolution of the conflicting requirements of drainage versus flood control and water quality objectives.

7.10.5 *Severity of flooding*

In conclusion, it is of interest to note that it is largely the human factors which have been discussed above that are believed to be responsible for the apparent increased severity of flooding during recent times. Floods, which may be defined as unusually high rates of discharge often leading to the inundation of land adjacent to the streams, are nearly

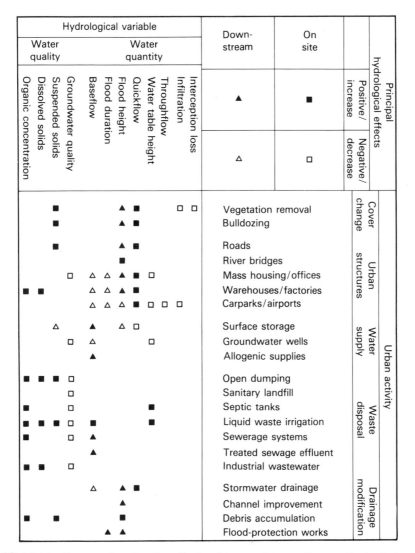

Figure 7.39 *Matrix diagram showing the effects of human intervention on the hydrology of urban areas (from an original diagram by Chandler* et al., *1976, with acknowledgement to* The Geographical Journal).

always the result of quickflow, rather than baseflow, and are thus usually caused by intense or prolonged rainfall or snowmelt, or a combination of these two.

Any increase in the severity of floods, therefore, is likely to be caused by increased rainfall intensity or duration, reduced infiltration capacity or the changed efficiency of the drainage network. The disastrous, widespread flooding which affected about two-thirds of Bangladesh in September 1988, was attributed partially to large-scale forest removal in mountainous headwaters but the principal cause, as always in the case of the Ganges and Brahmaputra, was undoubtedly the heavy monsoon rains. There is some evidence that, in many areas, the flood situation may indeed be deteriorating. Climatic change is undeniable. Small changes in the atmospheric circulation and subtle changes in the delicate interrelationships between ocean temperatures and the overlying atmosphere may well cause *local* alterations in the frequency of intense or prolonged precipitation, and thus in flood magnitude and frequency. The available evidence is, however, largely circumstantial and often difficult to interpret. It is equally undeniable that land use is continually changing. The spread of urbanization, of forest clearance and of agricultural underdrainage, and the ploughing up of natural grasslands have increased flood potential. There is, moreover, some evidence of the changed character of flooding itself, particularly in an intensively developed country like Britain. In early historical times rivers probably flooded much more frequently than today, responding to what would now be regarded as comparatively minor precipitation events. Because of the primitive, poorly drained character of catchments and floodplains the floodwaters once rose comparatively gradually to the peak and drained equally gradually back into the channel. Land-use changes and drainage improvement quickened the movement of water into stream channels which accordingly, over the centuries, have been slowly dredged, straightened and embanked. Now minor floods are contained and only major floods overtop the banks, although when they do so the inundation takes place rapidly and the floodwaters are unable to drain easily back into the channels.

7.11 Modelling runoff

From the detailed discussion of runoff processes in this chapter and indeed from a consideration of the complex interactions of the hydrological processes discussed in the preceding chapters, it will be clear that the development of a successful runoff model, even for a small and relatively simple catchment, will be a daunting task. Inevitably the challenge has been taken up and there now exist in the literature numberless runoff models, few of which can be applied indiscriminately beyond the conditions for which they have been developed, and none of which are totally satisfactory, i.e. in all cases modelled runoff deviates from observed runoff. These models have been developed for widely differing purposes, ranging from real-time flow forecasting for a particular river reach to long-term flow predictions for large river basins or groups of basins. Moreover, they have been developed to accommodate an equally wide range of hydrological data availability, from the large ungauged catchment with no precipitation data, at one extreme, to the continuous monitoring of many hydrological variables in a small, intensively instrumented, experimental catchment, at the other.

It is clearly not possible to review runoff modelling, even superficially, in a few paragraphs, and no attempt will be made to do so here. Instead a small number of selected examples will be used to illustrate contemporary and potentially promising developments in this field.

7.11.1 *Types of runoff model*

Broadly speaking, runoff models may be regarded as *deterministic* or *stochastic*. Deterministic models simulate the physical processes operating in the catchment to transform precipitation into runoff, whereas stochastic (probabilistic) models take into consideration the chance of occurrence or probability distribution of the hydrological variables. Models may also be described as *conceptual* or *empirical* depending on how much consideration is given to the physical processes acting on the input variables to produce the output of runoff, although, as Clarke (1973) demonstrated, this distinction is almost entirely artificial. Models may also be *linear* or *non-linear* in either the systems theory or statistical regression sense. Finally, they may be *lumped* or *distributed* depending on whether or not the spatial distribution of hydrological variables within the catchment is considered either in a probabilistic or in a geometrical sense.

In recent years there has been increasing emphasis on the development of physically based distributed models. The rationale behind this development is that a model that treats the catchment as a spatially variable physical system is intrinsically better and has significant theoretical advantages that make it more useful, over a wider range of applications, than other types of model.

However, such models are both complex and demanding in terms of data input and computer capacity so that simpler, lumped models, which average and aggregate for the entire catchment, may often be preferred. Indeed, Chow (1967) questioned the value of investigating distributed models, other than for '. . . the satisfaction of our intellectual curiosity', if lumped models can serve the purpose. Two decades later, as was emphasized in Chapter 1, hydrologists continue to be exercised by the apparent irreconcilability of the simplicity of large-scale system response (cf. stream hydrographs resulting from a given catchment rainfall) and the small-scale spatial complexity which defies description and experimental study at the catchment scale (Beven, 1987).

7.11.2 *The Système Hydrologique Européen (SHE) model*

An important example of a new generation of physically based distributed models is the SHE model which was developed collaboratively by the UK Institute of Hydrology (IH), the Danish Hydraulic Institute (DHI) and the Société Grenoblois d'Étude et d'Applications Hydrauliques (SOGREAH). IH developed the interception, evaporation and snowmelt components, DHI the saturated and unsaturated flow components and SOGREAH the overland and channel flow components.

The operation of the model, which was described in IH (1981, 1984), is based on dividing the catchment into a grid of squares, for each of which the appropriate data are specified individually. These data include, for example, surface elevation, vegetation and soil properties, meteorological data, channel and surface flow resistance data, impermeable bed levels and phreatic surface levels. Surface and subsurface flow conditions are developed by the model for each grid square, thereby permitting investigation of local hydrological variations.

Because the model uses catchment parameters, which have real physical significance and can be estimated from catchment characteristics, agreement between modelled and observed hydrographs is good from the outset. Moreover, the distributed nature of the model permits sophisticated calibration for the entire catchment based on comparisons between modelled and observed hydrographs at several points within it (IH, 1984). The model has been applied successfully to catchments in a wide range of environments, including New Zealand (New Zealand Ministry of Works and Development, 1985).

7.11.3 *Institute of Hydrology Distributed Model (IHDM)*

Like SHE this is also a physically based distributed model. The IHDM described by Beven *et al.* (1987), however, is based on dividing the catchment into hillslope and channel elements, and incorporates procedures to allow for the expansion and contraction of the hillslope area contributing quickflow to the channels. The hillslope flow processes are also linked to routines that permit distributed prediction of precipitation, snowmelt and evaporation rates by reference to topography and vegetation. Rigorous sensitivity analysis and calibration were reported by Rogers *et al.* (1985).

7.11.4 *Stochastic runoff models*

The hydrological processes that cause runoff from a catchment area are clearly both deterministic and stochastic. Most of the discussion in this book has focused on the deterministic linkages between precipitation, evaporation, storage and runoff since, it has been argued, only by understanding such linkages can the drainage basin hydrological system be adequately analysed. And yet the precipitation inputs which result in streamflow derive from atmospheric processes, still not fully understood, and show apparently random variations in timing and magnitude of individual, and especially of extreme, falls. In addition, variations in the receptivity of a catchment to precipitation, characterized by variations in interception, infiltration, storage, etc., may also be considered to contain a random component. Both the randomness of precipitation and of the catchment conditions on which it falls will be reflected in variations of runoff, together with the longer-term component of persistence, the Hurst phenomenon (see Section 7.7).

Ideally all of these stochastic components should be incorporated into models of the runoff process. Some of them have been, so that, for example, in one of the best-known 'deterministic' models, the Stanford watershed model (Crawford and Linsley, 1966), infiltration capacity is assumed to vary randomly over the catchment. For the most part, however, stochastic runoff models have been developed in a conceptual vacuum with little reference to their physical linkages with hydrological 'reality'. Attention has been concentrated on the descriptive mimicking of variations of runoff with time, i.e. the generation of synthetic runoff series, and excellent reviews have been provided by Beaumont (1979, 1982) and Lawrance and Kottegoda (1977). Most such models attempt to accommodate three main assumptions:

(a) The recorded historical sequence of streamflows is extremely unlikely to recur.
(b) It is unlikely that the maximum possible flood for a given stream is included within the historical record.
(c) Streamflow exhibits persistence.

Both short-memory and long-memory models have been used. Markovian or first-order autoregressive models, in which the present value of the process depends on its own past and on a random component independent of the past, were developed by Thomas and Fiering (1962), but because they are short-memory models they do not represent the Hurst phenomenon very satisfactorily. Another type of short-memory model approaches daily flow generation by considering the shape of the runoff hydrograph, e.g. the shot-noise model (Weiss, 1977). A more satisfactory treatment of persistence was suggested by Mandlebrot and van Ness (1968) who described a family of long-memory models, the fractional Gaussian noise (FGN) models, involving a complex correlation structure in which the current value of the process depends on the entire history of the process. Subsequently, another family of long-memory models, based on the broken line process, which will also preserve the Hurst phenomenon, was suggested by Rodriguez-Iturbe *et al.* (1972) and Meija *et al.* (1972). Alternatively, some hydrologists have successfully used the family of autoregressive integrated moving average (ARIMA) models outlined by Box and Jenkins (1970).

In the final analysis, however, the potential of stochastic modelling techniques will only be developed fully when stochastic runoff models incorporate an adequate physical conception of the catchment from which the runoff derives. Until that has been achieved we must heed the warning of Klemes (1978) that '. . . enquiry into the stochastic aspects of hydrology provides no prerogative for ignoring hydrology itself'.

References

Abdul, A. S. and R. W. Gillham (1984) Laboratory studies of the effects of the capillary fringe on streamflow generation, *WRR*, **20**: 691–8.

Acreman, M. C. (1985) The effects of afforestation on the flood hydrology of the upper Ettrick valley, *Scottish Forestry*, **39**: 89–99.

Agassi, M., J. Morin and I. Shainberg (1985) Effect of raindrop impact energy and water salinity on infiltration rates of sodic soils, *Proc. SSSA*, **49**: 186–90.

Akan, A. O. (1984) Simulation of runoff from snow-covered hillslopes, *WRR*, **20**: 703–13.

Anderson, D. G. (1968) Effects of urban development on floods in northern Virginia, *USGS Open File Report*, 26 pp.

Anderson, M. G. and T. P. Burt (1977) A laboratory model to investigate the soil moisture conditions on a draining slope, *J. Hydrol.*, **33**: 383–90.

Anderson, M. G. and T. P. Burt (1978) Toward a more detailed field monitoring of variable source areas, *WRR*, **14**: 1123–31.

Anderson, M. G. and T. P. Burt (1982) The contribution of throughflow to storm runoff: an evaluation of a chemical mixing model, *Earth Surf. Proc. and Landforms*, **7**: 565–74.

Archer, D. R. (1983) Computer modelling of snowmelt flood runoff in north-east England, *Proc. ICE*, **75**: 155–73.

ASCE-TASK (1980) Characteristics of low flow, *Trans. ASCE, J. Hydraul. Div.*, **106**(HY5): 717–31.

Atkinson, B. W. (1980) Climatic regions within the drought area, in *Atlas of drought in Britain 1975–76*, J. C. Doornkamp and K. J. Gregory (eds), Institute of British Geographers, London.

Ayers, H. D. (1965) Effect of agricultural land management on winter runoff in the Guelph, Ontario region, *Research Watersheds: Proceedings of Hydrology Symposium*, vol. 4, University of Guelph, Ontario, pp. 167–82.

Barnes, B. S. (1938) Contribution to the discussion of O. H. Meyer on analysis of runoff characteristics, *Trans. ASCE*, **103**: 83–141.

Barnes, B. S. (1939) The structure of discharge-recession curves, *Trans. AGU*, **20**: 721–5.

Barry, R. G. (1981) Trends in snow and ice research, *EOS*, **62**: 1139–44.

Barry, R. G. (1983) Research on snow and ice, in *Contributions in hydrology, US National Report 1979–1982, 18th General Association, IUGG, Hamburg*, American Geophysical Union, Washington, D.C., pp. 765–76.

Bauer, W. J. (1969) Urban hydrology, in *The progress of hydrology*, vol. 2, University of Illinois, Urbana, Ill., pp. 605–37.

Beaumont, C. D. (1979) Stochastic models in hydrology, *Progress in Physical Geography*, **3**: 363–91.

Beaumont, C. D. (1982) The analysis of hydrological time series, *Progress in Physical Geography*, **6**: 60–99.

Beaumont, P. (1978) Man's impact on river systems: a world-wide view, *Area*, **10**: 38–41.

Beckinsale, R. P. (1969) River regimes, in *Water, earth and man*, R. J. Chorley (ed.), Methuen, London, pp. 455–71.

Beran, M. A. and A. Gustard (1977) A study into the low-flow characteristics of British rivers, *J. Hydrol.*, **35**: 147–52.

Berndt, H. W. (1964) Inducing snow accumulation on mountain grassland watersheds, *J. Soil and Water Cons.*, **19**, 196–8.

Betson, R. P. (1964) What is watershed runoff?, *JGR*, **69**: 1541–52.

Beven, K. (1978) The hydrological response of headwater and sideslope areas, *Hydrol. Sci. Bull.*, **23**, 419–37.

Beven, K. (1987) Towards a new paradigm in hydrology, in *Water for the future: hydrology in perspective*, IAHS Publ. 164, pp. 393–403.

Beven, K. and E. F. Wood (1983) Catchment geomorphology and the dynamics of runoff contributing areas, *J. Hydrol.*, **65**: 139–58.

Beven, K., A. Calver and E. M. Morris (1987) *The Institute of Hydrology Distributed Model*, Report 98, The Institute of Hydrology, Wallingford, 33 pp.

Bezinge, A. (1987) Glacial meltwater streams, hydrology and sediment transport: the case of the Grande Dixence hydroelectricity scheme, in *Glacio-fluvial sediment transfer*, A. M. Gurnell and M. J. Clark (eds), Chichester, Wiley, pp. 473–98.

Bonell, M. and D. A. Gilmour (1978) The development of overland flow in a tropical rainforest catchment, *J. Hydrol.*, **39**: 365–82.

Bonell, M. and J. Williams (1986) The generation and redistribution of overland flow on a massive oxic soil in a eucalypt woodland within the semi-arid tropics of north Australia, *Hydrological Processes*, **1**: 31–46.

Bonell, M., D. A. Gilmour and D. S. Cassells (1983a) Runoff generation in tropical rainforests of northeast Queensland, Australia, and the implications for land use management, in *Hydrology of humid tropical regions*, Proceedings of Hamburg Symposium, August 1983, IASH Publ. 140, pp. 287–97.

Bonell, M., D. S. Cassells and D. A. Gilmour (1983b) Vertical soil water movement in a tropical rainforest catchment in northeast Queensland, *Earth Surf. Proc. and Landforms*, **8**: 253–72.

Bonell, M., M. R. Hendriks, A. C. Imeson and L. Hazelhoff (1984) The generation of storm runoff in a forested clayey drainage basin in Luxembourg, *J. Hydrol.*, **71**: 53–77.

Bonell, M., D. S. Cassells and D. A. Gilmour (1987a) Spatial variations in soil hydraulic properties under tropical rainforest in northeastern Australia, *International Conference on Infiltration, Development and Application, Hawaii, January 1987*, Water Resources Research Center, University of Hawaii, Honolulu, pp. 153–65.

Bonell, M., D. Cassells, D. Gilmour and J. Williams (1987b) Hillslope hydrology in the humid and semi-arid tropics of north-east Queensland: a comparative study, in *Readings in Australian geography*, A. Conacher (ed.), Institute of Australian Geographers (WA Branch) and Department of Geography, University of Western Australia, Nedlands, WA 6009, pp. 379–91.

Bosch, J. M. and J. D. Hewlett (1982) A review of catchment experiments to determine the effect of vegetation changes on water yield and evapotranspiration, *J. Hydrol.*, **55**: 3–23.

Boughton, W. C. and D. M. Freebairn (1985) Hydrograph recession characteristics of some small agricultural catchments, *Australian Journal of Soil Research*, **23**: 373–82.

Box, G. E. P. and G. M. Jenkins (1970) *Time series analysis, forecasting and control*, Holden-Day, San Francisco.

Burt, T. P. and D. P. Butcher (1985) On the generation of delayed peaks in stream discharge, *J. Hydrol.*, **78**: 361–78.

Calles, U. M. (1985) Deep groundwater contribution to a small stream, *Nordic Hydrology*, **16**: 45–54.

Cassells, D. S., D. A. Gilmour and M. Bonell (1985) Catchment response and watershed management in the tropical rainforests in north-eastern Australia, *Forest Ecology and Management*, **10**: 155–75.

Chandler, T. J., R. U. Cooke and I. Douglas (1976) Physical problems of the urban environment, *Geog. J.*, **142**: 57–80.

Chorley, R. J. (1978) The hillslope hydrological cycle, in *Hillslope hydrology*, M. J. Kirkby (ed.), John Wiley and Sons, New York, pp. 1–42.

Chow, V. T. (1964) Runoff, Sec. 14 in *Handbook of applied hydrology*, V. T. Chow (ed.), McGraw-Hill, New York.

Chow, V. T. (1967) General report of Technical Session No. 1, Proceedings of International Hydrology Symposium, vol. 2, Fort Collins, Colo., pp. 50–65.

Clarke, R. T. (1973) A review of some mathematical models used in hydrology, with observations on their calibration and use, *J. Hydrol.*, **19**: 1–20.

Colbeck, S. C. (1975) A theory for water flow through a layered snowpack, *WRR*, **11**: 261–6.

Colbeck, S. C. (1979) Water flow through heterogeneous snow, *Cold Regions Science Technology*, **3**: 37–45.

Colbeck, S. C., E. A. Anderson, V. C. Bissell, A. G. Crook, D. H. Male, C. W. Slaughter and D. R. Wiesnet (1979) Snow accumulation, distribution, melt and runoff, *EOS*, **60**, 464–71.

Costin, A. B., L. W. Gay, D. J. Wimbush and D. Kerr (1961) Studies in catchment hydrology in the Australian Alps, III. Preliminary snow investigation, *Division of Plant Industry Technical Paper 15*, CSIRO.

Crawford, N. H. and R. K. Linsley (1966) Digital simulation in hydrology: Stanford watershed model IV, *Technical Report 39*, Department of Civil Engineering, 210 pp.

Day, D. G. (1978) Drainage density changes during rainfall, *Earth Surf. Proc.*, **3**: 319–26.

Delleur, J. W. (1983) Urban hydrology, in *Contributions in Hydrology, US National Report 1979–1982, 18th General Association of IUGG, Hamburg*, American Geophysics Union, Washington, D.C., pp. 730–40.

De Vries, J. J. (1976) The groundwater outcrop-erosion model: evolution of the stream network in The Netherlands, *J. Hydrol.*, **29**: 43–50.

De Vries, J. J. (1977) The stream network in the Netherlands as a groundwater discharge phenomenon, *Geology Mijnbouw*, **56**: 103–122.

De Zeeuw, J. W. (1966) *Hydrograph analysis of areas with prevailing groundwater discharge*, Veenman and Zonen, Wageningen (with English summary).

Dincer, T., B. R. Payne, T. Florkowski, J. Martinec and E. Tongiori (1970) Snowmelt runoff from measurements of tritium and oxygen-18, *WRR*, **6**: 110–24.

DOE (1978) *Surface Water: United Kingdom 1971–73*, HMSO, London.

Doornkamp, J. C., K. J. Gregory and A. S. Burn (eds) (1980) *Atlas of drought in Britain 1975–6*, Institute of British Geographers, London.

Dragoun, F. J. (1969) Effects of cultivation and grass on surface runoff, *WRR*, **5**, 1078–83.

Dunne, T. and R. D. Black (1970) Partial area contributions to storm runoff in a small New England watershed, *WRR*, **6**: 1296–311.

Duysings, J. J. H. M., J. M. Verstraten and L. Bruynzeel (1983) The identification of runoff sources of a forested lowland catchment: a chemical and statistical approach, *J. Hydrol.*, **64**: 357–75.

Elliston, G. R. (1973) Water movement through the Gornergletscher, Symposium on the Hydrology of Glaciers, Cambridge, 1969, *IASH Publ.*, **95**: 79–84.

Engman, E. T. and A. S. Rogowski (1974) A partial area model for storm flow synthesis, *WRR*, **10**: 464–72.

Ernst, L. F. (1978) Drainage of undulating sandy soils with high groundwater tables, *J. Hydrol.*, **39**: 1–30, 31–50.

Espey, W. H., C. W. Morgan and F. D. Masch (1966) A study of some effects of urbanization on storm runoff from a small watershed, *Development Board Report 23*, Texas Water, 110 pp.

Farquharson, F. A. K., D. Mackney, M. D. Newson and A. J. Thomasson (1978) Estimation of run-off potential of river catchments from soil surveys, *Special Survey 11*, Soil Survey, Harpenden, 29 pp.

Ferguson, R. I. (1985a) High densities, water equivalents, and melt rates of snow in the Cairngorm mountains, Scotland, *Weather*, **40**: 272–7.

Ferguson, R. I. (1985b) Runoff from glacierized mountains: a model for annual variation and its forecasting, *WRR*, **21**: 702–8.

Ferguson, R. I. and E. M. Morris (1987) Snowmelt modelling in the Cairngorms, NE Scotland, *Trans. Roy. Soc. Edinburgh: Earth Sciences*, **78**: 261–7.

Fernow, B. E. (1902) Forest influences, *USFS Bull. 7*, US Department of Agriculture, Washington, D.C.

Fountain, A. G. and W. Tangborn (1985) Overview of contemporary techniques, in *Techniques for prediction of runoff from glacierized areas*, G. J. Young (ed.), IAHS Publ. 149, pp. 27–41.

Francis, J. R. D. (1973) Rain, runoff and rivers, *QJRMS*, **99**: 556–68.

Freeze, R. A. (1967) Quantitative interpretation of regional groundwater flow patterns as an aid to water balance studies, *Int. Assoc. Hydrol. Sci. Publ.*, **76**: 154–73.

Freeze, R. A. (1972) Role of subsurface flow in generating surface runoff, 1. Upstream source areas, *WRR*, **8**: 609–23.

Freeze, R. A. (1980) A stochastic-conceptual analysis of rainfall-runoff processes on a hillslope, *WRR*, **16**: 391–408.

Gillham, R. W. (1984) The capillary fringe and its effect on water-table response, *J. Hydrol.*, **67**: 307–24.

Gorgens, A. H. M. and D. A. Hughes (1982) Synthesis of streamflow information relating to the semi-arid karoo biome of South Africa, *South African Journal of Science*, **78**: 58–68.

Guilcher, A. (1965) *Précis d'Hydrologie, marine et continentale*, Masson, Paris.

Gurnell, A. M. (1981) Heathland vegetation, soil moisture and dynamic contributing area, *Earth Surf. Proc. and Landforms*, **6**: 553–70.

Gurnell, A. M. and M. J. Clark (eds) (1987) *Glacio-fluvial sediment transfer*, John Wiley and Sons, Chichester, 524 pp.

Gurnell, A. M., K. J. Gregory, S. Hollis and C. T. Hill (1985) Detrended correspondence analysis of heathland vegetation: the identification of runoff contributing areas, *Earth Surf. Proc. and Landforms*, **10**: 343–51.

Hall, D. G. (1968) The assessment of water resources in Devon, England, using limited hydrometric data, *Int. Assoc. Hydrol. Sci. Publ.*, **76**: 110–20.

Hall, M. J. (1985) *Urban hydrology*, Elsevier Applied Science, Barking.

Hanna, J. E. and D. N. Wilcock (1984) The prediction of mean annual flood in Northern Ireland, *Proc. ICE*, **77**: 429–44.

Harrison, G. (1985) *Drought '84*, Water Authorities Association, London, 27 pp.

Hertzler, R. A. (1939) Engineering aspects of the influence of forests on mountain streams, *Civil Engineering*, **9**: 487–9.

Hestnes, E. (1985) A contribution to the prediction of slush avalanches, *Annals of Glaciology*, **6**: 1–4.

Hewlett, J. D. (1961a) Watershed management, in *Report for 1961 Southeastern Forest Experiment Station*, US Forest Service, Ashville, N.C.

Hewlett, J. D. (1961b) Soil moisture as a source of baseflow from steep mountain watersheds, *Southeastern Forest Experiment Station, Paper 132*, US Forest Service, Ashville, N.C.

Hewlett, J. D. (1969) Tracing storm base flow to variable source areas on forested headwaters, *Technical Report 2*, School of Forest Resources, University of Georgia, Athens, Ga.

Hewlett, J. D. (1982) Personal communication.

Hewlett, J. D. and J. D. Helvey (1970) Effects of forest clear-felling on the storm hydrograph, *WRR*, **6**: 768–82.

Hewlett, J. D. and A. R. Hibbert (1963) Moisture and energy conditions within a sloping soil mass during drainage, *JGR*, **68**: 1081–7.

Hewlett, J. D. and A. R. Hibbert (1967) Factors affecting the response of small watersheds to precipitation in humid areas, in *Forest hydrology*, W. E. Sopper and H. W. Lull (eds), Pergamon, Oxford, pp. 275–90.

Hewlett, J. D. and W. L. Nutter (1969) *An outline of forest hydrology*, University of Georgia Press, Athens, Ga.

Hewlett, J. D. and W. L. Nutter (1970) The varying source area of streamflow from upland basins, in *Proceedings of Symposium on Watershed Management*, American Society of Civil Engineers, New York, pp. 65–83.

Hewlett, J. D. and C. A. Troendle (1975) Non-point and diffused water sources: a variable source area problem, in *Proceedings of Symposium on Watershed Management*, American Society of Civil Engineers, New York, pp. 21–46.

Hewlett, J. D., J. C. Fortson and G. B. Cunningham (1977) The effect of rainfall intensity on storm flow and peak discharge from forest land, *WRR*, **13**: 259–66.

Hewlett, J. D., J. C. Fortson and G. B. Cunningham (1984) Additional tests on the effect of rainfall intensity on storm flow and peak flow from wild-land basins, *WRR*, **20**: 985–9.

Hibbert, A. R. (1967) Forest treatment effects on water yield, in *Forest hydrology*, W. E. Sopper and H. W. Lull (eds), Pergamon, Oxford, pp. 527–43.

Hindley, D. R. (1973) The definition of dry weather flow in river flow measurement, *JIWE*, **27**: 438–40.

Hornbeck, J. W. (1973) Storm flow from hardwood-forested and cleared watersheds in New Hampshire, *WRR*, **9**: 346–54.

Horton, J. H. and R. H. Hawkins (1965) Flow path of rain from the soil surface to the water table, *Soil Sci.*, **100**: 377–83.

Horton, R. E. (1933) The role of infiltration in the hydrologic cycle, *Trans. AGU*, **14**: 446–60.

Huff, D. D., R. V. O'Neill, W. R. Emanuel, J. W. Elwood and J. D. Newbold (1982) Flow variability and hillslope hydrology, *Earth Surf. Proc. and Landforms*, **7**: 91–4.

Hursh, C. R. (1936) Storm-water and absorption. Contribution to Report of the Committee on Absorption and Transpiration, 1935–36, *Trans. AGU*, **17**: 296–302.

Hursh, C. R. (1944) Report of the sub-committee on subsurface flow. *Trans. AGU*, **25**, 743–6.

Hursh, C. R. and E. F. Brater (1941) Separating storm hydrographs from small drainage areas into surface and subsurface flow, *Trans. AGU*, **22**, 863–70.

IAHS (1973) *Symposium on the hydrology of glaciers*, IAHS Publ. 95.

IAHS (1975) *Snow and ice*, IAHS Publ. 104.

IAHS (1982) *Hydrological aspects of alpine and high mountain areas*, IASH Publ. 138.

IH (1980) *Low flow studies report*, Institute of Hydrology, Wallingford.

IH (1981) *Institute of Hydrology Research Report 1978–81*, NERC, Wallingford.

IH (1984) *Institute of Hydrology Research Report 1981–84*, NERC, Wallingford.

IH (1987) Report for 1985/86, in *The Natural Environment Research Council Report for 1985/86*, NERC, Swindon, pp. 61–78.

IH/BGS (1985) *Hydrological Data UK, 1982 Yearbook*, Institute of Hydrology/British Geological Survey, Wallingford.

IH/BGS (1988) *Hydrometric Register and Statistics 1981–85*, Institute of Hydrology/British Geological Survey, Wallingford.

Jones, J. A. A. (1971) Soil piping and stream channel initiation, *WRR*, **7**: 602–10.

Jones, J. A. A. (1979) Extending the Hewlett model of stream runoff generation, *Area*, **11**: 110–14.

Jones, J. A. A. (1981) *The nature of soil piping—a review of research*, BGRG Research Monograph 2, GeoBooks, Norwich, 301 pp.

Jones, J. A. A. (1986) Some limitations to the a/s index for predicting basin-wide patterns of soil water drainage, *Zeitschrift Geomorphologie, N.F.*, **60**: 7–20.

Jones, J. A. A. (1987) The effects of soil piping on contributing areas and erosion patterns, *Earth Surf. Proc. and Landforms*, **12**: 229–48.

Jones, J. A. A. and F. G. Crane (1984) Pipeflow and pipe erosion in the Maesnant experimental catchment, in *Catchment experiments in fluvial geomorphology*, T. P. Burt and D. E. Walling (eds), GeoBooks, Norwich, pp. 55–72.

Jungerius, P. D. (1985) Soils and geomorphology, *Catena, Suppl. 6*, 18 pp.

Keller, R. (1970) Water balance in the Federal Republic of Germany, in *Symposium on world water balance*, IAHS Publ. 93, vol. II, pp. 300–14.

Kibler, D. F. (ed.) (1982) *Urban stormwater hydrology*, Water Resources Monograph 7, American Geophysical Union, Washington, D. C., 271 pp.

Kidder, E. H. and W. F. Lytle (1949) Drainage investigations in the plastic till soils of north-eastern Illinois, *Agricultural Engineering*, **30**: 384–9.

Kirkby, M. J. and R. J. Chorley (1967) Throughflow, overland flow and erosion, *Int. Assoc. Hydrol. Sci. Bull.* **12**: 5–21.

Kirkham, D. (1947) Studies of hillside seepage in the lowan drift area, *Proc. SSSA*, **12**: 73–80.

Klemes, V. (1978) Physically based stochastic hydrologic analysis, *Advances in Hydroscience*, **11**: 285–356.

Klute, A., E. J. Scott and F. D. Whisler (1965) Steady-state water flow in a saturated inclined slab, *WRR*, **1**: 287–94.

Kunkle, G. R. (1968) A hydrogeologic study of the groundwater reservoirs contributing base runoff to Four Mile Creek, east-central Iowa, *USGS Wat. Sup. Pap.*, 1839–O.

Lane, L. J., Ferreira, V. A. and E. D. Shirley (1980) Estimating transmission losses in ephemeral stream channels, in *Hydrology and water resources in Arizona and the Southwest*, American Water Research Association and Arizona–Nevada Academy of Science, Las Vegas, pp. 193–202.

Larson, L. W. (1975) An application of the aerial gamma monitoring technique for measuring snow cover water equivalent on the Great Plains, *Proceedings of Symposium on Snow Management on the Great Plains. Great Plains Agriculture Council Publ. 73*, Agriculture Experimental Station, University of Nebraska, Lincoln.

Lawrance, A. J. and N. T. Kottegoda (1977) Stochastic modelling of riverflow time series, *Journal of Royal Statistical Society*, **140**: 1–47.

Lawrence, S. (1981) The New Etal drainage experiment, *Soil and Water*, **9**: 36–7.

Lazaro, T. R. (1979) *Urban hydrology: a multidisciplinary perspective*, Ann Arbor Science.

Linsley, R. K., M. A. Kohler and J. L. H. Paulhus (1958) *Hydrology for engineers*, McGraw-Hill, New York.

Lloyd, J. G. (1968) River authorities and their work, *JIWE*, **22**: 343.

Loganathan, G. V., C. Y. Kuo and T. C. McCormick (1985) Frequency analysis of low flows, *Nordic Hydrology*, **16**: 105–28.

Lvovitch, M. I. (1958) Streamflow formation factors, *Proc. IASH Gen. Assoc. of Toronto*, **3**: 122–32.

Lvovitch, M. I. (1973) The global water balance, United States, International Hydrological Decade Bulletin 23, in *Trans. AGU*, **54**: 28–42.

Lvovitch, M. I. (1980) Soil trend in hydrology, *Hydrol. Sci. Bull.*, **25**: 33–45.

McIntyre, D. S. (1958) Permeabillity of soil crusts formed by raindrop impact, *Soil Sci.*, **85**: 185–9.

Mandlebrot, B. B. and J. W. van Ness (1968) Fractional Brownian motions, fractional noises, and their applications, *SIAM Review*, **10**: 422–37.

Marsh, P. and M. K. Woo (1984) Wetting front advance and freezing of meltwater within a snow cover, 1. Observations in the Canadian Arctic, *WRR*, **20**: 1853–64.

Marsh, P. and M. K. Woo (1985) Meltwater movement in natural heterogeneous snow covers, *WRR*, **21**: 1710–16.

Marsh, T. J. (1988) Personal communication.

Marsh, T. and M. Lees (1985) *Hydrological data UK, the 1984 drought*, Institute of Hydrology/British Geological Survey, Wallingford.

Marsh, T. J. and I. G. Littlewood (1978) An estimate of annual runoff from England and Wales, 1728–1976, *Hydrol. Sci. Bull.*, **23**: 131–42.

Martinec, J., H. Siegenthaler, H. Oescheger and E. Tongiorgi (1974) New insight into the runoff mechanism by environmental isotopes, *Proceedings of Symposium on Isotope Techniques in Groundwater Hydrology*, vol. 1, International Atomic Energy Agency, Vienna, pp. 129–43.

Martinelli, M. (1964) Watershed management in the Rocky Mountain alpine and subalpine zones, *USFS Research Note RM-36*, 7 pp.

Meija, J. M., I. Rodriguez-Iturbe and D. R. Dawdy (1972) Streamflow simulation, 2. The broken line process as a potential model for hydrologic simulation, *WRR*, **8**: 931–41.

Micklin, P. P. (1981) A preliminary systems analysis of impacts of proposed Soviet river diversions on Arctic sea ice, *EOS*, **62**: 489–93.

Morin, J. and H. S. Jarosch (1977) Runoff rainfall analysis for bare soils, *Soil Erosion Research Station Pamphlet 164*, Israel Ministry of Agriculture, Bet Dagan, 22 pp.

Morin, J., Y. Benyamini and A. Michaeli (1981) The effect of raindrop impact on the dynamics of soil surface crusting, *J. Hydrol.*, **52**: 321–35.

Mosley, M. P. (1979) Streamflow generation in a forested watershed, New Zealand, *WRR*, **15**: 795–806.

Nakamura, R. (1971) Runoff analysis by electrical conductance of water, *J. Hydrol.*, **14**: 197–212.

NERC (1975) *Flood studies report*, Natural Environment Research Council, London.

Newson, M. D. and M. Robinson (1983) Effects of agricultural drainage on upland streamflow: case studies in mid-Wales, *Journal of Environmental Management*, **17**: 333–48.

New Zealand Ministry of Works and Development (1985) SHE—model of the '80s, *Streamland*, No. 39, The Ministry of Works and Development, Wellington.

Nortcliffe, S. and J. B. Thornes (1984) Floodplain response of a small tropical stream, in *Catchment experiments in fluvial geomorphology*, T. P. Burt and D. E. Walling (eds), GeoBooks, Norwich, pp. 73–86.

O'Brien, A. L. (1977) Hydrology of two small wetland basins in eastern Massachusetts, *Water Research Bulletin*, **13**: 325–40.

O'Loughlin, E. M. (1981) Saturation regions in catchments and their relations to soil and topographic properties, *J. Hydrol.*, **53**: 229–46.

Onesti, L. J. (1985) Meteorological conditions that initiate slushflows in the central Brooks Range, Alaska, *Annals of Glaciology*, **6**: 23–5.

Packman, J. C. (1979) The effect of urbanization on flood magnitude and frequency, in *Man's impact on the hydrological cycle in the United Kingdom*, G. E. Hollis (ed.), GeoAbstracts, Norwich, pp. 153–72.

Pardé. M. (1939) Hydrologie fluviale des Iles Britanniques, *Annales de Géographie*, **48**: 369–84.

Pardé, M. (1955) *Fleuves et Rivières*, 3rd edn, Armand Colin, Paris.

Parkinson, R. J. and I. Reid (1986) Effect of local ground slope on the performance of tile drains in a clay soil, *Journal of Agricultural Engineering Research*, **34**: 123–32.

Pearce, A. J., M. K. Stewart and M. G. Sklash (1986) Storm runoff generation in humid headwater catchments, I: Where does the water come from? *WRR*, **22**: 1263–72.

Petts, G. E. and J. Lewin (1979) Physical effects of reservoirs on river systems, in *Man's impact on the hydrological cycle in the United Kingdom*, G. E. Hollis (ed.) GeoAbstracts, Norwich, pp. 79–92.

Pierson, T. C. (1980) Piezometric response to rainstorms in forested hillslope drainage depressions, *J. Hydrol. (N.Z.)*, **19**: 1–10.

Pinder, G. F. and J. F. Jones (1969) Determination of the groundwater component of peak discharge from the chemistry of total runoff, *WRR*, **5**: 438–45.

Pirt, J. (1983) Low flow estimation in ungauged catchments, *Occasional Paper 6*, Department of Geography, University of Technology, Loughborough.

Pirt, J. and J. R. Douglas (1982) A study of low flows using data from the Severn and Trent catchments, *JIWES*, **36**: 299–308.

Pirt, J. and C. M. Simpson (1982) A study of low flows using data from the Severn and Trent catchments—Part II: Flow frequency procedures, *JIWES*, **36**: 459–69.

Price, A. G. and L. K. Hendrie (1983) Water motion in a deciduous forest during snowmelt, *J. Hydrol.*, **64**, 339–56.

Ragan, R. M. (1968) An experimental investigation of partial area contributions, *Proceedings of Symposium of Berne*, IASH Publ., 76, pp. 241–9.

Ramser, C. E. (1927) Run-off from small agricultural areas, *Journal of Agricultural Research*, **34**: 797–823.

Rango, A. (1985) An international perspective on large-scale snow studies, *Hydrol. Sci. J.*, **30**: 225–38.

Rawitz, E., E. T. Engman and G. D. Cline (1970) Use of the mass balance method for examining the role of soils in controlling watershed performance, *WRR*, **6**: 1115–23.

Reid, I. and R. J. Parkinson (1984) The nature of the tile-drain outfall hydrograph in heavy clay soils. *J. Hydrol.*, **72**: 289–305.

Renard, K. G. (1979) Transmission losses, Paper prepared for SCS Unit Hydrologist Meeting, Brainerd, Minn. (cyclostyled).

Robinson, M. (1985) The hydrological effects of moorland gripping: a re-appraisal of the Moor House research, *Journal of Environmental Management*, **21**: 205–11.

Robinson, M. (1986) Changes in catchment runoff following drainage and afforestation, *J. Hydrol.*, **86**: 71–84.

Robinson, M. and K. J. Beven (1983) The effect of mole drainage on the hydrological response of a swelling clay soil, *J. Hydrol.*, **64**: 205–23.

Robinson, M., J. Mulqueen and W. Burke (1987) On flows from a clay soil—seasonal changes and the effect of mole drainage, *J. Hydrol.*, **91**: 339–50.

Robinson, M., E. L. Ryder and R. C. Ward (1985) Influence on streamflow of field drainage in a small agricultural catchment, *Agric. Water Manag.*, **10**: 145–58.

Rodda, J. C., R. A. Downing and F. M. Law (1976) *Systematic hydrology*, Newnes-Butterworth, London.

Rodda, J. C., A. V. Sheckley and P. Tan (1978) Water resources and climatic change, *JIWES*, **31**: 76–83.

Rodhe, A. (1981) Spring flood, melt wa... or ground water?, *Nordic Hydrology*, **12**: 21–30.

Rodriguez-Iturbe, I., J. M. Meija and D. R. Dawdy (1972) Streamflow simulation, 1. A new look at Markovian models, fractional Gaussian noise, and crossing theory, *WRR*, **8**: 921–30.

Rogers, C. C. M., K. J. Beven, E. M. Morris and M. G. Anderson (1985) Sensitivity analysis, calibration and predictive uncertainty of the Institute of Hydrology distributed model, *J. Hydrol.*, **81**: 179–91.

Roots, E. F. and J. W. Glen (1982) Preface, in *Hydrological aspects of alpine and high mountain areas*, J. W. Glen (ed.), IAHS Publ. 138, pp. v–vi.

Rothlisberger, H. and H. Lang (1987) Glacial hydrology, in *Glacio-fluvial sediment transfer*, A. M. Gurnell and M. J. Clark (eds), John Wiley and Sons, Chichester.

Savini, J. and J. C. Kammerer (1961) Urban growth and the water regimen, *USGS Wat. Sup. Pap.*, 1591-A, 43 pp.

Scoging, H. M. and J. B. Thornes (1979) Infiltration characteristics in a semiarid environment, in *The hydrology of areas of low precipitation*, Proceedings of Canberra Symposium, December 1979, IAHS Publ. 128, pp. 159–68.

Searcy, J. K. (1959) Flow-duration curves, *USGS Wat. Sup. Pap.*, 1542-A.

Sheaffer, J. R., K. R. Wright, W. C. Taggart and R. M. Wright (1982) *Urban storm drainage management*, Marcel Dekker Inc., New York.

Sklash, M. G. and R. N. Farvolden (1979) The role of groundwater in storm runoff, *J. Hydrol.*, **43**: 45–65.

Sklash, M. G., M. K. Stewart and A. J. Pearce (1986) Storm runoff generation in humid headwater catchments, II. A case study of hillslope and low order stream response, *WRR*, **22**: 1273–82.

Smith, K. (1968) Some thermal characteristics of two rivers in the Pennine area of northern England, *J. Hydrol.*, **6**: 405–16.

Storr, D. (1972) Estimating effective net radiation for a mountainous watershed, *Boundary-Layer Met.*, **3**: 3–14.

Strahler, A. N. (1964) Quantitative geomorphology of drainage basins and channel networks, Sec. 4–11 in *Handbook of applied hydrology*, V. T. Chow (ed.), McGraw-Hill, New York.

Tanaka, T. (1982) The role of subsurface water exfiltration in soil erosion processes, *Hydrol. Sci. J.*, **27**: 233 (abstract).

Thomas, H. A. and M. B. Fiering (1962) Mathematical synthesis of streamflow sequences for the analysis of river basins by simulation, in *Design of water resource systems*, A. Maass *et al.* (eds), Harvard University Press, Cambridge, pp. 459–93.

Toler, L. G. (1965) Use of specific conductance to distinguish two base-flow components in Econfina Creek, Florida, *USGS Prof. Pap.* 525-C, pp. 206–8.

Toth, J. (1962) A theory of groundwater motion in small drainage basins in central Alberta, Canada, *JGR.*, **67**: 4375–87.

Trimble, S. W., F. H. Weirich and B. L. Hoag (1987) Reforestation and the reduction of water yield on the southern piedmont since circa 1940, *WRR*, **23**: 425–37.

Tsukamoto, Y., T. Ohta and H. Noguchi (1982) Hydrological and geomorphological studies of debris slides on forested hillslopes in Japan, *Hydrol. Sci. J.*, **27**: 234 (abstract).

Vaidhianathan, V. I. and C. Singh (1942) A new phenomenon in the movement of the free water-level in a soil and its bearing on the measurement of the water table, *Proceedings of Indian Academy of Science*, **15**: 264–280.

van de Griend, A. A. (1981) *A weather type hydrologic approach to runoff phenomena*, Rodopi, Amsterdam.

van de Griend, A. A. and E. T. Engman (1985) Partial area hydrology and remote sensing, *J. Hydrol.*, **81**: 211–51.

van der Molen, W. H. (1983) Personal communication.

Voronkov, P. P. (1963) Hydrochemical bases for segregating local runoff and a method of separating its discharge hydrograph, *Meteorologiya i Gidrologiya*, **8**: 21–8.

Walling, D. E. (1971) Streamflow from instrumented catchments in south-east Devon, in *Exeter essays in geography*, K. J. Gregory and W. Ravenhill (eds.), University of Exeter, pp. 55–81.

Walling, D. E. and T. P. Burt (1983) *Field experiments in fluvial geomorphology*, GeoBooks, Norwich.

Wankiewicz, A. C. (1978) Water pressure in ripe snowpacks, *WRR*, **14**: 593–600.

Ward, R. C. (1968) Some runoff characteristics of British rivers, *J. Hydrol.*, **6**: 358–72.

Ward, R. C. (1981) River systems and river regimes, in *British rivers*, J. Lewin (ed.), Allen and Unwin, London, pp. 1–33.

Ward, R. C. (1984) Some aspects of river flow in northern New South Wales, Australia, *J. Hydrol.*, **71**, 31–51.

Washburn, A. L. (1980) *Geocryology: a survey of periglacial processes and environments*, 2nd edn, John Wiley and Sons, New York, 406 pp.

Water Resources Board (1973) *Water resources in England and Wales*, HMSO.

Weiss, G. (1977) Shot noise models for the generation of synthetic streamflow data, *WRR*, **13**: 107–8.

Whipple, W., L. S. Tucker, N. S. Grigg, T. Grizzard, C. W. Randall and R. Shubinski (1982) *Stormwater management in urbanizing areas*, Prentice-Hall.

Williams, J. and M. Bonell (1987) Computation of soil infiltration properties from the surface hydrology of large field plots, *Proceedings of International Conference on Infiltration, Development and Application, Hawaii, January 1987*, Water Resources Research Centre, University of Hawaii, Honolulu, pp. 272–81.

Wilson, W. T. (1941) An outline of the thermodynamics of snowmelt, *Trans. AGU*, **22**: 182–95.

WMO (1986) *Intercomparison of models of snowmelt runoff*, Operational Hydrology Report 23, World Meteorological Organisation, Geneva.

Wolf, P. O. (1952) Forecast and records of floods in Glen Cannich in 1947, *JIWES*, **6**: 298–324.

Woodruff, J. F. and J. D. Hewlett (1970) Predicting and mapping the average hydrologic response for the eastern United States, *WRR*, **6**: 1312–26.

Young, G. J. (1985) *Techniques for the prediction of runoff from glacierized areas*, IAHS Publ. 149, 149 pp.

Zaltsberg, E. (1987) Evaluation and forecasting of groundwater runoff in a small watershed in Manitoba, *Hydrol. Sci. J.*, **32**: 69–84.

Zaslavsky, D. and G. Sinai (1981) Surface hydrology, I. Explanation of phenomena, *Proc. ASCE, J. Hydraul. Div.*, **107**(HY1): 1–16.

Zon, R. (1927) Do forests prevent floods?, *American Forests for Life*, **33**: 387–92, and 432.

8. Water quality

8.1 Introduction and definitions

The preceding chapters of this book have dealt with the storages and fluxes of the substance 'water' through the hydrological cycle. In dealing with the individual components of precipitation, interception, evaporation, soil water, groundwater and river flow, little mention has been made of the nature of the water and indeed, until comparatively recently, except in specific areas, hydrologists gave little attention to the chemical characteristics of water (Hem, 1985). There is now, however, a growing awareness among hydrologists of the importance of water *quality*, including both *chemical* characteristics, due to dissolved material, and *physical* characteristics, such as temperature, taste and suspended solids. Given the increasing usage and pollution of water sources by man's activities, it can be argued that the problems of water quality are now more difficult and demanding than those of water quantity (Stott, 1979).

Information on, and control of, water quality is of great importance for a wide range of purposes, including water supply and public health, agricultural and industrial uses. There is also a need to monitor and control the ever-increasing impact of man on water chemistry through various forms of pollution. Nowadays no waters are free from human influences—even Arctic precipitation contains constituents discharged into the atmosphere (Galloway *et al.*, 1982). The hydrologist studying the rates and pathways of water movement is thus placed in a unique position to help find solutions to these problems.

In principle, evaporation provides a source of pure (distilled) water for precipitation, which then becomes increasingly concentrated with dissolved material as it moves through the atmosphere and in subsequent stages of the hydrological cycle as it comes into contact with organic matter, soil and rock material. There is, however, still a great deal to be learned about the chemistry of these natural processes, as well as the mechanisms and pathways of solute fluxes. Water quality research is now probably the most rapidly expanding aspect of hydrology (Andrews and Webb, 1987).

This chapter provides a brief overview of a number of selected aspects of this rapidly developing area of research, but before turning to the water quality processes at different stages of the hydrological cycle, the properties of water and the nature of chemical reactions are reviewed.

8.1.1 *Properties of water*

Water occupies a central role in the transport of chemicals around the surface of our planet. Although its appearance is bland and pure water is almost colourless, tasteless and odourless, it has certain properties that make it unique. Water is a chemical *compound* of two commonly occurring *elements*, hydrogen (H) and oxygen (O), but differs in behaviour from most other compounds to such an extent that it has been called a 'maverick' compound (Leopold and Davis, 1970). One of the most unusual characteristics is that under normal climatic conditions it is commonly found in all three phases: solid (ice), liquid and gas (water vapour). The water *molecule* (H_2O) is strongly attracted to most inorganic substances (including itself), and its changes of phase (melting and evaporation, condensation and freezing) absorb or release more heat than most other common substances and act as a cushion against extremes of temperature. Many of its physical and chemical properties are unusual, but of most importance for water quality studies is the fact that virtually all substances are *soluble* to some extent in water. It is, in fact, practically impossible to produce and store absolutely pure water (Lamb, 1985).

These properties of water can be accounted for by its *molecular structure*. Each water molecule comprises two *atoms* of hydrogen attached to one oxygen atom by a very strong and stable mechanism, involving sharing a pair of electrons, known as a *covalent bond* (Franks, 1983). The two hydrogen atoms are not on diametrically opposite sides of the oxygen atom but at an angle of 105° apart (Fig. 8.1a). This produces a bipolar molecule, equivalent in effect to a bar magnet, with an unbalanced distribution of *electrical charge*: the oxygen atom on one side of the molecule has a negative charge while the side with the two hydrogen atoms has a positive charge. As a result of this *electrostatic* effect, adjacent water molecules tend to interact by a process known as *ionic* or *hydrogen bonding* (Fig. 8.1b). It is the combined strength of these two types of bond that accounts for the unusually large latent and specific heat capacity of water. It is also responsible for the water's cohesive nature and large surface tension which enables it to 'wet' surfaces and move through materials such as soils and plant stems by capillarity.

These properties account for the solubility of many materials in water. The atoms in many substances are held together not by strong covalent bonds but by weaker

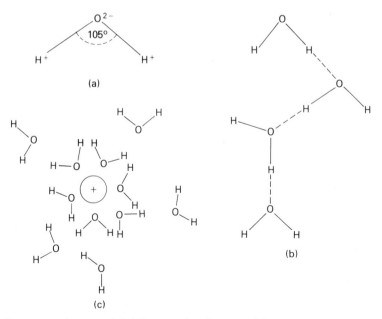

Figure 8.1 *Structure of water. (a) Water molecule comprising one oxygen atom and the two hydrogen protons. (b) Hydrogen bonding between water molecules. (c) Hydration 'shell' of water molecules surrounding a cation in solution.*

electrostatic attraction. These bonds may be weakened further by the bipolar water molecules which act to cancel out some of the electrostatic attraction and enable the atoms to move apart as separate electrostatically charged atoms or groups called *ions*— positively charged *cations* and negatively charged *anions*. When these ions become surrounded by water molecules and have little direct influence on each other, the substance is said to be *dissolved* (Fig. 8.1c). The liquid (in this case, water) is called the *solvent* and the dissolved solid is called the *solute*. The mixture of solvent and solute is called a *solution*. This chapter is concerned with *aqueous solutions*, in which water is the solvent. Most inorganic compounds dissociate into ions when they dissolve in water, although there may be some interaction between oppositely charged ions to form *complex ions*. Organic compounds occur in solution as uncharged molecules (Hem, 1985). The powerful solvent action of water is vital for plant and animal life as it provides the medium for the transport of chemicals and nutrients. In that sense it may be considered to be the original 'elixir of life' (Lamb, 1985). However, this same solvent action also works to transport harmful pollutants and toxic substances.

8.1.2 *Water quality characteristics*

The notion of water quality comprises consideration of many different factors. Commonly quoted *determinands* include physical characteristics such as colour, temperature, taste and odour, as well as chemical characteristics such as acidity, hardness, and the concentrations of various constituents such as nitrates, sulphates and dissolved oxygen and man-made pollutants including pesticides and herbicides. There is not a simple measure of the purity of water, and the term 'quality' only has meaning when related to some specific use of water. Thus, according to Tebbutt (1977), the concentration of total dissolved material in raw sewage is similar to that in many groundwater supplies used for drinking water. By that single criterion both are about 99.9 per cent pure water, but they are obviously very different in other respects!

There are many texts that detail analytical procedures for the evaluation of different aspects of water quality to which the interested reader may refer (cf. DOE, 1972; WPCF, 1980). It is outside the scope of this chapter to deal with them in detail, but a limited number of aspects will be outlined which are of importance for discussions in subsequent sections in which the data and results from the completed analyses are discussed. For many purposes it is not the total amount of a particular element that may be of interest but the chemical form in which it occurs. Thus, for example, nitrogen may occur in a number of *species*, including organic nitrogen, ammonia (NH_3), nitrite (NO_2) and nitrate (NO_3), and these may have very different effects on the suitability of the water for different uses.

Methods of chemical analysis are available to identify and measure the concentrations in water of many elements and compounds. The most commonly used unit for expressing the concentration of dissolved constituents is as the weight of solute per unit volume of water, e.g. milligram per litre (mg/l). For some purposes the weight of solute per unit weight of solution is used, measured most often as parts per million (p.p.m.). For most practical purposes the two systems yield the same numbers; however, for highly mineralized water with solute concentrations greater than 7000 mg/l a density correction should be used to convert between the two. For the calculation of the masses of substances involved in chemical reactions the molar concentrations are used (these are concentrations expressed in *moles* per litre, where a mole of a substance is its atomic or molecular weight in grams). For thermodynamic calculations, described later, chemical *activities* rather than concentrations are used. A correction factor, or activity coefficient, usually represented as γ_i, is applied to the concentration values to allow for non-ideal behaviour of ions in solution. Its value is unity for ideal conditions; for dilute solutions (less than 50 mg/l of dissolved ions) the coefficient is generally $\geqslant 0.95$, which is similar to the measurement error of concentrations, but at high concentrations (e.g. 500 mg/l) for ions with a large charge, or *valency*, it may be as low as 0.7 (Hem, 1985).

For many purposes it is useful to express chemical species by their *equivalent* weight. This is the concentration in moles per litre multiplied by the ionic charge, and is usually expressed in units of milliequivalents per litre (meq/l). Table 8.1 gives the chemical formulae of many of the common ions in solution, which are discussed in this chapter, and conversion factors between the different methods of expressing concentration. In any solution the overall number of positive and negative electrical charges must be equal to maintain electrical neutrality, i.e. the total meq/l of cations must equal the total meq/l of anions. This requirement of *electroneutrality* is useful for checking the accuracy of the determination of ionic concentrations and for ensuring that all of the significant ionic species in a solution have been accounted for. Some ions, however, such as silica (SiO_2), do not have a charge, and an equivalent weight cannot be computed.

Table 8.1 Names and formulae of some common chemical species showing their electrical charge (valency), formula weight and conversion factor to equivalent weight units. The concentration (mg/l) divided by formula weight gives micromoles per litre. Concentration (mg/l) multiplied by final column (valency/formula weight) gives milliequivalents per litre (based on data in Hem, 1985).

Name	Species	Formula weight (approx.)	mg/l → meq/l
Aluminium	Al^{3+}	26.9	0.111 19
Ammonium	NH_4^+	18.0	0.055 44
Bicarbonate	HCO_3^-	61.0	0.016 39
Calcium	Ca^{2+}	40.1	0.049 90
Carbonate	CO_3^{2-}	60.0	0.033 33
Chloride	Cl^-	35.4	0.028 21
Hydrogen	H^+	1.0	0.992 16
Hydroxide	OH^-	17.0	0.058 80
Iron (ferrous)	Fe^{2+}	55.8	0.035 81
(ferric)	Fe^{3+}	55.8	0.053 72
Magnesium	Mg^{2+}	24.3	0.082 29
Nitrate	NO_3^-	62.0	0.016 13
Nitrite	NO_2^-	46.0	0.021 74
Phosphate	PO_4^{3-}	95.0	0.031 59
Orthophosphate $\{$	HPO_4^{2-}	96.0	0.020 84
	$H_2PO_4^-$	97.0	0.010 31
Potassium	K^+	39.1	0.025 58
Silica	SiO_2	60.1	—
Sodium	Na^+	23.0	0.043 50
Sulphate	SO_4^{2-}	96.1	0.020 82

An aspect of water quality which is of great importance, since it affects many chemical reactions, is the *acidity* of the water. Whether or not solutes are present in water, some of the water molecules will dissociate into hydrogen (H^+) and hydroxyl (OH^-) ions. Since the resulting concentrations of H^+ ions are very low, they are expressed in terms of the *pH*, or negative log_{10} of the H^+ ion activity, i.e.

$$pH = log_{10} \frac{1}{[H^+]} \tag{8.1}$$

The square brackets denote chemical activities in moles per litre. Values of pH less than 7 (10^{-7} moles/litre of H^+ ions) are said to be acidic, while those above 7 are *alkaline*. A pH of 7 at 25°C is said to be *neutral* but as hydrogen ion behaviour is temperature dependent, this value decreases somewhat with increasing temperature. Waters that are uninfluenced by pollution generally have pH values of between 6 and 8.5 (Hem, 1985). This may appear a small variation but it should be remembered that since pH has a logarithmic scale, a change of one unit corresponds to a ten-fold change in H^+ ion concentration.

The term 'acidity' applied to aqueous solutions may also be defined as the ability to react with OH^- ions. This may be determined by titration with an alkali (Stumm and Morgan, 1981). It is a function of a number of solute species (including, for example, iron) and is not simply related to the H^+ concentration. In contrast, the 'alkalinity' of water (i.e. its ability to react with H^+ ions) can usually be identified with the concentration of CO_3^{2-} and HCO_3^- ions (Hem, 1985). The 'strength' of an acid refers to the extent to which it dissociates in solution.

To understand the chemical processes in natural waters that affect the composition of water and to make quantitative statements about them requires the application of certain fundamental concepts, of which some of the most useful are the principles of chemical thermodynamics.

8.2 Processes controlling the chemical composition of water

Chemical processes in natural waters are principally concerned with reactions in relatively dilute aqueous solutions; these are usually *heterogeneous* systems comprising a liquid phase with either or both a solid and a gaseous phase. Many reactions are *reversible*, being able to proceed in both directions, and in practice a dynamic equilibrium will be established between the two opposing reactions. The behaviour of such reversible reactions may be studied using the principles of chemical *thermodynamics*. This enables the likely direction of a reaction over time to be determined and the final equilibrium solute concentrations in the water to be predicted (Lewis and Randall, 1961; Sposito, 1981). The final products of an *irreversible* reaction will be determined by the quantities of the reactants available.

The solution of gaseous carbon dioxide in water is a reversible reaction producing carbonic acid (H_2CO_3), and may also form the ions HCO_3^- and CO_3^{2-}:

$$CO_2(g) + H_2O \leftrightharpoons H_2CO_3(aq) \tag{8.2}$$
$$\updownarrow$$
$$HCO_3^- + H^+ \tag{8.3}$$
$$\updownarrow$$
$$CO_3^{2-} + 2H^+ \tag{8.4}$$

The second and third steps produce hydrogen ions (H^+) and will alter the pH of the solution. Subscript letters, used in this chapter, indicate the physical state of the substance; g = gaseous, aq = aqueous species occurring in solution as written and c = crystalline solid.

Similarly, a solid may dissolve in water; an example of this is calcite ($CaCO_3$) which occurs in many carbonate rocks:

$$CaCO_3(c) + H^+ \rightleftharpoons HCO_3^- + Ca^{2+} \qquad (8.5)$$

Depending upon the pH of the water there may be subsequent interactions between the dissolved carbonate species, i.e.

$$HCO_3^- \rightleftharpoons CO_3^{2-} + H^+ \qquad (8.6)$$

or

$$HCO_3^- + H^+ \rightleftharpoons H_2CO_3 \qquad (8.7)$$

The *equilibrium constant* (K) of a reversible reaction has a constant value for a given combination of reactants and products at a given temperature. Experimentally obtained values at standard temperature (usually 25°C) are available in the chemical literature (e.g. Sillen and Martell, 1964). Alternatively, the equilibrium constant of a reaction may be calculated from the Gibbs free energy (Drever, 1982), using published values (e.g. Woods and Garrels, 1987).

The solution of $CaCO_3$ in water may be used to illustrate the use of these principles to give the final equilibrium values of a set of reactants. The equilibrium constant is calculated from the ratio of the activities of the products divided by the activities of the reactants, i.e. for the reaction given in Eq. (8.5),

$$K = \frac{[Ca^{2+}][HCO_3^-]}{[CaCO_3(c)][H^+]} \qquad (8.8)$$

where K for this reaction has a published value (Jacobson and Langmuir, 1974) of 81. The activity of a solid (here $CaCO_3$) is taken as unity, so the equation becomes

$$81 = \frac{[Ca^{2+}][HCO_3^-]}{[H^+]} \qquad (8.9)$$

Therefore, given measurements of the pH, solution temperature and concentrations of calcium (Ca) and bicarbonate (HCO_3^-) it is possible to say whether the system is in equilibrium. If the quotient is less than K the water may dissolve more calcite (assuming the solid is present); if it equals K the water is at equilibrium (the concentrations will not change unless outside influences alter, e.g. temperature); or if it is greater than K the solution is supersaturated and will precipitate calcite. The fact that the final equilibrium condition depends upon the amounts of the reactants and products is known as the law of *mass action*. It does not provide quantitative information on the rate of a reaction, although in general the further it is from equilibrium the faster it may be.

For a chemical in gaseous form the *partial pressure* is used in such calculations. This is the proportion (by volume) of the particular gas, multiplied by the total pressure (measured in atmospheres).

More complex reactions may be dealt with by combining several equilibrium equations. For example, the dissolution of CO_2 in water produces H^+ and HCO_3^- ions, which are a reactant and product respectively of the dissolution of calcite. Adding the equation for these two reactions (Garrels and Christ, 1965; Drever, 1982) enables the solubility of calcite to be expressed as a function of the partial pressure of CO_2 (Fig. 8.2).

In practice there are many limitations to the application of thermodynamic procedures since in real world situations, outside of the chemistry laboratory, there are likely to be exchanges of energy and reactants with the surrounding environment and equilibrium is not attained. The extent to which natural water systems are in chemical equilibrium is not well known. It is more likely to be attained in aquifer systems, where the rate of movement is relatively slow and residence times are long, than for near soil surface flows.

Nevertheless, the principles have proved very useful for indicating the direction and the maximum extent of reactions and are widely adopted. Thermodynamics has been called 'one of the most useful tools in physical chemistry' (Alberty, 1987). A number of computer programs are available to facilitate the calculations of equilibrium conditions (Nordstrum *et al.*, 1979), of which one of the best known is the WATEQ program (Truesdell and Jones, 1974).

The rate of different chemical reactions can vary enormously: some reactions are so rapid that equilibrium can be attained almost immediately, while others are so slow that an equilibrium may not be achieved before environmental conditions alter. While thermodynamics deals with equilibrium states, chemical *kinetics* is concerned with the mechanism

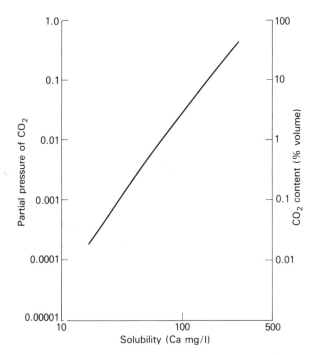

Figure 8.2 *Solubility of $CaCO_3$ in water at 25°C (expressed as mg/l of Ca) for different atmospheric concentrations of CO_2.*

and rate of operation of chemical changes and with the factors controlling the reaction rate. For a reversible reaction the ratio of the forward and backward reaction rates equals the equilibrium constant. Many reversible reactions involve a sequence of intermediate steps, some rapid, some slow. Kinetics can identify the slowest or limiting change that determines the overall rate of the reaction (Drever, 1982; Stumm and Morgan, 1981). Thus kinetics is a more fundamental science than thermodynamics, but due to the great complexities it is much less well understood, and thermodynamic principles are much more generally used.

In addition to these chemical considerations of the thermodynamics and kinetics of reactions, hydrology plays an important role in determining solute composition and concentrations. Apart from snow and ice, and with the exception of some deep groundwater systems, water is generally in continual movement. Its velocity, and hence residence time, will strongly influence whether or not the water attains a chemical equilibrium for a particular reaction. Many reactions are 'diffusion controlled' (Alberty, 1987), i.e. their rate constants are controlled by the physical speed at which reactants can diffuse together, rather than by the rate of chemical reaction at a point. Furthermore, the dynamic nature of flow, particularly in the soil zone, means that between storms pore water solute concentrations may increase as minerals are dissolved but are then flushed out by new waters in the next storm. Thus the flushing frequency and interstorm period may be important variables.

Succeeding sections of this chapter will now deal in turn with particular aspects of water quality pertaining to different components of the hydrological cycle. These will discuss first the composition of precipitation and then look at the water quality behaviour and changes as the water passes through soils and groundwater, ending with the mixture of chemicals found in rivers and lakes. Both natural and man-made sources will be considered, as it may often be difficult or impossible to separate the two. Finally, the relation between water quality and the characteristics of the region or catchment area will be discussed.

8.3 Atmospheric solutes

At the moment that a droplet is formed in the atmosphere the water is very pure, but its chemistry will alter rapidly due to conditions both within the cloud and in the atmosphere between the cloud layer and the earth's surface. Particulate material may act as nuclei for raindrop formation (Section 2.2), determining its initial chemical composition, and as the precipitation moves through the atmosphere it will accumulate further particulates by entrainment and various gases in the atmosphere will dissolve in the droplets. The particulates in the atmosphere originate from a wide variety of sources including ash from volcanoes and power stations and wind-blown dust. Of particular importance as cloud condensation nuclei are *aerosols* (Charlson et al., 1978). These are very small particles (less than 1 μm) which may be liquid or solid material and originate from the land or sea, or from chemical reactions in the atmosphere.

The removal of gases and particulates is a very complex process, which is discussed in detail elsewhere (e.g. Fowler, 1984). Note, however, that this natural 'scrubbing' of the atmosphere by precipitation is a major means by which it is purged of materials that might otherwise accumulate to reach dangerous concentrations (Lamb, 1985). The removal of

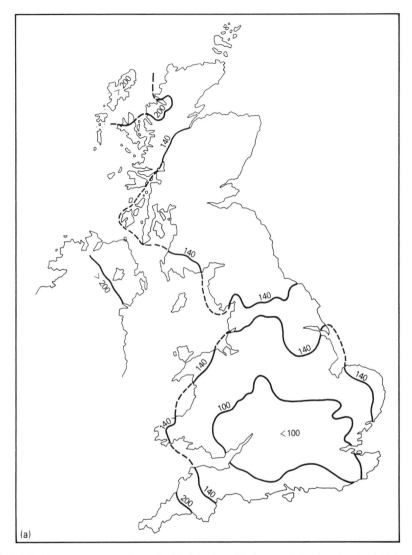

(a)

Figure 8.3 (a) *Mean concentration of chloride (μeq/l) in precipitation in the UK in 1986 (data supplied by Warren Spring Laboratory); (b) ratio of Na/Cl concentrations (mg/l) in precipitation over the conterminous USA, July 1955–June 1956 (redrawn from an original diagram by Junge and Werby*, Journal of Meteorology, *1958. By permission of the American Meteorological Society.*

particulates from the atmosphere by heavy rain or snowfall often gives rise to a period of much improved visibility.

The oceans comprise about 70 per cent of the earth's surface and sea salts are a major source of dissolved material in precipitation. Seawater droplets become entrained in the atmosphere when waves break and are carried upwards by turbulence, becoming increasingly concentrated as their water evaporates. This may continue until just a solid

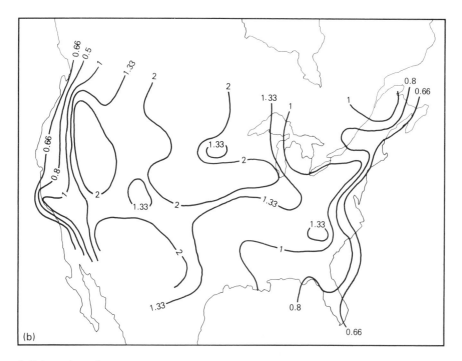

(b)

Figure 8.3 (*continued*)

aerosol particle is left, carried in the wind until it is dissolved in rain. The supply of sea salt to the atmosphere will vary with meteorological conditions and the state of the sea surface. Skartveit (1982), for example, found a close relation between the concentration of sea salt in the precipitation of a coastal area and the windspeed over the sea.

A number of studies have mapped the concentrations of the elements found in precipitation (Junge and Werby, 1958; Stevenson, 1968; Munger and Eisenreich, 1983) and have demonstrated a decline with distance inland from the coast in those elements, including Na^+, Cl^-, Mg^{2+} and K^+, which are derived from marine sources (Fig. 8.3a). In contrast, the solutes in precipitation falling in inland areas are derived predominantly from terrestrial sources, and include Ca^{2+}, NH_4^+, SO_4^{2-}, HCO_3^- and NO_3^- (Cryer, 1986). These comprise substances from natural sources such as gases from plants and soils and wind-blown dust and, in addition, oxides of sulphur and nitrogen produced by the burning of fossil fuels and from industrial and vehicle emissions. As a consequence of the difference between terrestrial and marine solutes, there are differences in the relative amounts of ions, e.g. the ionic ratio Na^+/Cl^- increases with distance from the ocean due to Na enhancement from terrestrial sources, e.g. dust (Fig. 8.3b). The ratio of Na/Cl concentrations (mg/l) in sea water is about 0.56 (or about 0.85 using equivalent weight units).

The relative concentrations of different marine-derived ions in precipitation are not necessarily the same as in sea water due to a number of processes causing *fractionation* and enrichment (Bloch *et al.*, 1966; Chesselet *et al.*, 1972). Certain substances in the sea, such as iodide, are attracted to the organic microlayer on the ocean surface and are then lost in greater proportions to the atmosphere. Rising bubbles tend to retain ions with larger charge/mass ratios, ejecting them into the atmosphere on bursting at the surface. Sodium, chloride and sulphate, in contrast, occur naturally in similar proportions in precipitation as in the oceans, although concentrations are smaller than in sea water by at least a factor of 1000. The ratio of Cl^- to SO_4^{2-} in rainfall may be compared to that in the oceans to determine the 'excess' input of sulphate to the atmosphere (i.e. above that from natural marine sources), assuming that all the Cl^- in a rainwater sample is of the marine origin. Table 8.2 shows data, assembled by Meybeck (1983) from various sources, to compare the chemistry of ocean water with that of precipitation under different influences. The concentrations of ions (in equivalent weight units) are compared to that of Cl as the reference element. While there is enrichment of some species in the oceanic precipitation, major changes are evident for the inland sites (>100 km from the coast). The calcareous rocks of the French pre-Alps, for example, provide Ca^{2+} and Mg^{2+}, and soil dust in arid areas provides ions such as Ca^{2+} and SO_4^{2-}. The industrial town of Rouen shows the effect of pollution, including SO_2 gas, resulting in very high SO_4^{2-} levels in the rain.

In recent years considerable concern has arisen about the possible effects of the large number of polluting substances emitted into the atmosphere from man's activities. These *anthropogenic* materials comprise both particulates and gases, and while many are naturally occurring constituents even in 'unpolluted' atmospheres, some of the substances may be harmful if present in high enough concentrations. Appreciable amounts of pollution in precipitation began to occur in the Industrial Revolution. Peat soil profiles near industrial towns in northern England contain widespread accumulations of soot and heavy metals within the layers formed in the last two hundred years. This pollution and the associated gases (which are not retained in the peat record) are thought to have been responsible for the major changes in natural vegetation species that occurred at that time over extensive areas of the Pennine uplands of Britain (Ferguson and Lee, 1983).

Not surprisingly, early concern about atmospheric pollution was centred on the clearly visible depositions of particulates, rather than the dissolved material in precipitation,

Table 8.2 Chemical composition of ocean water and precipitation showing the ratio of the concentrations of different ions to that of Cl^- (equivalents/equivalent), chloride concentrations and mean annual precipitation.

	Ca^{2+}	Mg^{2+}	Na^+ (ratio to Cl^-)	K^+	SO_4^{2-}	Cl (mg/l)	Annual rain (mm)
(a) World average							
Ocean water	0.037	0.19	0.85	0.018	0.10	19 300	—
Ocean precipitation	0.160	0.24	0.86	0.021	0.30	4.0	—
(b) Inland sites							
French pre-Alps	7.4	0.95	1.15	0.9	4.9	0.6	1380
Arid central USSR	1.1	0.24	0.80	—	0.68	39.0	150
Western Ontario	2.2	0.94	0.83	0.33	3.0	0.35	790
Central Amazonia	0.16	0.17	0.90	0.07	0.74	0.49	2250
(c) Polluted site							
Rouen, north France	—	—	0.74	0.13	13.7	5.0	450

leading many countries to introduce smoke abatement legislation such as 'Clean Air' Acts. As long ago as the mid-nineteenth century, however, Smith (1852) discovered that higher concentrations of sulphuric acid were found in precipitation nearer to the industrial town of Manchester in northern England. The term 'acid rain' was probably used for the first time in the 1870s (Smith, 1872). In the 1950s Gorham (1958a, 1958b) showed evidence of links between atmospheric pollution and the acidity of precipitation and small pools, but the significance of this work was not recognized by other workers at that time.

The problem of 'acid rain' causing damage to the environment was first identified in the late 1960s by the work of Odén, who pointed to declining fish populations in Scandinavia as a consequence of the acidification of freshwater rivers and lakes (Havas, 1986). Since that time acidification of fresh waters has been observed in other parts of Europe (Flowers and Battarbee, 1983; Paces, 1985) and in North America (Harris and Verry, 1985), and it has been found to affect forests, crops and soils (Cresser and Edwards, 1987; Innes, 1987).

Sulphuric acid accounts for about 60–70 per cent of the mean annual acidity of precipitation in north-west Europe and North America, and most of the rest is due to nitric acid (Fowler et al., 1982; Seip and Tollan, 1985). The primary chemical pollutants are sulphur dioxide (SO_2) and nitrogen oxides NO and NO_2 (often referred to together as NO_x) which are produced by the burning of fossil fuels. These undergo oxidation in the atmosphere, to sulphuric acid (H_2SO_4) and nitric acid (HNO_3), in a number of complex reactions, involving sunlight, moisture, oxidants and catalysts, that are still not well understood. Many of the reactions involve photochemical oxidants, including ozone (O_3), the hydroxyl radical (OH) and hydrogen peroxide (H_2O_2).

The reactions in very simplified form may be viewed as:

$$NO + O_3 \rightarrow NO_2 + O_2 \tag{8.10a}$$

$$NO_2 + OH^- \rightarrow HNO_3 \rightarrow NO_3^- + H^+ \tag{8.10b}$$

$$SO_2 + O_2 \rightarrow SO_3 \overset{H_2O}{\rightarrow} H_2SO_4 \rightarrow SO_4^{2-} + 2H^+ \tag{8.10c}$$

Sulphuric acid is the main cause of acid deposition because sulphur is emitted in much larger quantities than nitrogen, and the sulphuric acid molecule in solution releases two H^+ ions whereas nitric acid releases one (Swedish Ministry of Agriculture, 1982). Although there is a certain amount of natural acid deposition, for example from volcanic gases, these are of minor importance in comparison with anthropogenic sources.

Although the term 'acid rain' is widely used (e.g. Watt Committee on Energy, 1984; Delleur, 1986; Mohnen, 1988) it is a misleading name (Seip and Tollan, 1985), and phrases such as acid input or acid deposition are more accurate and are to be preferred. Even pure water in equilibrium with atmospheric CO_2 has an 'acidic' pH of about 5.6 (i.e. well below the neutral pH 7). This value, pH 5.6, has been used as the reference level for distinguishing 'natural' from polluted 'acid' rain (Barrett and Broden, 1955). However, other natural materials, including dissolved aerosols and volcanic gases, can also influence the pH, resulting in still lower values. Furthermore, while rain is usually the most important mechanism by which atmospheric water transfers pollutants to the ground it is not the only one. In some areas snow is an important component of precipitation, while in upland and coastal areas frequent cloud or mist can provide a significant contribution to acidification. The fine water droplets of mist and cloud may contain much higher concentrations of acid than the drops of rain (Sadasivan, 1980; Castillo *et al.*, 1985; Fowler *et al.*, 1988), with the result that the so-called *occult deposition* of these fine droplets on vegetation may cause a proportionately much more important chemical input than the quantity of water deposited would suggest. At an upland site in northern England, for example, Dollard *et al.* (1983) estimated that cloudwater deposition could amount to 20 per cent of the chemical input to a bulk precipitation collector.

These forms of 'wet' deposition are efficient processes for removing material from the atmosphere, but are restricted to the times when condensation and precipitation occur. Wet deposition may be highly 'episodic'. For example, about 30 per cent of the total deposition of H^+ ions in 1980 at Goonhilly in Cornwall took place on only five rainy days (Watt Committee on Energy, 1984). As well as large total rainfall on these days, the concentrations of pollutants were very high, the air masses having moved to the site after stagnating for some time over areas of high emissions. Appreciable acid deposition also takes place by means of the 'dry' deposition of particles and aerosols directly onto the surfaces of soils, plants and water bodies. The main process is by the absorption of gases, such as SO_2 and NO_2, rather than particles (Garland, 1978; Cape *et al.*, 1987), and will depend upon the chemical and physical affinity of each gas for a particular surface (Fowler, 1984). In contrast to the episodic nature of wet deposition, this is a continuous process, although the rate of deposition may fall as the collecting surface approaches 'saturation' (Fowler and Cape, 1984). Subsequent oxidation to SO_4^{2-} and NO_3^- takes place on the soil and vegetation surfaces when they are wetted by rain or dew. In addition, gases may also pass into the plant stomata and be metabolized.

The relative importance of dry and wet deposition varies with factors such as geographical location and season (due to differences in the amounts of rainfall and of artificial emissions). In general, dry deposition dominates close to emission sources and wet deposition is more important at greater distances. Before discussing deposition rates in more detail it is necessary to consider some of the difficulties encountered in the measurement or estimation of the different processes.

The different forms of acid deposition have created great problems in the design of measurement devices (Cryer, 1986; Barrett, 1987). The majority of the available measurements are from *bulk* collectors—storage raingauges collecting precipitation, and any dry deposited material on the gauge funnel that gets washed in. Apart from the problems of assessing the point depth and areal variability of the rain, which were discussed in Chapter 2, there are quality problems. These include not just obvious sources of contamination, such as bird droppings, but also the dry deposition rates of gases and aerosols which may be very different on the raingauge surface compared to that occurring on adjacent soils and vegetation. For this reason 'wet only' collectors have been designed that have a movable cover which is opened when a sensor detects rainfall. They are, however, expensive and prone to miss the early part of the rain (which often has the highest solute concentrations).

Dry deposition is extremely difficult to measure. Deposited particulate matter has been measured by the use of air filters, and the deposition rate of gases may be inferred from their decrease in concentration near to the ground level (Garland, 1978). Gas deposition depends on the physical and chemical properties of the gas, and the type and roughness of the surface, and may vary with temperature and the presence of moisture on the surface (Fowler, 1984).

It has been found that pollutants may be carried many hundreds of kilometres in the atmosphere before being deposited. Dry deposition usually occurs within two or three days and is greatest close to the source of emission, while if the pollutants remain longer in the atmosphere there is greater opportunity of being oxidized to sulphuric and nitric acids. These acids are then dissolved in precipitation. This may be demonstrated by comparing the nature of deposition occurring in an industrial country such as Britain with a

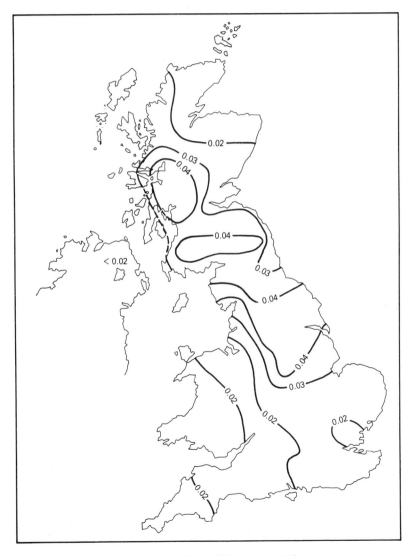

Figure 8.4 *Wet deposited acidity in the UK in 1986 (gm of $H^+/(m^2 \cdot year)$) (data supplied by Warren Springs Laboratory).*

predominantly rural country such as Sweden. In Britain as a whole, dry deposition of acidity exceeds that by wet deposition, and 75 per cent of the deposited sulphur originates from British pollutant sources (Watt Committee on Energy, 1984; Barrett, 1987). Sweden, in contrast, has much less industry of its own, but it is downwind of a number of industrialized countries. Consequently, it is found that wet deposition of acidity predominates, and only 20–25 per cent of the sulphur deposited derives from Swedish sources (Swedish Ministry of Agriculture, 1982).

The pattern of acid deposition is, however, not simply related to the distribution of sources. In Britain, for example, while the largest inputs by dry deposition are in the industrial parts of the country, the largest loads of wet deposition are in fact upwind of the emission sources (Fowler et al., 1985). These are the remote uplands of the north and west country where, despite low concentrations of H^+ ions in the rain, the total input is greatest due to the high rainfall amounts (Fig. 8.4; compare with Fig. 2.1). In remote areas such as these and parts of Scandinavia where wet deposition predominates and precipitation is high, there may be an appreciable contribution to acid deposition from very distant sources, either from Europe or from North America (Watt Committee on Energy, 1984).

The relative importance of wet and dry deposition will be influenced by local factors including topography and the prevailing meteorological conditions, but in general terms the ratio of dry/wet deposition declines systematically from about 10 close to pollution sources to $\leqslant 1$ for areas over 300 km away (Fowler, 1984). Even in remote areas, however, dry deposition is still a significant contributor to the total deposition.

The sensitivity of a particular catchment to acid deposition depends on a variety of factors, including the geology, soils and land use, and this is discussed further in Sections 8.5 and 8.6. The areas where damage is most severe are not necessarily those receiving the greatest deposition of acidity.

8.4 Interception and evaporation

It is known that different land uses affect the water budget of an area. Chapter 3 presents the results of studies showing much higher evaporation losses from forests than from grassland due to the greater interception capacity of the trees. This loss will act to increase the solute concentrations of the remaining water reaching the forest floor as throughfall or stemflow, although it will not affect the solute loads. Vegetation can also directly influence the total amount of solutes reaching the ground in several ways. It has already been shown (Section 8.3) that trees provide large collecting surfaces for the deposition of fine droplets, which may have much higher concentrations of solutes than rainfall. They may also receive deposition of particulate materials, which are subsequently dissolved in rain to reach the ground as *washoff*. In addition, trees may absorb gases into their leaves by stomatal uptake and this material, together with nutrients translocated from their roots and exuded on the leaves, may be transferred to rainwater running over them, as *crown leaching*. In a study of water chemistry changes in the canopy of a coniferous forest in northern Britain, Cape et al. (1987) found that the sulphate loads in the rainfall above the trees amounted to only 30 per cent of that reaching the ground via throughfall and stemflow. They concluded that the bulk of this gain was due to leaching of SO_4^{2-} from the foliage and that this material originated from gaseous SO_2 taken up by stomata and from particles containing SO_4^{2-} which had been deposited externally on the vegetation.

The view that trees may filter pollutants from the air was developed by Mayer and Ulrich (1974) who suggested that the enhancement in chemical fluxes between the rain above the tree canopy and the throughfall (and stemflow) beneath was equal to the dry deposition of particles and gases as well as occult deposition by mists. It is very difficult, however, to separate the net gain of atmospheric material by wet deposition and the washoff of dry deposition, on the one hand, from the recycling of materials by crown leaching, on the other hand (Miller and Miller, 1980). While increases in acidity under coniferous forests have been noted in many studies (Cape *et al.*, 1987; Skeffington, 1987; Stevens *et al.*, 1987), this is not always the case, and some investigators have found little difference in acidity between precipitation and throughfall (Miller, 1984; Reynolds *et al.*, 1986). Several studies of throughfall under different tree species have found greater acidity under conifers than under hardwoods, which may be due to more efficient 'filtering' of atmospheric pollutants by the former and the greater cation exchange capacity of the latter (Clesceri and Vasudevan, 1980; Joslin *et al.*, 1987). In a study of the water chemistry changes as rainwater passed through the canopy of a decidous forest in the north-eastern USA, Likens *et al.* (1977) found a large increase in solute concentrations and that much of the acidity was neutralized. Increasing acidity of rainfall will tend to accelerate the leaching of many cations from the foliage and this exchange of H^+ for cations can, in some cases, result in the rainwater becoming progressively less acid as it passes through the tree canopy (Watt Committee on Energy, 1984). Such neutralization of acid inputs will, nevertheless, still result in acidification of the overall soil–plant system as these cations are ultimately derived from the root zone and will be lost in drainage water.

The pattern of chemical input to the ground surface under trees is likely to be very spatially variable. If the canopy is discontinuous there will be direct incident precipitation between the tree crowns and enhanced input of leached material under the canopy. Stemflow may provide high solute concentrations in water to a very localized zone immediately surrounding the base of each tree trunk (Nicholson *et al.*, 1980). The role of vegetation in water chemistry does not end when the water reaches the ground due to the intimate role of vegetation in soil chemical systems, involving organic matter and nutrient cycling, as well as in physical processes including soil structure that affect soil water movement.

8.5 Soil water and groundwater

Before discussing the water quality processes operating in these subsurface zones it is necessary to give some attention to the nature of the media in which the water resides and through which it passes, since they may provide important sources of, and sinks for, solutes. The relative abundance of the elements in the surface layers of the earth is determined by the composition of the earth's crust; the materials in the rocks and soils derive directly or indirectly from rock minerals formed originally under conditions of extreme heat and temperature, and which are found in *igneous* and some *metamorphic* rocks. Cooling magma formed *primary* minerals such as feldspars, quartz and micas. Apart from quartz, which is very resistant, these are however, unstable at the earth's surface, and are prone to chemical alteration to more stable *secondary* minerals, such as clays and iron oxides. In addition, the operation of biochemical processes forms new minerals such as calcite. About 75 per cent of the land surface of the globe comprises these reworked *sedimentary* rocks (Leopold *et al.*, 1964), which are much more important for holding and transmitting water than the relatively impermeable and low-porosity igneous and metamorphic rocks.

8.5.1 *Weathering of rocks*

It has been estimated that 99 per cent by weight of the earth's crust comprises just eight elements: 47% oxygen (O), 28% silica (Si), 8% aluminium (Al), 5% iron (Fe), 3.5% calcium (Ca), 3% sodium (Na), 2.5% potassium (K) and 2% magnesium (Mg). The chemicals are combined into minerals, which have a definite chemical composition. The concern of the hydrologist centres on the weathering of these minerals to make substances available to go into solution and the behaviour of the soil and rock systems in retaining, cycling and leaching these chemicals.

The most effective mechanism of chemical weathering is the action of rainwater, containing dissolved acids, on rock minerals. The main source of natural acidity in the environment is provided by the solution of CO_2 in water to form carbonic acid (H_2CO_3). This dissociates in water (Eqs 8.2 to 8.4) to form bicarbonate, and to a lesser extent carbonate, ions and generates H^+ ions:

$$H_2 + CO_2 \rightleftharpoons H_2CO_3 \rightleftharpoons H^+ + HCO_3^-$$
$$\updownarrow$$
$$CO_3^{2-} + H^+ \qquad (8.11)$$

The CO_2 is dissolved from the atmosphere, but due to plant root respiration and the decay of organic matter, the concentration of CO_2 in the air in soil pores may be 100 times greater than that in the atmosphere (Bolt and Bruggenwert, 1978). This results in a much higher concentration of carbonic acid in the soil water than is found in surface water such as rivers and lakes. In most humid areas other, stronger acids may also be important, including very dilute H_2SO_4 and HNO_3, as well as organic acids formed from decaying vegetation (Brady, 1984).

The major mechanism of mineral weathering is by acid *hydrolysis*, whereby H^+ ions replace cations in the mineral, leading to an expansion and decomposition of its silicate structure. An example of such chemical action is the weathering of the mineral orthoclase feldspar, found in igneous rocks, to the clay mineral, kaolinite:

$$\underset{\substack{\text{Feldspar}\\\text{(primary mineral)}}}{2KAlSi_3O_8(c)} + \underset{\text{Acid}}{2H^+(aq)} + 9H_2O \rightleftharpoons$$

$$\underset{\substack{\text{Kaolinite}\\\text{(secondary mineral)}}}{Al_2Si_2O_5(OH)_4(c)} + \underset{\text{(In solution)}}{4H_4SiO_4(aq)} + 2K^+(aq) \quad (8.12)$$

Although chemically a reversible reaction, in practice it is essentially irreversible because the feldspar cannot be reconstituted to any significant extent without imposing very great temperatures and pressures. The silica and potassium are removed in solution to groundwater and to streams, pushing the reaction to right, and kaolinite clay accumulates as part of the soil mantle. Such secondary minerals do not necessarily represent the final weathering products since they are only stable within certain pH limits, and may under suitable conditions undergo further chemical weathering to even more stable chemical forms. Another example of mineral weathering is given by Eq. (8.5) which describes the chemical solution of calcite (a major constituent of limestones). The rate of weathering, and the resulting concentrations of solutes in streams and underground waters for a given reaction, will depend on a number of factors including the temperature and the flux of water. In general, reaction rates speed up with increasing temperature, and rates of weathering and leaching in the tropics are several times greater than in temperate areas (Young, 1976). Tropical soils have much higher clay contents (often 60 per cent or more) than temperate areas (e.g. 35 per cent clay is considered high in Britain), and some authors have invoked this much stronger weathering history to explain why the solute loads of streams in the tropics are much lower than those in temperate regions (Nortcliffe, 1988). Geographical differences in rock type may, however, be more important in accounting for these differences (Walling and Webb, 1983). The rate of weathering will also be affected by the flux of water, removing products in solution and bringing new water into contact with the minerals. In some cases the rate of dissolution may reach an upper limit, governed by surface chemical processes (Berner, 1978).

The rate of supply of minerals by chemical weathering of rocks is fairly slow, and if there was no mechanism for retaining the substances that are in solution they would be quickly washed out in drainage water. Such mechanisms do exist, however, and are closely related to the operation of biological processes and have an important control over short-term solute dynamics.

8.5.2 *Adsorption and exchange reactions*

When plants colonize the weathered rock debris they have a direct physical effect by controlling the removal of particulate weathering products by erosion, and they also result in a number of chemical changes. Plants take atmospheric gases into their foliage and dissolved minerals into their root systems and return chemicals to the soil as crown leaching (see Section 8.4) or as leaf litter and other partially decomposed plant remains, known as *humus*. The humus combines with the soil clays to form colloidal complexes which have extremely large areas per unit weight. Their surfaces have electrical charges which enable them to attract and *adsorb* a 'swarm' of dissolved ions (Fig. 8.5). This electrostatic attachment of adsorbed ions is sufficiently weak for them to be easily exchanged for other ions in solution. The exchange of ions between the soil exchange surfaces and the soil solution is a continuous process. The rate of exchange is generally rapid and, following a change in the composition of the soil solution, requires only a few minutes for a new equilibrium to be established between the adsorbed ions and those in solution (Leopold *et al.*, 1964). The amount of a given ion that is adsorbed depends upon the abundance of the different ions in solutions, the ion exchange capacity of the clay or humus and the relative strength of adsorption of the various ions.

With the exception of kaolinite (which has the fewest surface charges), the clay and humus surfaces have more negative than positive charges and consequently attract more cations than anions (Plant and Raiswell, 1983). The *cation exchange capacity* (CEC) measured in meq/100 g or centimole/kg (cmol/kg) varies from about 10 for kaolinite to 100–150 for montmorillonite, and organic colloids have capacities of 200 or even more. Values of CEC for topsoils typically range from about 5 meq/100 g for sandy soils up to 50 meq/100 g for a heavy clay with a high organic matter content. The relative strength of adsorption of the different ionic species increases with their charge and decreases with their hydrated ion radius (i.e. the charged ion surrounded by a 'shell' of polar water molecules). Thus for *equal* concentrations of ions in solution (in equivalent weight units) the relative strength of adsorption of cations is:

$$Al^{3+} > Ca^{2+} > Mg^{2+} > K^+ > NH_4^+ > Na^+$$

The capacity of a given exchange surface for holding cations is not constant, but varies with the pH of the solution. As acidity increases, e.g. due to the accumulation of organic matter, there is an increase in the number of H^+ ions in solution. These H^+ ions are strongly adsorbed onto the exchange surfaces and consequently will displace some of the previously adsorbed cations, other than Al which is too strongly held. Increasing acidity also increases the rate of Al supply to solution from mineral weathering. Since there is a dynamic equilibrium between the adsorbed Al and that in solution some of this Al is adsorbed on the surfaces as Al^{3+} or Al-hydroxy ($Al(OH)_x$) ions. The H and the Al ions tend to dominate acid soils, and both contribute to H^+ concentrations in solution—the H^+ ions directly and the Al ions indirectly by hydrolysis, releasing H^+ ions:

Increasing pH (↓)

$$Al^{3+}(aq) + H_2O \rightleftharpoons Al(OH)^{2+}(aq) + H^+ \qquad (8.13a)$$

$$Al(OH)^{2+}(aq) + H_2O \rightleftharpoons Al(OH)_2^+(aq) + H^+ \qquad (8.13b)$$

$$Al(OH)_2^+(aq) + H_2O \rightleftharpoons Al(OH)_3(c) + H^+ \qquad (8.13c)$$

Gibbsite

Under low pH condition Al becomes soluble in the form of Al^{3+} and Al hydroxy cations which are very strongly adsorbed on exchange surfaces. At higher pH conditions these Al ions react with OH^- ions to form insoluble $Al(OH)_3$, making the exchange surfaces available to other cations. In contrast to Al and H these *base cations*, which principally comprise Ca, Mg, K and Na, act to neutralize acidity and dominate the CEC in neutral and alkaline soils. In acid soils, Al and H tend to be the dominant cations due to their greater adsorption by the soil, while Ca, Mg, K and Na are leached out in solution. The percentage of the CEC accounted for by base cations is known as the *base saturation*, and acid soils which are therefore poor in these plant nutrients have a low base saturation (White, 1987). The leaching of these cations depends both upon the equilibrium between the cations in solution and on exchange surfaces, and also on the presence of a *mobile* anion, such as SO_4^-, which is not itself readily retained on soil surfaces or taken up by plants and can transport the released cations from the soil in solution (e.g. as $CaSO_4$ or $MgSO_4$) in drainage water (Seip, 1980). Anions may also be adsorbed on soil particles, but to a much lesser extent than cations. The 'mobile' sulphate anion, for example, may be adsorbed by

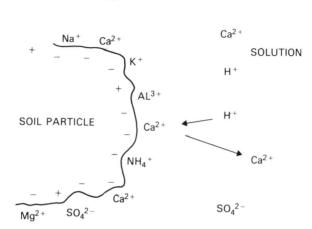

Figure 8.5 *Ion exchange between cations in solution and cations adsorbed on the soil particles.*

some soils but there is evidence that under high atmospheric deposition rates of sulphate, SO_4 'saturation' of the soil may result (Cresser and Edwards, 1987). Once this happens, further sulphate inputs can percolate freely through the soil.

The ions adsorbed on the exchange surfaces are, like those in solution, largely available to plants and microorganisms. Although the exact biochemical processes are not yet fully understood, it is well established that plants can take up, or absorb, ions selectively, and that different plant species have different nutrient requirements. The uptake or these ions must not disturb the overall charge neutrality within the plant, however, so the roots must excrete H^+ ions (if there is an excess cation absorption) or HCO_3^- or OH^- ions (if anion absorption predominates). Plants can thus directly influence the pH and the ionic composition of the soil water around their roots. Important ions for plant nutrition include the cations K, Ca, Mg and S, and the anions phosphorus (P), nitrogen (N) and sulphur (S). The concentration of these ions in the soil solution consequently varies systematically over a growing season, becoming depleted as the plants take up nutrients, and then increasing again over the winter period, or when there is no crop (Gregory, 1988).

From the preceding discussion it will be evident that the pH of the soil solution has a very important effect on the way in which substances are gained (by mineral weathering), retained (by ion adsorption or plant uptake) or lost from the soil (dissolved in drainage water). The soil solution 'acidity', defined in terms of its ability to neutralize OH^- ions, is related to the concentration of H and Al ions. This can be considered to comprise two forms: an *active* acidity due to the H^+ (and Al) ions in solution and a much larger *reserve* or *exchange* acidity comprising the H and Al ions adsorbed on exchange surfaces. The reserve acidity is very much larger than the active form, being about 10^3 times greater in sandy soils and about 10^5 times greater for a clay soil with a high organic matter content (Brady, 1984). Since the active and reserve acidities are in a dynamic equilibrium any change in the concentration of the H^+ ions in solution (e.g. an increase caused by acid deposition from the atmosphere or a reduction due to adding lime to farmland) will tend to be balanced by the adsorption or release of adsorbed H^+ ions. Thus any pH change in the soil solution following the addition of an acid or base will be negligibly small until there has been a significant change in the (much larger) reserve of adsorbed ions. Most natural waters exhibit this resistance to change of the pH which is known as *buffering*. It is therefore important to distinguish between intensity factors (pH) and capacity factors, i.e. the total acid or base neutralizing capacity (Stumm and Morgan, 1981). The buffering capacity of a soil is related to its cation exchange capacity and is important in determining the effect on a soil of external inputs of water and solutes. Thus, for example, the soils and surface waters of Scandinavia and eastern North America have been greatly affected by atmospheric acidification because they have lime-poor bedrocks that are very resistant to weathering and provide only small amounts of base cations (Swedish Ministry of Agriculture, 1982). Similarly, in Britain the upland soils of the north and west, which have developed on igneous rocks, low in cations, are more susceptible to acidification by atmospheric deposition than the soils of lowland, eastern Britain where, although the total deposition is greater, the soils have a much higher content of exchangeable cations (Catt, 1985). By the same token, of course, the pH of these poorly buffered upland soils could in theory be relatively easily increased by the application of lime (Bache, 1980). Different buffering mechanisms correspond to broad soil pH ranges. In neutral and slightly acid soils $CaCO_3$ is an efficient buffer [Eqs (8.5) to (8.7)]. Aluminium is a major source of buffering in the pH range 4–5 [Eqs (8.13a) to (8.13c)] while in very acid soils, $<$ pH 3.5, iron oxides control the pH.

Once in solution the movement of solutes may be influenced by a number of processes, both physical and chemical, and these are discussed below.

8.5.3 *Solute movement in soils and groundwater*

Materials disssolved in water will be transported by the flow of that water, but they will rarely travel at exactly the same rate. Movement of a solute species depends on three mechanisms: *advection*, *dispersion* and *reaction* (Freeze and Cherry, 1979). The mass flow of water will carry solutes by *advection* (sometimes also called *convection*) which, in the absence of other processes, results in chemicals being transported at the same rate as the macroscopic velocity of the water. The velocity of solutes will, however, in practice vary from this rate due to *hydrodynamic dispersion*, leading to a range of velocities. This results from two processes: mechanical dispersion and molecular diffusion, which are similar in effect, and in consequence are generally considered together. Mechanical dispersion is a consequence of the complexities of the pore system of the medium through which the water is moving. Flows are faster through large pores than through small pores, and across the middle of pores than near the pore walls. Flows also vary due to the tortuosity of pore networks, with flows in some pores at an angle to the mean direction of water flow (Fig. 8.6). Molecular diffusion is much smaller and results from the random, thermal-kinetic motion of molecules, and occurs regardless of whether or not there is net water movement. Both mechanical dispersion and molecular diffusion cause a movement of solute from areas of high to low concentration, and this response to concentration gradients makes the solute species become more diffuse with time. The resulting intermixing of chemicals between the moving 'mobile' water in the larger pores and the

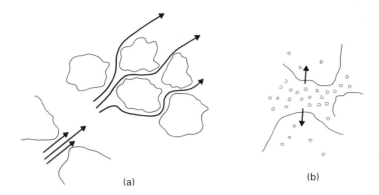

Figure 8.6 *Dispersion of a solute in porous media: (a) mechanical disperson, (b) molecular diffusion.*

largely 'immobile' water held in finer pores of the matrix can influence the overall rate of transport of solutes (van Genuchten and Wierenga, 1976).

Changes in solute concentrations may also take place due to chemical *reactions*. These may be with the solid matrix of the medium or between dissolved substances or, in the unsaturated zone, between the solution and the gas phase. Many different chemical and biochemical reactions can occur, including oxidation, reduction, solution and precipit-ation, but of the most general importance is the ion exchange process of *adsorption*.

The effects of these processes on the movement of a solute species through a porous medium may be presented in graphical form as a breakthrough curve (BTC). Figure 8.7 shows such curves for the simplest case of a non-reactive tracer (of concentration C_o) continuously added to steady, saturated, downward flow through a column of a porous, uniform medium. As the tracer replaces an existing solution (tracer concentration zero), there is an increase in the concentration of the tracer in the outflow (C). Advection alone results in a simple 'piston flow' displacement of the existing solution by the new solution

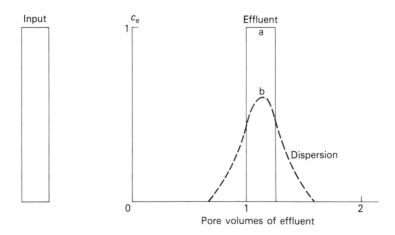

Figure 8.7 *Schematic breakthrough curves resulting from piston-type displacement of a pulse input of solute.*

with a sudden, step-like change (line a) in the proportion of new solution in the outflow $C_e(=C/C_o)$ (Fig. 8.7). This moves at the average linear velocity of the water. Hydro-dynamic dispersion (line b) results in some of the tracer moving faster, and some slower, than this velocity. With increasing flow distance from the tracer input this spread of the concentration profile due to dispersion becomes greater.

For groundwater systems the adsorption reactions are normally very rapid relative to the flow velocity and an equilibrium state is usually assumed between the solute species in solution and adsorbed on solid particles (Freeze and Cherry, 1979). In this situation the relation between the concentration of a solute species adsorbed per unit weight of a particular medium and the concentration in solution at a given temperature can be described by the 'adsorption isotherm' (Stumm and Morgan, 1981). From this 'partition-ing' of the chemical between solid and liquid phases a retardation factor can be calculated to describe the delay in solute movement due to adsorption (van Genuchten and Wierenga, 1976; Smettem, 1986). The effect of greater adsorption is to alter the time dis-tribution of a pulse of tracer, reducing the average rate of movement (shifting the BTC to the right) and making the concentration curve more asymmetrical (Fig. 8.8).

The displacement process may be described by the convective–dispersion equation for one-dimensional, steady-state, uniform flow in a homogeneous saturated medium as

$$\frac{\delta C}{\delta t} = D \frac{\delta^2 C}{\delta Z^2} - V \frac{\delta C}{\delta Z} \left[+ \frac{\rho}{n} \frac{\delta S}{\delta t} \right] \qquad (8.14)$$

$$\underbrace{\phantom{D\frac{\delta^2 C}{\delta Z^2}}}_{\text{Dispersion}} \quad \underbrace{\phantom{V\frac{\delta C}{\delta Z}}}_{\text{Convection}} \quad \underbrace{\phantom{\frac{\rho}{n}\frac{\delta S}{\delta t}}}_{\text{Sorption}}$$

where the third term is an extension to account for the gain or loss of the solute by reaction. C is the concentration of the solute species under consideration, Z is the distance along the flow line, t is the time, D is the coefficient of hydrodynamic dispersion and V is the average flow velocity; ρ and n are the bulk density and porosity of the medium and S is

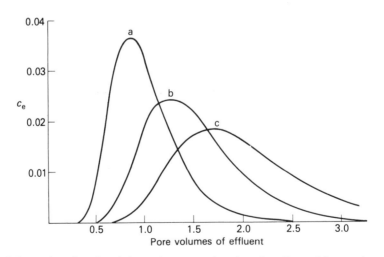

Figure 8.8 *Schematic solute breakthrough curves showing the effect of increasing adsorption of the solute (curves a to c) by the medium. (Redrawn from an original diagram by Smettem, 1986. By permission of John Wiley and Sons Ltd.)*

the mass of the chemical adsorbed per unit mass of solid. If the adsorption rate is slow compared to the flow rate, $\delta S/\delta t$ can be estimated from reaction kinetics (Lapidus and Amundson, 1952).

In some soils and rocks macrovoids can act as conduits for the rapid movement of water and solutes. Such channels may include shrinkage cracks and root channels in soils (see Section 5.5.2 under 'Macropores') and fracture lines and jointing in rocks (see Section 6.4.1), and, under certain circumstances, enable flows to effectively bypass the soil or rock matrix.

Schematic breakthrough curves are shown in Fig. 8.9 for media with different ranges of pore sizes. For simplicity, adsorption and dispersion effects are ignored. The rate of inflow is sufficiently great for preferential flow to be generated in the larger pores. This pattern

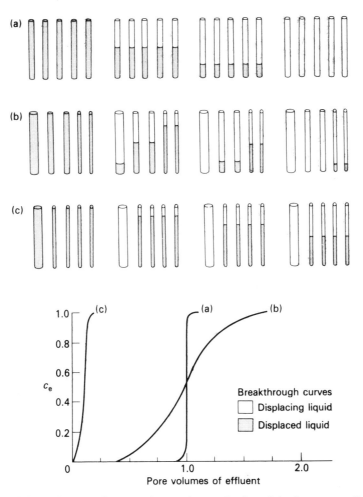

Figure 8.9 *Breakthrough curves for a continuous input of solute: (a) piston-type displacement, (b) a medium with a limited range of pore sizes, (c) a medium with bypass flow through continuous macropores (from an original diagram by Bouma, 1981).*

may be contrasted with the classical infiltration equations developed by soil physicists (Section 5.4.2) which deal with ideal situations of a homogeneous medium (uniform pore size) and a uniform initial moisture content. In the case of preferential or bypass flow, the water does not come into chemical equilibrium with the soil as it can only react with the soil mass by lateral diffusion. The obvious consequence is that greater amounts of dissolved chemicals will pass unchanged through the medium to underlying layers and to receiving water bodies. Macropores may be important in the contamination of ground-water since, under high surface inputs of water, they enable solutes to bypass the natural purification actions in the soil matrix and penetrate beyond the reach of plant roots. In areas subject to 'acid rain', bypass flow through the soil will react less completely with the soil buffering processes, and this may result in greater acidity of the water in rivers and lakes.

In a study of nitrate losses from a sandy loam soil under grass, Barraclough *et al.* (1983) observed high concentrations early in the winter period which could not be accounted for by normal convective–dispersive flow, and they attributed this to bypass flow with nitrate moving in the larger, water-filled pores. A similar conclusion was reached by Smettem *et al.* (1983) in a study of nitrate losses under arable (winter wheat) cultivation. Very high stream nitrate losses in runoff were recorded immediately after the application of nitrogen fertilizer (due to bypassing), but only a month later the nitrate had relocated in the immobile water in small pores in the soil peds and was 'protected' from leaching by subsequent bypass flow.

Further work is needed, however, to elucidate the factors controlling the generation of macropore flow and the resulting solute travel time distributions in such situations (White, 1985). Macropore flow results in an early rise in the BTC while adsorption on the solid phase and solute diffusion from mobile to immobile water zones will cause a delay. Subsequent desorption and diffusion back to the mobile flow zone will both give rise to a long tail to the BTC. Due to the complexity of these effects and to the generally unknown extent of the spatial variability of soil structure at the field scale, a 'black-box' transfer function, analogous to the 'unit hydrograph' approach in catchment rainfall-runoff studies, has been used to model the solute travel time distribution BTC (Jury, 1982). In a number of studies the pore water velocities have been found to be approximated by a log-normal frequency distribution.

In field situations it is often unlikely than an equilibrium will be reached between the concentration of the solute and the soil matrix due to the sorption rate being much slower than the rate of change of the solute concentrations. This will be the case for situations with (a) rapid and varying soil water fluxes, (b) varying solute concentrations and (c) slow adsorption rates. The relative importance of these factors will vary between sites where different hydrological processes are operating as well as between different chemicals at the same site.

Much of the available information on adsorption rates has been derived from laboratory experiments, often dealing with homogeneous solutions or well-stirred suspensions, and may give very different results to those encountered in field situations. The soil solution is not always well mixed, resulting in local concentration gradients, and it may in practice be very difficult to separate out the effects of the chemical processes from those of diffusion (Skopp, 1986). In addition to adsorption, other processes which may be occurring at the same time include precipitation or solution, degradation, volatilization and oxidation. The great complexity of the solute –water–soil system creates great problems in investigating solute movement and concentration changes, and Travis and Etnier (1981), in a review paper, suggested that we may never achieve a full understanding and ability to make quantitative predictions. Nevertheless, significant progress has been achieved and some broad generalizations regarding surface and shallow subsurface solute processes in storm events can be made.

Subsurface flows will have a much greater opportunity for reactions with the solid phase than flow over the ground surface. Consequently, subsurface flows may carry much higher concentrations of solutes than overland flow (Barnett et al., 1972; Hubbard and Sheridan, 1983). Overland flow will, however, take up organic chemicals from leaf litter and vegetation matter (Bache, 1984), and on poorly vegetated sloping sites the erosion of soil particles (and any adsorbed chemicals) may occur (Foster, 1982). The distinction between subsurface and surface flow chemistry need not be clear cut, however, since infiltrating water must pass through the surface organic layers, and subsurface lateral flow may subsequently reemerge at the ground surface further downslope as 'return flow' (Dunne, 1978). The chemical and physical properties of different soil horizons may also give rise to different solute concentrations in the water following different flow paths. Whitehead et al. (1986) used a simple two-compartment soil model, with an upper layer representing a thin acidic organic soil overlying a more alkaline mineral subsoil, to investigate the effect of changes in flow paths on stream chemistry. Storm runoff was very acid, but the baseflow in dry weather periods was much less acid and had lower Al levels since a greater proportion of the flow was moving through the subsoil and undergoing acid buffering. In such areas artificial drainage to encourage percolation into the deeper, less acid soil horizons may help to improve the water quality of the streams. Such a management practice will not necessarily be effective at other sites and indeed in some situations deep drainage may increase stream Al and H^+ concentrations (Bache, 1984; Cresser and Edwards, 1987). Knowledge of soil processes, including adsorption, can be used to limit the loss of agricultural chemicals into the streams by discouraging their application close to streams. The intervening 'buffer' strips act to filter out the chemicals (Hall et al., 1983). Similarly, by applying liquid chemicals at low intensities it will be less likely that bypass flow in macropores will be generated, so that chemicals will pass slowly through the soil matrix and have a greater opportunity to be adsorbed (Miller et al., 1965).

8.5.4 Tracers

The direction and speed of water movement may be studied by 'labelling' the water by adding a tracer. This should move as part of the water flow and it is therefore important to select one which is unreactive. Thus, cations are not good tracers due to exchange reactions. Certain anions are also unsuitable, e.g. phosphate (PO_4^{2-}) undergoes a variety of reactions. Sulphate (SO_4^{2-}), nitrate (NO_3^-) and chloride (Cl^-) are generally considered to be 'mobile' (or 'conservative'), being adsorbed little on solids and travelling at more nearly the same speed as the water. Of these, Cl^- is the most widely used although even it is not ideal. For example, in a study in Israel, Gvirtzman et al. (1986) found that Cl^- tracer moved faster than the average water movement, given by tritiated water, THO, due to *anion exclusion*. This results from the Cl^- anions being repelled by the negatively charged soil surfaces, and so being concentrated in the faster moving flow in the middle of the pores (Thomas and Swoboda, 1970).

A wide range of substances has been used as tracers, including salts, fluorescent dyes and radioisotopes. A detailed discussion of their relative advantages and disadvantages has been given by various authors (e.g. Atkinson and Smart, 1981; Moser et al., 1986).

Tracers can be used in a number of ways. Some of the earliest work, for example, used dyes to study conduit flow paths in limestone terrain. One important group of tracers are *isotopes* which are atoms of a given element that have a different atomic mass but the same chemical properties, and may be used to 'label' chemical species in solution. They may be either *stable isotopes* or *radioisotopes*. Wright et al. (1982) compared the ratio of the radioisotope ^{14}C, produced naturally by cosmic radiation interaction with atmospheric nitrogen, to the stable isotope ^{12}C in the carbonate dissolved in groundwater and rainfall. They used the known rate of decay of ^{14}C to date the large groundwater reserves under the Libyan desert and concluded that these were largely derived from recharge in a much more rainy period some 15 000–35 000 years ago.

Tracers have also been used in a long-standing debate concerning the relative importance of matrix and fissure flow in the unsaturated zone of chalk aquifers. Since its intergranular pores are exceedingly small it was widely assumed that downward flow is dominated by movement in the fairly frequent joints and fissures. In a classic study, Smith et al. (1970) examined the tritium profile of pore water in the unsaturated zone. Tritium T, or ^3H, is an isotope of hydrogen that occurs naturally in small concentrations in rainfall, but very much larger concentrations occurred due to thermonuclear weapon tests in the atmosphere after 1952 with peak values in 1963–4. They concluded that about 85 per cent of the percolating water was as intergranular seepage moving by piston displacement at about 0.9 m/year. This, at first, surprising result has important water supply implications with potentially large quantities of recent man-made pollutants in slow but steady movement downwards to the water table. Subsequently, however, Foster (1975) argued that the observed tritium profile could alternatively have been produced by flow through the fissures if allowance is made for diffusion of tritium from the joint water into the pore water in the matrix. Later work by Wellings and Cooper (1983) and Price (1987), using measurements of both matric potentials and solute movement, suggested that different processes may be operating at different locations and in different strata. Where the matrix permeability is sufficiently high fissure flow does not occur; where matrix permeability is much lower, or infiltration rates are very large, water cannot enter the matrix fast enough and fissure flow can occur.

8.5.5 *Chemical evolution of groundwater*

Since the kinetics of mineral weathering are often slow, it is unlikely that thermodynamic equilibrium with the flowing water will be reached in the soil zone. In large groundwater systems, however, residence times are much longer and as water moves through the medium equilibrium may be progressively established.

In areas of groundwater recharge there is a net transfer of mineral matter from the soil zone to the underlying saturated zone. As this groundwater moves along flow lines to discharge areas, its chemistry will be altered by the minerals with which it comes into contact.

As groundwater moves through the system it will normally dissolve further material from the matrix rocks and the concentration of total dissolved solids will increase. Some materials may be precipitated out due to changes in temperature and pressure affecting their solubility or to reactions with ions dissolved from the matrix forming new insoluble compounds.

In a classic study of the chemical changes in groundwater as it moves from areas of recharge to areas of discharge, Chebotarev (1955) found a progressive evolution in the dominant anion species with increasing travel distance and age:

$$HCO_3^- \rightarrow SO_4^{2-} \rightarrow Cl^-$$

This sequence is determined by mineral availability and solubility (Freeze and Cherry, 1979). In broad terms, minerals which dissolve to release Cl^- (e.g. halite) are more readily soluble than those releasing SO_4^{2-} (e.g. gypsum and anhydrite), which are in turn more soluble than those releasing HCO_3^- (e.g. calcite and dolomite). Thus Cl^- and, to a lesser extent, SO_4^{2-} may have been largely leached from the recharge zone of a groundwater system. Furthermore, HCO_3^- enters the groundwater system in recharge zones from the dissolution of CO_2 from the atmosphere and the soil air.

While this chemical sequence is a useful conceptual model, these changes cannot be defined quantitatively to particular ages or distances, and it is rarely observed in its entirety due to the often dominant effects of the local physical environment. Thus some waters may not evolve past the HCO_3^- or SO_4^{2-} stage, while if the water comes into contact with a highly soluble mineral such as halite it may evolve directly to the Cl^- stage. Many sedimentary deposits are not homogeneous, but comprise assemblages of minerals, and one of the most important factors controlling the chemistry of their groundwater is the order in which the minerals are encountered by the flowing water. Thus the chemical reactions in one stratum, changing the ionic composition and the pH of the water, will influence the reactions in subsequent strata (Freeze and Cherry, 1979). In a heterogeneous aquifer system with strata of widely varying permeability, flow will occur predominantly in the more permeable strata, and the mineral composition of these layers, rather than the composition of the whole aquifer, will be the main factor influencing water chemistry. Furthermore, there may be great differences between the solutes in the different strata. Cation changes also occur along groundwater flow lines, although due to cation exchange reactions, the pattern is much less clear than that outlined by Chebotarev for the anions, but as a broad generalization Ca and Mg are replaced by Na (Price, 1985).

8.5.6 *Presentation of water chemistry data*

It is often useful to summarize and display data on the chemical composition of groundwater, for example to study changes along a flow line, or to detect mixing of waters of different compositions (Walton, 1970). There are many different ways of presenting the data, and some of the most important and commonly used were summarized by Hem (1985). The most widely adopted system is the trilinear diagram attributed to Piper (1944) and used for the major ion composition. Groundwater is characterized by three cation constituents: Ca^{2+}, Mg^{2+} and Na^+ plus K^+, and three anion constituents: HCO_3^-, SO_4^- and Cl^-. The trilinear diagram (Fig. 8.10) combines three plotting fields. The cation and anion compositions are plotted in the left- and right-hand triangles respectively,

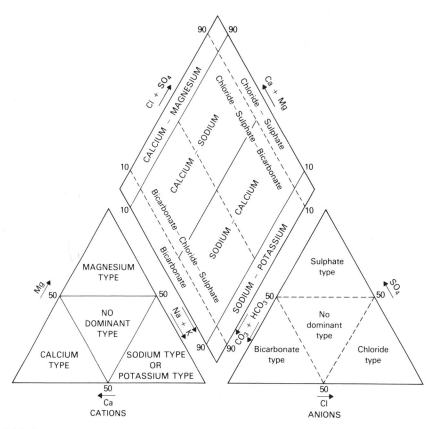

Figure 8.10 *Trilinear diagram used to represent the chemical composition of groundwater. See text for explanation (from an original diagram by Davis and De Wiest, 1966).*

expressed in equivalents per litre. The intersection of the rays projected from the cation and anion plotting points onto the central field represents the overall major-ion character of the groundwater. The values given are the relative rather than absolute concentrations, and since the latter are important in many situations, it is common to plot in the central field not a point but rather a circle centred on that point whose area is proportional to the total dissolved concentration of the water (Hem, 1985; Walton, 1970). This type of presentation is useful for describing differences in chemistry of water samples and it can be extended to give identifiable categories or *facies* (Back, 1966), based on the concept of the composition of groundwater tending towards equilibrium with the matrix rocks through which it is flowing. The hydrochemical facies represent the observed spatial pattern of solute concentrations in the groundwater which results from the chemical processes operating, the rock types and the flow paths in the region.

8.6 Runoff

Surface waters, comprising rivers, streams and lakes, are almost always a mixture of waters from different sources. They receive varying inflows of overland flow, throughflow and baseflow, which due to their very different flow paths and residence times, have different solute and sediment loads. These waters are then mixed together in the channel and routed downstream. In addition, there are point sources of pollution entering the river system as industrial waste discharges and effluent from sewage treatment works.

The following sections deal with the water quality changes and processes occurring within stream channels and lakes and then look at the water quality of a water course in the context of the overall catchment, in terms of its precipitation chemistry, mineral weathering, vegetation and land use.

8.6.1 *Processes in stream channels*

A number of changes may occur when water from the soil zone or groundwater enters a stream channel. These include physical, chemical and biological processes which result, perhaps most obviously, in changes in water temperature, dissolved oxygen and the ability to transport solid particles. Solute concentrations may increase due to the solution of materials in the stream bed and banks. In semi-arid areas, in particular, solute concentration increases as stream water volume is depleted by evaporation.

The solute composition of water flowing into stream channels will be related to the flow paths and residence times of that water in the soil and groundwater systems. Thus baseflow in streams during periods of dry weather is likely to be composed of water that has been in contact with mineral material for some time and as a result has a higher solute concentration than flows in storm periods which have had a much shorter period of contact with the vegetation and soil material. As a consequence, solute concentrations in river water are generally found to be inversely related to flow rates (Walling and Webb, 1986). In fact some studies have used this to develop chemical means of apportioning a stream flow hydrograph into 'baseflow' and 'stormflow' components (see Section 7.4.5).

As noted in Section 8.5, subsurface waters may contain much higher levels of CO_2 than atmospheric levels due to biological activity. When such waters enter stream channels there is a rapid 'degassing' of CO_2 to levels that are much closer to atmospheric values. This leads to the consumption of H^+ ions and a consequent raising of the pH [Eqs (8.2) to (8.4)] and may result in the precipitation of $CaCO_3$ [Eqs (8.5) to (8.7)] as well as certain ions that become less soluble in less acid conditions. An example of the latter is the increasing insolubility of aluminium in less acid water.

There may also be interactions between the solutes entering the channels and sediments in the channel (Bencala *et al.*, 1984). Ions may be adsorbed by stream sediments; this may act as a buffer against changes in the concentrations of ions in the streamflow and in the short term reduce the dissolved concentration of a pollutant in the water. Pollutants, including pesticides, organic mercury compounds and radioactive isotopes, may enter into ion exchange reactions with river and lake sediments. In some cases these reactions may be irreversible, e.g. cesium-137 becomes fixed between the lattice plates of illite, while in other cases the effect is simply to introduce a time delay before the adsorbed materials are released. For example, inorganic mercury in sediments may be subsequently converted by bacterial action into methyl mercury. This can accumulate in living tissue and may be magnified in the food chain, reaching toxic concentrations. The study of sediment movement and storage may therefore be important not just in its own right, in relation to problems of land erosion and sedimentation of channels and lakes, but also for the fate of adsorbed hazardous or toxic substances—whether they are buried and immobilized deep in floodplains, reservoirs and estuaries, on the one hand, or continue to be available to aquatic organisms, on the other.

A crude, but convenient, classification of the sediment movement in a river is into *suspended* load, comprising the finer particles which are held in the flow by turbulence, and *bed* load, comprising the coarser particles which move by sliding, hopping or rolling along the stream bed (Nordin, 1985). In practice there is no clear division between the two modes of transport, and particles that may move as bed load in times of low flow may be carried as suspended sediment in times of high flow.

The mechanism of sediment movement and the transporting capacity of given flow conditions are discussed in various texts on hydraulics and fluvial geomorphology (e.g. Graf, 1971; Bogardi, 1974; Richards, 1982). Rating curves between flow and sediment discharge (Campbell and Bauder, 1940) are commonly used in sediment yield studies as a means to interpolate between the often infrequent sediment measurements. Hydraulic factors control the sediment transport *capacity* of a stream and so are most appropriate for

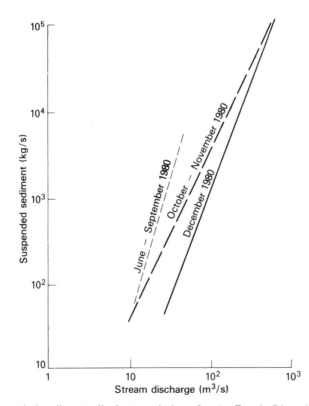

Figure 8.11 *Suspended sediment–discharge relations for the Toutle River, Washington, showing the effect of the enormous quantity of easily eroded sediment made available by the Mount St Helens eruption in May 1980. Previously the stream sediment loads had been negligible. (Redrawn from an original diagram by Nordin, 1985. By permission of John Wiley and Sons Ltd.)*

alluvial rivers, which have beds and banks formed of river deposited material that can be transported by high flows. Many rivers are non-alluvial and the discharge of sediment loads, and in particular that of the finer suspended load, is controlled more by the quantity and timing of the supply of sediment into the stream than by the capacity of the flow to transport it. In such rivers, although there may be a positive relation between flow and sediment load (the higher flows can move larger particles and storm erosion supplies new sediment to the channels), the relationship between flow and sediment discharge may be very poorly defined (Fig. 8.11). Improved estimates of sediment loads may be obtained by the use of separate rating curves for different seasons and for rising and falling river levels (Walling, 1977). Ideally, however, sediment sampling should be sufficiently frequent to define the variations over time. Turbidity meters provide one means of achieving a detailed picture of temporal variations without excessive expense and sampling frequency, but are not suitable for all situations.

The situation is further complicated by the fact that solute and sediment travel at different speeds in a river and both move more slowly than the rate of propagation of a flood wave (Heidel, 1956; Glover and Johnson, 1974). Furthermore, the distributed nature of river networks results in the complex mixing of inputs from various tributaries.

The natural pattern of sediment and solute behaviour in a river is therefore best understood in terms of the characteristics of, and processes operating in, the drainage basin as a whole. This is discussed in Section 8.6.3, but first attention must be paid to the study of point source inputs into a river network which may be viewed in terms of the processes operating within the channels. Point inputs to stream channels may have some important consequences; an example, which is of fundamental biological importance, is the role of certain types of organic waste in reducing the concentrations of dissolved oxygen in streams. When large amounts of biodegradable materials, such as sewage and animal wastes, enter a stream or lake, their chemical and microbial breakdown consumes appreciable amounts of oxygen. This lowers the dissolved oxygen content of the water and may damage the aquatic ecosystem. Low concentrations (less than 3 mg/l) may be harmful to fish and, if the dissolved oxygen becomes exhausted, further decomposition will occur by means of anaerobic processes, which generally produce noxious odours (Lamb, 1985).

The pollutant 'strength' of an effluent, i.e. its potential for removing dissolved oxygen, is measured in terms of its *biological oxygen demand* (BOD). This is a laboratory-derived measure of the consumption of oxygen under standard conditions, usually at 20°C over a period of five days' duration. Details of the method of determination are given in standard texts. While five days is too short for some resistant chemicals, such as those found in wood pulp wastes, for which a 20-day BOD is often used, it is sufficiently long for the majority of the oxygen demand for domestic sewage and many industrial wastes (Dunne and Leopold, 1978). Even so, the necessity of at least a five-day delay may be unacceptable for some purposes, such as monitoring the performance of a treatment plant, and the quicker measure of *chemical oxygen demand* (COD) can provide a value in a couple of hours, which can then be related by an empirical relation for the site to BOD (Lamb, 1985). Figure 8.12 illustrates a typical 'dissolved oxygen sag' curve due to the deoxygenation and subsequent recovery after the addition of organic effluent to a stream from a single outfall point. This curve represents the balance between the rate of consumption of oxygen (which progressively declines as the waste is broken down) and

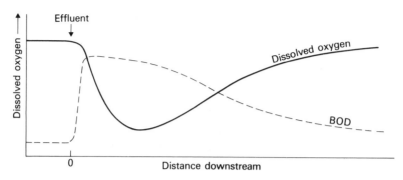

Figure 8.12 *Dissolved oxygen 'sag' curve and biological oxygen demand (BOD) in a river resulting from the introduction of an organic effluent.*

the replenishment of oxygen, mainly by solution from the atmosphere. The rate of reoxygenation increases with the severity of oxygen depletion and the turbulence of the water and reduces with increasing water temperatures. Since the rate of reoxygenation is much slower than that of depletion, the minimum levels of dissolved oxygen may occur some distance, perhaps many kilometres, downstream of the effluent outfall. The magnitude of oxygen sag will depend on the flow in the river, to dilute the effluent, so that any problems of low oxygen levels will be particularly critical at times of low flow in warm weather.

8.6.2 *Lakes*

The discussion of water quality changes in open water bodies has, so far, centred on changes in channels containing flowing water. It is appropriate at this point to briefly discuss the properties of lakes, which generally tend to reduce the wide variations in water quality of the rivers entering them. Due to their reduced water velocity and turbulence, lakes may cause important differences in water quality. Much of the sediment and organic matter entering with river water may settle out, reducing turbidity, and allowing sunlight to penetrate further, which in turn allows more photosynthesis and a richer growth of plants and algae. Under natural processes this gradual accumulation of organic matter will eventually result in the infilling of the lake and creation of a peat bog. The rate of this aquatic plant growth may be limited by a shortage of dissolved nutrients, of which phosphorus and nitrogen seem to be the most important (OECD, 1986). Increased levels of these nutrients are released into streams by man from agricultural fertilizers and domestic sewage, and by removing the natural limits to aquatic plant production this may result in excessive growth of algae, called *algal blooms*. These may cause problems for water supply abstractions as they affect the taste and smell and cause filtration problems. They may, in some cases, cause severe oxygen depletion when the algae die, and can lead to the death of fish and other aquatic life.

One important aspect of some lakes is *thermal stratification*. This is a consequence of water being at its maximum density at 4°C. In deep lakes or reservoirs there may be little mixing of the surface layer of water with that below during most of the year. In summer the surface waters are warmer and less dense than the water below and in winter the surface water may be colder and less dense. As a result, the surface layer (or epilimnion), which may be some 5–8 metres deep, has a high dissolved oxygen content, abundant sunlight and a rich plant life, while in contrast the deeper water (or hypolimnion) is insulated from direct contact with atmospheric oxygen and so may have little or no dissolved oxygen and receives less sunlight. Under these conditions the accumulation of decaying organic matter, sinking down from the surface, leads to stagnant conditions. In the deep water of lakes, phosphate and silica may be released into solution, and ammonia and other gases produced from the biological decay. These two layers of the lake water may remain distinct until with the onset of wintry weather the surface water cools and becomes denser, allowing mixing of the layers by wind action. This 'overturn' of the lake water may occur over only a few hours, and the outflow of the bottom stagnant water may cause serious reductions in the dissolved oxygen levels both within the lake and in streams flowing from it. The sudden upwelling of nutrients from the bottom to the surface can also result in algal blooms.

8.6.3 *Catchments*

The foregoing sections discussed the main processes controlling the water chemistry of the different components of the hydrological cycle. The following sections give a few examples of how the water quality at the outlet of a catchment can be interpreted as the integration of these processes. These discuss the influence on water quality of geology, climate and man's activities, including the use of chemicals such as fertilizers and pesticides, and detail the main water quality characteristics of a number of British rivers.

Geology
Since weathering reactions and their soluble products will be largely determined by the minerals available, it is to be expected that there will be a broad relationship between the composition of stream solutes and the type of rocks underlying a catchment. On a global scale, about 60 per cent of the total natural dissolved load of rivers is derived from rock weathering (Walling and Webb, 1986). The rest comes mainly from the dissolution of atmospheric CO_2 (to form HCO_3^-) and from sea salts.

A number of studies have shown the importance of lithology on stream chemistry. Miller (1961) found that solute concentrations in streams draining sandstone catchments were 10 times higher than in those draining quartzite, and he attributed much of the difference to the dissolution of thin limestone beds in the sandstones. Johnson and Reynolds (1977) found lower solute concentrations in streams draining granite catchments than in those draining mixed metamorphic sedimentary rocks. In a study of major ions and total dissolved solids in 56 single bedrock-type catchments across the USA, Peters (1984) concluded that lithology was a major factor determining dissolved loads, with the highest values in streams draining limestone basins.

Table 8.3 shows typical stream solute concentrations that have been reported in the literature, arranged into broad classes by rock type. Igneous rocks are generally rather impervious and their minerals are resistant to weathering; they generally have the lowest solute concentrations while sedimentary rocks commonly give higher values (but this is very dependent upon their mineral composition). Streams on metamorphic rocks tend to have solute concentrations that are intermediate between igneous and sedimentary rocks,

Table 8.3 Concentrations of dissolved solids in stream water draining catchments underlain by different rock types.

Rock type	Total dissolved solids (mg/l)	Principal ions
Igneous and metamorphic	<100	Na, Ca, HCO$_3$
Sedimentary (detrital)*	50–250	Variable
Limestone and dolomite	100–500	Ca, Mg, HCO$_3$
Evaporites	<10 000	Na, Ca, SO$_4$, Cl

* Very variable depending on composition.

depending on the original material and the degree of alteration. Precipitates such as limestone (mainly $CaCO_3$) and dolomite ($CaMg(CO_3)_x$) generally have much higher concentrations, while evaporites such as halite (NaCl) and anhydrite ($CaSO_4$), which are derived from soluble minerals deposited by evaporation of water, may produce the highest concentrations of dissolved solids. In areas where rock weathering is very slow the stream solutes may be strongly dependent upon atmospheric inputs (Likens *et al.*, 1977; Reid *et al.*, 1981).

Rates of chemical weathering cannot be considered in isolation from physical weathering since the latter is necessary to expose fresh rock material to attack (Drever, 1982). Thus, solute loads may be higher from steep or poorly vegetated areas where physical erosion continually removes the solid weathering products. In many situations the subsurface waters pass through several different rock strata, so the controls on the resulting chemical composition of the surface waters may be very complex.

The water quality of streams depends on a number of interrelated environmental factors, but in general terms the most important natural factor, along with geology, is climate.

Climate
The most important control on the speed of chemical weathering is the availability of liquid water. Temperature is less crucial, and while weathering reactions are faster at higher temperatures this may be offset by less rapid exposure of bedrock by physical erosion due to a denser vegetation cover. In a classic study, Langbein and Schumm (1958) demonstrated that mechanical erosion declined with increasing mean annual rainfall as a result of the greater vegetation cover protecting the ground surface from erosion.

In a study of annual flow and solute data for nearly 500 catchments in which pollution was relatively limited, Walling and Webb (1983) found clear evidence that *loads* increased with annual runoff, although this was at a slower rate than the increase in flows, so solute *concentrations* actually decreased. There was a great deal of scatter in the solute/runoff relationships, due largely to differences in catchment lithologies. They could not make 'positive' suggestions regarding the climatic controls on solute loads due to differences in the predominant rock types in different climate zones. Thus, for example, temperate regions have a greater abundance of sedimentary rocks while tropical areas have predominantly crystalline (igneous and metamorphic) rocks.

Effects of man

In recent years the activities of man have had an increasing effect on the water quality of all the components of the hydrological cycle. This has been both through the addition of chemicals (whether into the atmosphere, onto the land or directly into water courses) and by the alteration of catchments by land-use changes. The following discussion is by no means comprehensive, since limitations of space prevent a full treatment of all the many influences and interactions. It is intended, therefore, to indicate some of the processes involved and some of the management options available, by reference to the examples of chemical fertilizers and pesticides and to the planting of land for commercial forestry.

Fertilizers are applied to the land to correct nutrient deficiencies limiting plant growth. They are particularly important for arable crops since nutrients are removed when the crops are harvested. The main fertilizers in use supply nitrogen (N), phosphorus (P), potassium (K) and sulphur (S).

In recent years there has been a great deal of concern regarding the possible links between nitrates in water and health risks to humans (including stomach cancer and methaemoglobinaemia). Over the 40-year period 1940–80 there was an eight-fold increase in the use of fertilizer nitrogen in the UK as part of an overall intensification of agricultural production (DOE, 1986), and this was one of the principal causes of increases in the nitrate levels in British rivers. The European Economic Community has specified a maximum admissible NO_3 concentration of 50 mg/l (= 11.3 mg/l N as NO_3) in drinking water.

Soils naturally contain a large amount of nitrogen, most of which is held in the humus-rich surface horizons. This is, however, largely unavailable to plants (and to leaching) and only slowly becomes available by a two-stage process comprising decomposition (mineralization) to ammonium (NH_4^+) and then oxidation (nitrification) to NO_3^- by soil bacteria. As noted earlier, NO_3^- is a 'mobile' anion which is not adsorbed by the soil and will move with the water flux, but in the upper soil layers it is taken up by plants. Nitrogen fertilizer is generally applied to crops in spring or early summer when plant growth is rapid, and the crops are subsequently harvested in late summer. This is reflected in the seasonal pattern of nitrogen concentrations in the stream water draining an intensively farmed arable catchment (Fig. 8.13). Nitrate leaching losses are greatest in the winter when plant uptake is minimal, and lowest in the summer when the plants are growing rapidly. The late spring peak values correspond to the 'flushing' of nitrogen by the first big storm after the fertilizer application, and this has also been observed in other studies (Webber and Wadsworth, 1976; Roberts, 1987).

In catchments where the dominant source of nitrate is sewage effluent the seasonal pattern of concentrations is reversed, with the fairly constant input of effluent undergoing greater dilution by the generally larger flows of river water in winter (Roberts and Marsh, 1987).

Agricultural practices will also affect the rate of loss of nitrate from farmland. Ploughing leads to aeration of the soil and greater mineralization and nitrification, and in the case of ploughing up of old grassland it can result in an increase in the release of nitrogen over a number of years. Similarly, artificial drainage will increase nitrate leaching and will also alter the flow paths by diverting some of the leachate from downward movement towards the groundwater to lateral flow in pipes discharging into streams (OECD, 1986).

There has also been concern regarding the increasing concentrations of nitrate found in

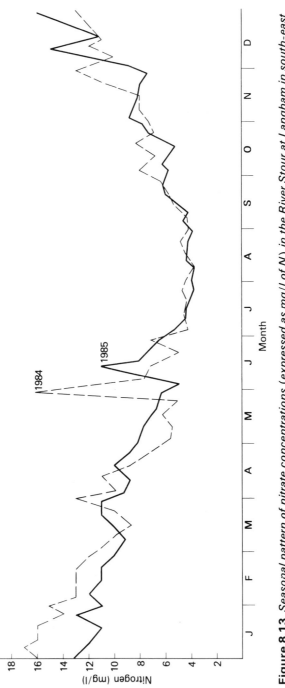

Figure 8.13 *Seasonal pattern of nitrate concentrations (expressed as mg/l of N) in the River Stour at Langham in south-east England in 1984 and 1985 (data supplied by Anglian Water Authority).*

groundwater supplies. Foster and Young (1981) studied pore water nitrate levels in the unsaturated zone of the chalk aquifer in Britain. Cores were drilled at 60 locations and the interstitial water was extracted by centrifuge. Nitrate concentrations were found to be closely related to the history of agricultural practice on the overlying land. Thus, *all* sites under arable farming had NO_3–N levels > 11.3 mg/l and many had over double this figure, while under permanent unfertilized vegetation, such as rough grassland and woodland, concentrations were much lower, i.e. generally less than 5 mg/l and often less than 1 mg/l. Values under fertilized grassland were intermediate between these land-use categories, i.e. generally in the range 5–10 mg/l. Water in the underlying saturated zone was generally 10–15 mg/l under arable crops (cf. 20–30 mg/l in the unsaturated zone above). This suggests that as this water moves downwards to the water table nitrate concentrations in the groundwater will continue to rise for some years, even if no further fertilizer is applied.

While much attention has been given in recent years to the loss of nitrates from agricultural land, much less is known about the leaching of *pesticides*. These comprise an enormous range of both naturally derived and synthetic chemical compounds, designed to be toxic and persistent (stable), to protect against pests, disease and weeds. There has been a dramatic increase in the use of pesticides since 1940 and, for example, in the UK virtually all agricultural and horticultural crops are treated (OECD, 1986).

In temperate climate areas the most commonly used pesticides are *herbicides* to kill weeds (CAB/BCPC, 1988) while in tropical areas *insecticides* have revolutionized health and life expectancy by controlling insects which are the principal vectors of disease. However, while it has been claimed that pesticides have saved over one hundred million lives in developing countries widespread concern has arisen, especially in developed countries, regarding their environmental effects. Pesticides may accumulate in the food chain and so pose a threat to fish, birds and mammals, and this has led to a long-standing debate about whether they are 'lifesaving chemicals' or the 'elixirs of death' (Harding, 1988).

Due to their potential environmental damage there has been a change from the persistent organochlorine pesticides to less persistent chemicals which will degrade, by chemical hydrolysis or bacterial oxidation, over a few months rather than years. These newer compounds are, however, generally more soluble in water, posing the risk of contamination of groundwater and surface water. The fate of an applied chemical in the soil depends upon its partitioning between the soil particles and the soil solution (McEwan and Stephenson, 1979; Lawrence and Foster, 1987). The partitioning of pesticide loss between leaching in solution and adsorption on soil particles is shown in Fig. 8.14 for a number of compounds with different solubilities.

As with fertilizers, pesticides are particularly prone to leaching in solution in the first couple of weeks after application, before processes such as degradation, adsorption and volatilization make the residues unavailable (Wauchope, 1978). In Britain the most commonly used herbicide is *mecoprop*, and in several studies concentrations greater than 10 μg/l have been recorded in streams during storms that occurred within a few days of its application (Williams, 1988).

There are many different types of land-use change and disturbance to catchments resulting from man's activities which may alter the water quality of streams. In Britain one of the largest land-use changes this century has been the planting of upland areas for

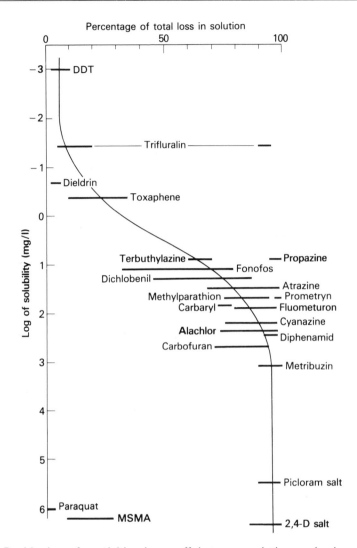

Figure 8.14 *Partitioning of pesticides in runoff between solution and adsorption on eroded sediments. Based on a large number of field studies (Reproduced from* Journal of Environmental Quality, **7** *(4), October–December 1978, pp. 459–72, by permission of the American Society of Agronomy, Inc.; Crop Science Society of America, Inc.; and Soil Science Society of America, Inc.)*

commercial forestry. Although the percentage of the country under forest (10 per cent) is low by comparison to many other countries in Europe, it represents a four-fold increase since 1920. The influence of forestry activities on water quality may be used to illustrate some of the aspects of chemistry and hydrology that have been discussed in this chapter.

Forests may alter water quality both directly through factors such as the increased 'capture' of atmospheric pollutants, alterations to throughfall chemistry and the acidic nature of the litter under conifers, as well as indirectly due to increased evaporation losses and hence an increase in the solute concentrations in the remaining water. Forestry management practices will also affect water quality. The drainage and cultivation of the land prior to tree planting will have a direct effect by releasing sediments into stream systems (Robinson and Blyth, 1982). Drainage of organic soils increases aeration and may result in substantial losses of nutrients including nitrate and ammonium (Kenttamies, 1980), and can lead to problems of eutrophication in lakes and reservoirs downstream. Chemicals are widely used in British forest plantations. Fertilizers such as phosphorus and potassium are generally applied from the air, and some will inevitably fall directly into streams and the forestry drainage ditches, leading to high losses of particulate and dissolved fertilizer for some months afterwards. Herbicides are also used to control competing weeds when the young trees are first becoming established. Felling of the forest for timber will also affect stream water quality. As a result of the decay of the organic debris, the soil disturbance caused by harvesting machinery and the lack of any plants to take up the released nutrients, there is likely to be a large increase in nitrate leaching. Likens *et al.* (1977) reported a large rise in nitrate concentrations after clear felling at Hubbard Brook in the eastern USA. Nitrate losses in the first two years after felling were about 100 times greater than from undisturbed land. Hornung and Newson (1986) found much smaller increases in nitrates in small plot-scale studies in Britain.

Water quality modelling and management
As demand for water resources increases and the variety of pollutants becomes more diverse, there is an increasing conflict between the use of rivers for water supply and as 'sewers' for disposing of industrial and domestic effluent. Considerable effort has been applied to developing techniques to improve decision making in water quality problems, and mathematical models can be used to assess alternative control measures to operate water and wastewater treatment facilities. Such models may be used to develop alternative operating rules in river basin management to take into account factors such as flood control, water supply, sewage disposal, fisheries, recreation and amenity activities. Examples of the types and application of mathematical models of water quality management were given by Whitehead (1980, 1984).

Information on common water quality determinands in British rivers is analysed and collated as part of a national network of 'harmonized monitoring' sites (Simpson, 1978). Table 8.4 gives data from a subset of 16 catchments showing median values of samples taken at approximately two-week intervals over a ten-year period. Differences between the sites can be related to differences in catchment characteristics. The Nene and Stour, for example, contain much intensively farmed (and fertilized) land, with high rates of nitrate–nitrogen loss. In contrast the Exe, Carron and Spey have only poor farmland, so the water is of high quality with low solute concentrations and a high level of dissolved oxygen. The Trent and Aire are examples of catchments with large inputs of industrial effluent and domestic sewage, and as a result have high concentrations of orthophosphate, chloride, NO_3–N and BOD. The Avon has an intensively farmed catchment and a high sewage effluent input which together give a median NO_3–N level close to the EEC limit of 11.3 mg/l.

8.7 In conclusion

The *quality* of water is of great importance in determining the uses to which the water can be put. Increasing sources of pollution (amounts and types) mean that there is greater

Table 8.4 Median values of determinands in river water 1975–85 (data from IH, 1988).

	pH	Conductivity ($\mu s/cm$)	Suspended solids (mg/l)	Dissolved oxygen (mg/l)	BOD (mg/l O)	Ammonia nitrogen (mg/l)	Nitrate (mg/l N)	Ortho-phosphate (mg/l P)	Chloride (mg/l)	Alkalinity (mg/l $CaCO_3$)
Ribble at Samlesbury	7.7	423	9	10.3	2.6	0.19	3.6	0.27	30	115
Trent at Nottingham	7.7	920	17	9.6	3.5	0.30	8.7	1.49	99	165
Avon at Evesham	7.9	930	18	10.2	2.7	0.15	10.1	1.40	73	197
Aire at Fleet	7.5	660	18	8.1	6.6	1.80	5.0	1.19	76	123
Nene at Wansford	8.8	895	13	10.8	3.0	0.19	9.0	0.98	69	210
Stour at Langham	8.1	920	10	10.8	2.3	0.08	8.0	0.59	66	250
Thames at Teddington	8.0	578	14	10.2	2.3	0.22	7.2	1.08	40	189
Great Stour at Horton	7.8	698	7	10.9	2.6	0.15	5.5	0.83	47	224
Itchen at Gatersmill	8.1	492	9	10.6	2.1	0.09	5.2	0.35	20	230
Axe at Whitford	7.9	390	5	10.9	1.7	0.06	3.3	0.22	22	139
Exe at Thorverton	7.5	162	6	11.2	1.6	0.05	2.3	0.08	17	37
Dee at Overton	7.2	163	3	11.2	1.1	0.02	1.0	0.03	18	25
Carron at New Kelso	6.7	47	1	11.3	0.8	0.01	0.1	0.01	10	5
Spey at Fochabers	7.2	77	2	11.2	0.9	0.03	0.3	0.01	11	25
Almond at Cragiehall	7.5	585	12	9.6	2.8	0.95	3.7	0.43	62	123
Leven at Linnbrane	7.1	70	4	11.0	1.6	0.02	0.3	0.01	10	18

potential for environmental damage and emphasize the need for public and political pressure to protect and manage our common resource.

In scientific terms the study of water quality is more than simply an 'add-on' to our existing knowledge of water quantity fluxes and stores. The use of tracers, both natural and artificial, can provide very revealing information on the flow paths of water, and the use of natural tracers such as ^{18}O has led to a questioning of some of the traditional hydrological theories of storm runoff generation (Section 7.4.5). Water quality can also affect the behaviour and movement of subsurface water and high solute concentrations can lower the evaporation from water bodies as a result of the reduced vapour pressure (Calder and Neal, 1984) (Section 4.3.2 under 'Salinity'). Increasingly the hydrologist must consider both the quantity and the quality of water, and in the future the latter may pose the greatest challenges.

References

Alberty, R. A. (1987) *Physical chemistry*, 7th edn, John Wiley and Sons, New York, 934 pp.

Andrews, E. D. and B. W. Webb (1987) Emerging issues in surface water quality research, *IASH Publ.*, **171**: 27–33.

Atkinson, T. C. and P. L. Smart (1981) Artificial tracers in hydrology, Ch. 13 in *A survey of British hydrogeology 1980*, C. R. Argent and D. J. H. Griffin (eds), The Royal Society, London, pp. 173–90.

Bache, B. W. (1980) The acidification of soils, in *Effects of acid precipitation on terrestrial ecosystems*, T. C. Hutchinson and M. Havas (eds), Plenum Press, New York, pp. 183–202.

Bache, B. W. (1984) Soil-water interactions, *Phil. Trans. Roy. Soc. Lond.*, B, **305**: 393–407.

Back, W. (1966) Hydrochemical facies and groundwater flow patterns in northern part of Atlantic Coastal Plain, *USGS Prof. Pap.*, 498-A, 42 pp.

Barnett, A. P., J. R. Carrekar, F. Abruna, W. A. Jackson, A. E. Dooley and J. H. Holladay (1972) Soil and nutrient losses in runoff with selected cropping treatments on tropical soils, *Agron. J.*, **64**: 391–5.

Barraclough, D., M. J. Hyden and G. P. Davies (1983) Fate of fertilizer nitrogen applied to grassland: 1. Field leaching results, *J. Soil Sci.*, **34**: 483–97.

Barrett, C. F. (Chairman) (1987) *Acid deposition in the United Kingdom 1981–1985*, 2nd Report of the UK Review Group on Acid Rain, Warren Spring Laboratory, Stevenage, 104 pp.

Barrett, E. and G. Broden (1955) The acidity of Scandinavian precipitation, *Tellus*, **7**: 251–64.

Bencala, K. E., V. C. Kennedy, G. W. Zellweger, A. P. Jackman and R. J. Avanzino (1984) Interactions of solutes and streambed sediment: 1. An experimental analysis of cation and anion transport in a mountain stream, *WRR*, **20**: 1797–803.

Berner, R. A. (1978) Rate controlled mineral dissolution under earth surface conditions, *American Journal of Science*, **278**: 1235–52.

Bloch, M. R., D. Kaplan, V. Kertes and J. Schnerb (1966) Ion separation in bursting air bubbles: an explanation for the irregular ion ratios in atmospheric precipitations, *Nature*, **209**: 802–3.

Bogardi, J. (1974) *Sediment transport in alluvial streams*, Akademiai Kiado, Budapest, 826 pp.

Bolt, G. H. and M. G. M. Bruggenwert (1978) *Soil chemistry—A. Basic elements*, Elsevier, Amsterdam, 281 pp.

Bouma, J. (1981) Soil morphology and preferential flow along macropores, *Agric. Water Manag.*, **3**: 235–50.

Brady, N. C. (1984) *The nature and properties of soils*, 9th edn, Macmillan, New York, 560 pp.

CAB/BCPC (1988) *The U.K. pesticide guide*, CAB International, Wallingford/British Crop Protection Council, Bracknell, 434 pp.

Calder, I. R. and C. Neal (1984) Evaporation from saline lakes: a combination equation approach, *Hydrol. Sci. J.*, **29**: 89–97.

Campbell, F. B. and H. A. Bauder (1940) A rating curve method for determining silt discharge of streams, *Trans. AGU*, **21**: 603–7.

Cape, J. N., D. Fowler, J. W. Kinnaird, I. A. Nicholson and I. S. Paterson (1987) Modification of rainfall chemistry by a forest canopy, in *Pollutant transport and fate in ecosystems*, P. J. Coughtrey, M. H. Martin and M. H. Unsworth (eds), British Ecological Society Special Publication 6, pp. 155–69.

Castillo, R. A., J. Kadlecek and S. McLaren (1985) Selected Whiteface Mountain cloud water concentrations, summers 1981 and 1982, *Water, Air and Soil Pollution*, **24**: 323–8.

Catt, J. (1985) Natural soil acidity, *Soil Use and Management*, **1**: 8–10.

Charlson, R. J., D. S. Covert, T. V. Larson and A. P. Waggoner (1978) Chemical properties of tropospheric sulphur aerosols, *Atmos. Environ.*, **12**: 39–53.

Chebotarev, I. I. (1955) Metamorphism of natural waters in the crust of weathering, *Geochima et Cosmochimica Acta*, **8**: 22–48, 137–70, 198–212.

Chesselet, R., J. Morelli and P. Buat-Menard (1972) Variations in ionic ratios between reference sea water and marine aerosols, *JGR*, **77**: 5116–31.

Clesceri, N. L. and C. Vasudevan (1980) Acid precipitation, throughfall chemistry and canopy processes, in *Ecological impact of acid precipitation*, D. Drablos and A. Tollan (eds), SNSF Conference, Oslo, pp. 258–9.

Cresser, M. and A. Edwards (1987) *Acidification of freshwaters*, Cambridge University Press, Cambridge, 136 pp.

Cryer, R. (1986) Atmospheric solute inputs, in *Solute processes*, S. T. Trudgill (ed.), John Wiley and Sons, New York, pp. 15–84.

Davis, S. N. and R. J. M. de Wiest (1966) *Hydrogeology*, John Wiley and Sons, New York, 463 pp.

Delleur, J. W. (1986) Hydrological response to acid rain, *Int. Assoc. Hydrol. Sci., Publ.*, **157**: 175–84.

DOE (1972) *Analysis of raw, potable and waste waters*, Department of the Environment, London, HMSO, 305 pp.

DOE (1986) Nitrate in water, *Pollution Paper 26*, Report by the Nitrate Coordination Group, Department of the Environment, London, 101 pp.

Dollard, G. J., M. H. Unsworth and M. J. Harvey (1983) Pollutant transfer in upland regions by occult precipitation, *Nature*, **302**: 241–3.

Drever, J. I. (1982) *The geochemistry of natural waters*, Prentice-Hall, Englewood Cliffs, N.J., 388 pp.

Dunne, T. (1978) Field studies of hillslope flow processes, in *Hillslope hydrology*, M. J. Kirkby (ed.), John Wiley and Sons, Chichester, pp. 227–93.

Dunne, T. and L. B. Leopold (1978) *Water in environmental planning*, Freeman, San Francisco, 818 pp.

Ferguson, P. and J. A. Lee (1983) Past and present sulphur pollution in the southern Pennines, *Atmos. Environ.*, **17**: 1131–7.

Flowers, R. J. and R. W. Battarbee (1983) Diatom evidence for recent acidification of two Scottish lochs, *Nature*, **305**: 130–3.

Foster, G. R. (1982) Modelling the erosion process, in *Hydrologic modelling of small watersheds*, C. T. Haan, H. P. Johnson and D. L. Brakensiek (eds), Monograph 5, American Society of Agricultural Engineers, pp. 297–382.

Foster, S. S. D. (1975) The chalk groundwater tritium anomaly—a possible explanation, *J. Hydrol.*, **25**: 159–65.

Foster, S. S. D. and C. P. Young (1981) Effects of agricultural land use on groundwater quality with special reference to nitrate, Ch. 4 in *A survey of British hydrogeology*, Royal Society, London, pp. 47–59.

Fowler, D. (1984) Transfer to terrestrial surfaces, *Phil. Trans. Roy. Soc. London, B*, **305**: 281–97.

Fowler, D. and J. N. Cape (1984) The contamination of rain samples by deposition on rain collectors, *Atmos. Environ.*, **18**: 183–9.

Fowler, D., J. N. Cape, I. D. Leith, I. S. Paterson, J. W. Kinnaird and I. A. Nicholson (1982) Rainfall acidity in northern Britain, *Nature*, **297**: 383–6.

Fowler, D., J. N. Cape and I. D. Leith (1985) Acid inputs from the atmosphere in the United Kingdom, *Soil Use and Management*, **1**: 3–5.

Fowler, D., J. N. Cape, I. D. Leith, T. W. Choularton, M. J. Gay and A. Jones (1988) The influence of altitude on rainfall chemistry at Great Dun Fell, *Atmos. Environ.*, **22**: 1355–62.

Franks, F. (1983) *Water*, Royal Society of Chemistry, London, 96 pp.

Freeze, R. A. and J. A. Cherry (1979) *Groundwater*, Prentice-Hall, Englewood Cliffs, N.J., 604 pp.

Galloway, J. N., G. E. Likens, W. C. Keen and J. M. Miller (1982) The composition of precipitation in remote areas of the world, *JGR*, **87**: 8771–86.

Garland, J. A. (1978) Dry and wet removal of sulphur from the atmosphere, *Atmos. Environ.*, **12**: 349–62.

Garrels, R. M. and C. Christ (1965) *Minerals, solutions and equilibria*, Harper and Row, New York, 450 pp.

Glover, B. J. and P. Johnson (1974) Variations in the natural chemical concentrations of river water during flood flows, and the lag effect, *J. Hydrol.*, **22**: 303–16.

Gorham, E. (1958a) Atmospheric pollution by hydrochloric acid, *QJRMS*, **84**: 274–6.

Gorham, E. (1958b) The influence and importance of daily weather conditions in the supply of chloride, sulphate and other ions to fresh waters from atmospheric precipitation, *Phil. Trans. Roy. Soc. London, B*, **241**: 147–78.

Graf, W. H. (1971) *Hydraulics of sediment transport*, McGraw-Hill, New York, 509 pp.

Gregory, P. J. (1988) Growth and functioning of plant roots, Ch. 4 in *Russell's soil conditions and plant growth*, A. Wild (ed.), Longman Scientific and Technical/Wiley, New York, pp. 113–67.

Gvirtzman, H., D. Ronen and M. Magaritz (1986) Anion exclusion during transport through the unsaturated zone, *J. Hydrol.*, **87**: 267–83.

Hall, J. K., N. L. Hartwig and L. D. Hoffman (1983) Application mode and alternate cropping effects on atrazine losses from a hillside, *J. of Environmental Quality*, **12**: 336–40.

Harding, D. J. L. (ed.) (1988) *Britain since 'Silent Spring': an update on the ecological effects of agricultural pesticides in the U.K.*, Proceedings of Symposium at Cambridge, 18 March 1988, Institute of Biology, London, 131 pp.

Harris, A. R. and E. S. Verry (1985) Wet deposition of sulphate and nitrate acids and salts in the United States, in *Hydrological and hydrogeochemical mechanisms and model approaches to the acidification of ecological systems*, I. Johansson (ed.), Nordic Hydrological Programme Report 10, Stockholm, pp. 57–65.

Havas, M. (1986) Effects of acidic deposition on aquatic ecosystems, in *Air Pollution*, vol. 6: *Air pollutants, their transformation, transport and effects*, A. C. Stern (ed.), 3rd edn., Academic Press, New York, pp. 351–89.

Heidel, S. G. (1956) The progressive lag of sediment concentration with flood waves, *Trans. AGU*, **37**: 56–66.

Hem, J. D. (1985) Study and interpretation of the chemical characteristics of natural water, *USGS Wat. Sup. Pap. 2254*, 3rd edn, 263 pp.

Hornung, M. and M. D. Newson (1986) Upland afforestation: influences on stream hydrology and chemistry, *Soil Use and Management*, **2**: 61–5.

Hubbard, R. K. and J. M. Sheridan (1983) Water and nitrate–nitrogen losses from a small, upland, coastal plain watershed, *J. of Environmental Quality*, **12**: 291–5.

IH (1988) *Hydrological data UK 1986*, Institute of Hydrology, Wallingford, 178 pp.

Innes, J. L. (1987) Air pollution and forestry, *Forestry Commission Bulletin 70* HMSO, London, 39 pp.

Jacobson, R. L. and D. Langmuir (1974) Dissociation constants of calcite and $CaHCO_3^+$ from 0°C to 50°C, *Geochima et Cosmochimica Acta*, **38**: 301–18.

Johnson, A. H. and R. C. Reynolds (1977) Chemical character of headwater streams in Vermont and New Hampshire, *WRR*, **13**: 469–73.

Joslin, J. D., P. A. Mays, M. H. Wolfe, J. M. Kelley, R. W. Garber and P. F. Brewer (1987) Chemistry of tension lysimeter water and lateral flow in spruce and hardwood stands, *J. of Environmental Quality*, **16**: 152–60.

Junge, C. E. and W. T. Werby (1958) The concentration of chloride, sodium, potassium, calcium and sulphate in rain water over the United States, *J. Meteorol.*, **15**: 417–25.

Jury, W. A. (1982) Simulation of solute transport using a transfer function model, *WRR*, **18**: 363–8.

Kenttamies, K. (1980) The effects on water quality of forest drainage and fertilisation in peatlands, *IAHS Publ.*, **130**: 277–83.

Lamb, J. C. (1985) *Water quality and its control*, John Wiley and Sons, New York, 384 pp.

Langbein, W. B. and S. A. Schumm (1958) Yield of sediment in relation to mean annual precipitation, *Trans. AGU*, **39**: 1076–84.

Lapidus, L. and N. R. Amundson (1952) Mathematics of adsorption in beds, *Journal of Physical Chemistry*, **56**: 984–95.

Lawrence, A. R. and S. S. D. Foster (1987) The pollution threat from agricultural pesticides and industrial solvents, *Hydrogeological Report, 87/2*, BGS, Wallingford, 29 pp.

Leopold, L. B. and K. S. Davis (1970) *Water*, Time-Life International, 191 pp.

Leopold, L. B., M. G. Wolman and J. P. Miller (1964) *Fluvial processes in geomorphology*, Freeman, San Francisco, 522 pp.

Lewis, G. N. and M. Randall (1961) *Thermodynamics*, 2nd edn, revised by K. S. Pitzer and L. Brewer, McGraw-Hill, New York, 723 pp.

Likens, G. E., F. H. Bormann, R. W. Pierce, J. S. Eaton and N. M. Johnson (1977) *Biogeochemistry of a forested ecosystem*, Springer-Verlag, New York, 146 pp.

McEwan, F. L. and G. R. Stephenson (1979) *The use and significance of pesticides in the environment*, John Wiley and Sons, New York, 538 pp.

Mayer, R. and B. Ulrich (1974) Conclusions on the filtering action of forests from ecosystem analysis, *Oecologia Plantarum*, **9**: 157–68.

Meybeck, M. (1983) Atmospheric inputs and river transport of dissolved substances, *IAHS Publ.*, **141**: 173–92.

Miller, H. G. (1984) Deposition–plant–soil interactions, *Phil. Trans. Roy. Soc. London, B*, **305**: 339–52.

Miller, H. G. and J. D. Miller (1980) Collection and retention of atmospheric pollutants by vegetation, in *Ecological impact of acid precipitation*, D. Drablos and A. Tollan (eds), SNSF Conference, Oslo, pp. 33–40.

Miller, J. P. (1961) Solutes in small streams draining single rock types, Sangre de Cristo Range, New Mexico, *USGS Surv. Wat. Sup. Pap.*, 1535F, 23 pp.

Miller, R. J., D. R. Nielsen and J. W. Biggar (1965) Chloride displacement in Panoche clay loam in relation to water movement and distribution, *WRR*, **1**: 63–73.

Mohnen, V. A. (1988) The challenge of acid rain, *Scientific American*, **259**: 14–22.

Moser, H., W. Rauert, G. Morgenschweis and H. Zojer (1986) Study of groundwater and soil moisture movement by applying nuclear, physical and chemical methods, in *Technical documents in hydrology*, UNESCO, Paris, 104 pp.

Munger, J. W. and S. J. Eisenreich (1983) Continental-scale variations in precipitation chemistry, *Environmental Science and Technology*, **17**: 32A–42A.

Nicholson, I. A., J. N. Cape, D. Fowler, J. W. Kinnaird and I. S. Paterson (1980) Effects of a Scots pine canopy on the chemical composition and decomposition patterns of precipitation, in *Ecological impact of acid precipitation*, D. Drablos and A. Tollan (eds), SNSF Conference, Oslo, pp. 148–9.

Nordin, C. F. (1985) The sediment loads of rivers, Ch. 7 in *Facets of hydrology*, J. C. Rodda (ed.), vol. II, John Wiley and Sons, Chichester, pp. 183–204.

Nordstrum, D. K. and 18 others (1979) A comparison of computerized chemical models for equilibrium calculations in aqueous systems, in *Chemical modelling in aqueous systems*, E. A. Jenne (ed.), ACS Symposium Series 93, American Chemical Society, Washington, D.C., pp. 857–92.

Nortcliffe, S. (1988) Soil formation and characteristics of soil profiles, Ch. 5 in *Russell's soil conditions and plant growth*, A. Wild (ed.), Longman Scientific and Technical/Wiley, New York, pp. 168–212.

OECD (1986) *Water pollution by fertilizers and pesticides*, Organisation for Economic Cooperation and Development, Paris, 144 pp.

Paces, T. (1985) Sources of acidification in Central Europe estimated from elemental budgets in small basins, *Nature*, **315**: 32–6.

Peters, N. E. (1984) Evaluation of environmental factors affecting yields of major dissolved ions in streams in the United States, *USGS Wat. Sup. Pap.*, 2228, 39 pp.

Piper, A. M. (1944) A graphic procedure in the geochemical interpretation of water analyses, *Trans. AGU*, **25**: 914–23.

Plant, J. A. and R. Raiswell (1983) Principles of environmental geochemistry, in *Applied environmental geochemistry*, I. Thornton (ed.), Academic Press, London, pp. 1–39.

Price, M. (1985) *Introducing groundwater*, Allen and Unwin, London, 195 pp.

Price, M. (1987) Fluid flow of the chalk of England, in *Fluid flow in sedimentary basins and aquifers*, J. C. Goff and B. P. J. Williams (eds), Geology Society Special Publ. 34, pp. 141–56.

Reid, J. M., D. A. McLeod and M. S. Cresser (1981) Factors affecting the chemistry of precipitation and river water in an upland catchment, *J. Hydrol.*, **50**: 129–45.

Reynolds, B., C. Neal, M. Hornung and P. A. Stevens (1986) Baseflow buffering of streamwater acidity in five mid-Wales catchments, *J. Hydrol.*, **87**: 167–85.

Richards, K. (1982) *Rivers: form and process in alluvial channels*, Methuen, London, 358 pp.

Roberts, G. (1987) Nitrogen inputs and outputs in a small agricultural catchment in the eastern part of the U.K., *Soil Use and Management*, **3**: 148–54.

Roberts, G. and T. J. Marsh (1987) The effects of agricultural practices on the nitrate concentrations in the surface water domestic supply sources of western Europe, *IAHS Publ.*, **164**: 365–80.

Robinson, M. and K. Blyth (1982) The effect of forestry drainage operations on upland sediment yields: a case study, *Earth Surf. Proc. and Landforms*, **7**: 85–90.

Sadasivan, S. (1980) Trace constituents in cloud water, rainwater and aerosol samples collected near the west coast of India during the southwest monsoon, *Atmos. Environ.*, **14**: 33–8.

Seip, H. M. (1980) Acidification of freshwater—sources and mechanisms, in *Ecological impact of acid precipitation*, D. Drablos and A. Tollan (eds), SNSF Conference, Oslo, pp. 358–9.

Seip, H. M. and A. Tollan (1985) Acid deposition, Ch. 3 in *Facets in hydrology*, J. C. Rodda (ed.), vol. II. John Wiley and Sons, Chichester, pp. 69–98.

Sillen, L. G. and A. E. Martell (1964) *Stability constants of metal ion complexes*, Chemical Society Special Publication 17, London, 754 pp.

Simpson, E. A. (1978) The harmonisation of the monitoring of the quality of inland fresh water, *JIWES*, **32**: 45–56.

Skartveit, A. (1982) Wet scavenging of sea salts and acid compounds in a rainy coastal area, *Atmos. Environ.*, **16**: 2715–24.

Skeffington, R. A. (1987) Do all forests act as sinks for air pollutants?, in *Acidification and water pathways*, vol. 2, Norwegian National Committee for Hydrology/UNESCO/WMO, pp. 85–94.

Skopp, J. (1986) Analysis of time-dependent chemical processes in soils, *J. of Environmental Quality*, **15**: 205–13.

Smettem, K. R. J. (1986) Solute movements in soils, in *Solute processes*, S. T. Trudgill (ed.), John Wiley and Sons, Chichester, pp. 141–65.

Smettem, K. R. J., S. T. Trudgill and A. M. Pickles (1983) Nitrate loss in soil drainage waters in relation to by-passing flow and discharge on an arable site, *J. Soil Sci.*, **34**: 499–509.

Smith, D. B., P. L. Wearn, H. J. Richards and P. C. Rowe (1970) Water movement in the unsaturated zone of high and low permeability strata by measuring natural tritium, *Isotope Hydrology 1970*, International Atomic Energy Authority, Vienna, pp. 73–81.

Smith, R. A. (1852) On the air and rain of Manchester, *Mem. Proceedings of Manchester Literary and Philosophical Society*, Series (2) 10, 207–17.

Smith, R. A. (1872) *Air and rain: the beginnings of chemical climatology*, Longmans, London, 600 pp.

Sposito, G. (1981) *The thermodynamics of soil solutions*, Oxford University Press, Oxford, 223 pp.

Stevens, P. A., M. Hornung and A. R. McLauchlin (1987) Acidification of water beneath a mature Sitka spruce plantation in Beddgelert forest, north Wales, *Acidification and water pathways*, vol. 2, Norwegian National Committee for Hydrology/UNESCO/WMO, pp. 107–14.

Stevenson, C. M. (1968) An analysis of the chemical composition of rainwater and air over the British Isles and Eire for the years 1959–64, *QJRMS*, **94**: 56–70.

Stott, P. (1979) Opening address: Conference on *Water resources—a changing strategy*, Institution of Civil Engineers, London.

Stumm, W. and J. J. Morgan (1981) *Aquatic chemistry*, John Wiley and Sons, New York, 780 pp.

Swedish Ministry of Agriculture (1982) *Acidification today and tomorrow*, Stockholm, 230 pp.

Tebbutt, T. H. Y. (1977) *Principles of water quality control*, 2nd edn, Pergamon, Oxford, 201 pp.

Thomas, G. W. and A. R. Swoboda (1970) Anion exclusion effects on chloride movement in soil, *Soil Sci.*, **110**: 163–6.

Travis, C. C. and E. L. Etnier (1981) A survey of sorption relationships for reactive solutes in soil, *J. of Environmental Quality*, **10**: 8–17.

Truesdell, A. H. and B. F. Jones (1974) WATEQ, a computer program for calculating equilibrium of natural waters, *USGS Journal of Research*, **2**: 233–48.

Van Genuchten, M. T. and P. J. Wierenga (1976) Mass transfer studies in sorbing porous media, I. Analytical solutions, *Soil Sci. Soc. Amer. Proc.*, **40**, 473–80.

Walling, D. E. (1977) Assessing the accuracy of suspended sediment rating curves for a small basin, *WRR*, **13**: 531–8.

Walling, D. E. and B. W. Webb (1983) The dissolved load of rivers: a global overview, *IAHS Publ.*, **141**: 3–20.

Walling, D. E. and B. W. Webb (1986) Solutes in river systems, Ch. 7 in *Solute processes*, S. T. Trudgill (ed.), John Wiley and Sons, New York, pp. 251–327.

Walton, W. C. (1970) *Groundwater resource evaluation*, McGraw-Hill, New York, 664 pp.

Watt Committee on Energy (1984) Acid rain, *Report 14*, London, 58 pp.

Wauchope, R. D. (1978) The pesticide content of surface water draining from agricultural fields—a review, *J. of Environmental Quality*, **7**: 459–72.

Webber, J. and G. A. Wadsworth (1976) Nitrate and phosphate in borehole, well and stream waters, in *Agriculture and water quality*, Bulletin 32, Ministry of Agriculture, Fisheries and Food, London, pp. 237–51.

Wellings, S. R. and J. D. Cooper (1983) The variability of recharge of the English chalk aquifer, *Agric. Water Manag.*, **6**: 243–53.

White, R. E. (1985) The influence of macropores on the transport of dissolved and suspended matter through soils, *Advances in Soil Science*, **3**: 95–120.

White, R. E. (1987) *Introduction to the principles and practice of soil science*, 2nd edn, Blackwell, Oxford, 244 pp.

Whitehead, P. G. (1980) Water quality modelling for design, *IAHS Publ.*, **130**: 465–75.

Whitehead, P. G. (1984) The application of mathematical models of water quality and pollution transport: an international survey, in *Technical documents in hydrology*, UNESCO, Paris, 50 pp.

Whitehead, P. G., C. Neal and R. Neale (1986) Modelling the effects of hydrological changes on stream water acidity, *J. Hydrol.*, **84**: 353–64.

Williams, R. J. (1988) Personal communication, Institute of Hydrology.

Woods, T. L. and R. M. Garrels (1987) *Thermodynamic values at low temperature for natural inorganic materials*, Oxford University Press, Oxford, 242 pp.

WPCF (1980) *Standard methods for the examination of water and wastewater*, 15th edn, Water Pollution Control Federation, Washington, D.C.

Wright, E. P., A. C. Benfield, W. M. Edmunds and R. Kitching (1982) Hydrogeology of the Kufra and Sirte basins, eastern Libya, *QJEG*, **15**: 83–103.

Young, A. (1976) *Tropical soils and soil survey*, Cambridge University Press, Cambridge, 468 pp.

Reference abbreviations

The following is a list of the abbreviations most commonly used in the alphabetical lists of chapter references:

Agric. Met.	Agricultural Meteorology
Agric. Water Manag.	Agricultural Water Management
Agron. J.	Agronomy Journal
Atmos. Environ.	Atmospheric Environment
Boundary-Layer Met.	Boundary-Layer Meteorology
Bull. Amer. Geol. Soc.	Bulletin of the American Geological Society
Bull. Amer. Met. Soc.	Bulletin of the American Meteorological Society
Bull. IASH, Bull. IAHS	Bulletin of the International Association of Scientific Hydrology (now the International Association of Hydrological Sciences)
Earth Surf. Proc. and Landforms	Earth Surface Processes and Landforms
Geog. J.	Geographical Journal
Hydrol. Sci. J./Bull.	Hydrological Sciences Journal/Bulletin
IASH Publ., IAHS Publ.	International Association of Scientific Hydrology Publication (now the International Association of Hydrological Sciences)
Int. Assoc. Hydrol. Sci. Bull./Publ.	International Association of Hydrological Sciences Bulletin/Publication
J. Agric. Sci.	Journal of Agricultural Science
J. Appl. Met.	Journal of Applied Meteorology
JAWA	Journal of American Waterworks Association
JGR	Journal of Geophysical Research
JIWE(S)	Journal of Institution of Water Engineers (and Scientists)
J. Hydrol.	Journal of Hydrology
J. Meteorol.	Journal of Meteorology
JMSJ	Journal of Meteorological Society of Japan
J. Soil and Water Cons.	Journal of Soil and Water Conservation
J. Soil Sci.	Journal of Soil Science
Met. Mag.	Meteorological Magazine
Mon. Wea. Rev.	Monthly Weather Review
Neth. J. Agric. Sci.	Netherlands Journal of Agricultural Science
Plant Physiol.	Plant Physiology
Proc. ASCE	Proceedings American Society of Civil Engineers
Proc. ICE	Proceedings Institution Civil Engineers
Proc. Phil. Trans. Roy. Soc.	Proceedings Philosophical Transactions of the Royal Society
Proc. SSSA	Proceedings Soil Science Society of America
Publ. in Climatol.	Publications in Climatology (New Jersey)
QJEG	Quarterly Journal of Engineering Geology
QJRMS	Quarterly Journal of the Royal Meteorological Society
Soil Sci.	Soil Science
Soviet Hydrol.	Soviet Hydrology

Trans. AGU	Transactions American Geophysical Union
Trans. ASAE	Transactions American Society of Agricultural Engineers
Trans. ASCE	Transactions American Society of Civil Engineers
Trans. IBG	Transactions Institute of British Geographers
USDA	United States Department of Agriculture
USFS	United States Forest Service
USGS Prof. Pap.	United States Geological Survey Professional Papers
USGS Wat. Sup. Pap.	United States Geological Survey Water Supply Papers
WRR	Water Resources Research

Simplified metric conversion tables

DISTANCE*

Inches		Millimetres
0.039	1	25.4
0.079	2	50.8
0.118	3	76.2
0.158	4	101.6
0.197	5	127.0
0.236	6	152.4
0.276	7	177.8
0.315	8	203.2
0.354	9	228.6

Feet		Metres
3.281	1	0.305
6.562	2	0.610
9.842	3	0.914
13.123	4	1.219
16.404	5	1.524
19.685	6	1.829
22.966	7	2.134
26.246	8	2.438
29.527	9	2.743

Yards		Metres
1.094	1	0.914
2.187	2	1.829
3.281	3	2.743
4.375	4	3.658
5.468	5	4.572
6.562	6	5.486
7.656	7	6.401
8.750	8	7.316
9.843	9	8.230

Miles		Kilometres
0.621	1	1.609
1.243	2	3.219
1.864	3	4.828
2.486	4	6.437
3.107	5	8.047
3.728	6	9.656
4.350	7	11.265
4.971	8	12.875
5.592	9	14.484

VOLUME*

Cu. feet		Cu. metres
35.315	1	0.028
70.629	2	0.057
105.943	3	0.085
141.258	4	0.113
176.572	5	0.142
211.887	6	0.170
247.201	7	0.198
282.516	8	0.227
317.830	9	0.255

Gallons (imp)		Litres
0.220	1	4.544
0.440	2	9.087
0.660	3	13.631
0.880	4	18.174
1.101	5	22.718
1.321	6	27.262
1.541	7	31.805
1.761	8	36.349
1.981	9	40.892

*AREA**

Sq. feet		Sq. metres
10.764	1	0.093
21.528	2	0.186
32.292	3	0.279
43.056	4	0.372
53.819	5	0.465
64.583	6	0.557
75.347	7	0.650
86.111	8	0.743
96.875	9	0.836

Sq. yards		Sq. metres
1.196	1	0.836
2.392	2	1.672
3.588	3	2.508
4.784	4	3.345
5.980	5	4.181
7.176	6	5.016
8.372	7	5.853
9.568	8	6.690
10.764	9	7.526

TEMPERATURE
(Centigrade to Fahrenheit)

°C	0	1	2	3	4	5	6	7	8	9
+40	104.0	105.8	107.6	109.4	111.2	113.0	114.8	116.6	118.4	120.2
+30	86.0	87.8	89.6	91.4	93.2	95.0	96.8	98.6	100.4	102.2
+20	68.0	69.8	71.6	73.4	75.2	77.0	78.8	80.6	82.4	84.2
+10	50.0	51.8	53.6	55.4	57.2	59.0	60.8	62.6	64.4	66.2
+ 0	32.0	33.8	35.6	37.4	39.2	41.0	42.8	44.6	46.4	48.2
− 0	32.0	30.2	28.4	26.6	24.8	23.0	21.2	19.4	17.6	15.8
−10	14.0	12.2	10.4	8.6	6.8	5.0	3.2	1.4	−0.4	−2.2
−20	−4.0	−5.8	−7.6	−9.4	−11.2	−13.0	−14.8	−16.6	−18.4	−20.2
−30	−22.0	−23.8	−25.6	−27.4	−29.2	−31.0	−32.8	−34.6	−36.4	−38.2
−40	−40.0	−41.8	−43.6	−45.4	−47.2	−49.0	−50.8	−52.6	−54.4	−56.2

* In the Distance, Area, and Volume conversion tables the figures in the central columns may be read as either the metric or the imperial unit, e.g., 1 cubic foot = 0.028 cubic metre; or 1 cubic metre = 35.315 cubic feet.

Index